CHEMILUMINESCENCE
and
BIOLUMINESCENCE

CHEMILUMINESCENCE
and
BIOLUMINESCENCE

EDITED BY
M. J. CORMIER, D. M. HERCULES, AND J. LEE

Departments of Biochemistry and Chemistry
University of Georgia
Athens, Georgia

PLENUM PRESS • NEW YORK - LONDON • 1973

Library of Congress Catalog Card Number 73-76169
ISBN 0-306-30733-2

© 1973 Plenum Press, New York
A Division of Plenum Publishing Corporation
227 West 17th Street, New York, N. Y. 10011

United Kingdom edition published by Plenum Press, London
A Division of Plenum Publishing Company, Ltd.
Davis House (4th Floor), 8 Scrubs Lane, Harlesden, London, NW10 6SE, England

All rights reserved

No part of this publication may be reproduced in any form without
written permission from the publisher

Printed in the United States of America

PREFACE

This volume is a collection of papers presented at the International Conference on Chemiluminescence held at the University of Georgia, October 10-13, 1972. This conference was a sequel to the first International Conference on Chemiluminescence held at Duke University in 1965. Between the two conferences considerable work had been done in all areas of chemiluminescence, and this second conference was an attempt to bring together the many workers in the field to share ideas about mutual problems.

The phenomenon of chemiluminescence covers a wide range of disciplines, and as a consequence the backgrounds and interests of scientists studying chemiluminescence are diverse. Topics ranging from the theory of atomic collisions to evolutionary biology have direct relevance to chemiluminescence, and in turn chemiluminescence offers a unique tool for their study. The objective of the symposium was to bring together workers having diverse backgrounds to discuss both the fundamental science relating to chemiluminescence as well as applications of chemiluminescence to important problems. A perusal of the table of contents will reveal the diversity of this volume. It covers chemiluminescence in the gas phase and in condensed media; chemiluminescence from radical reactions; reactions of singlet oxygen; analysis for trace elements; mechanisms in bioluminescence; mechanisms of organic chemiluminescent reactions; and lasers produced from chemiluminescent reactions.

This volume presents the invited lectures given at the conference as well as abstracts of the contributed papers. An edited version of the discussions following each invited lecture is also presented.

It is always difficult to express gratitude to all of those persons who have contributed so much to the success of a conference and a volume such as this. We believe a special note of thanks goes to our session chairmen for their willingness to undertake the difficult task of editing the remarks presented at each session. We also give a special note of thanks to our graduate students and postdoctoral fellows who aided in the editing and recording of these comments. We also extend our thanks to those agencies listed on the following page for the financial support that made this conference possible.

                                                Milton J. Cormier
                                                David M. Hercules
                                                John Lee

February 1, 1973

The organizers of the International Conference on Chemiluminescence wish to thank the following organizations for their sponsorship of the Conference:

    Environmental Protection Agency

    National Science Foundation

    Office of Naval Research

    United States Army Research Office - Durham

    United States Atomic Energy Commission

    University of Georgia

CONTENTS

List of Participants   xiii

INTRODUCTION

Mechanisms of Chemiluminescence and Bioluminescence   1
   John Lee, David M. Hercules, and Milton J. Cormier

THEORY AND GAS PHASE REACTIONS

Chemiluminescence in Gases   7
   Tucker Carrington

Transvibronic Reactions in Molecular Beams   29
   D. R. Herschbach

Studies of Vibrationally Excited Molecules by
         Infrared Chemiluminescence   43
   Ian W. M. Smith

Chemical Lasers Produced from $O(^3P)$ Atom Reactions. II.
        A Mechanistic Study of 5-μm CO Laser
        Emission from the $O + C_2H_2$ Reaction   61
   M. C. Lin

The Nitrogen Afterglow   73
   B. A. Thrush and M. F. Golde

The Air Afterglow Revisited   83
   Frederick Kaufman

Chemiluminescence Reaction Involving Metal Vapors   101
   C. J. Duthler and H. P. Broida

OXYGEN REACTIONS

The Quenching of Singlet Molecular Oxygen   111
   James A. Davidson and E. A. Ogryzlo

Chemiluminescence of Perhydroxyl- and Carbonate Radicals . 131
  J. Stauff, U. Sander, and W. Jaeschke

RADICAL IONS

Chemiluminescent Reactions of Radical-Ions 143
  Edwin A. Chandross

Cation—Anion Annihilation of Naphthalene,
        Anthracene, and Tetracene 147
  G. J. Hoytink

Chemiluminescence from Radical Ion Recombination. VI.
        Reactions, Yields, and Energies 169
  A. Weller and K. Zachariasse

Chemiluminescence from Radical Ion Recombination. VII.
        Hetero-Eximer Chemiluminescence Yields . . . 181
  A. Weller and K. Zachariasse

On the Efficiency of Electrogenerated Chemiluminescence . . 193
  Allen J. Bard, Csaba P. Keszthelyi,
  Hiroyasu Tachikawa, and Nurhan E. Tokel

ORGANIC REACTION MECHANISMS

Chemiluminescence of Diazaquinones and Related
        Compounds 209
  Karl-Dietrich Gundermann

The Chemiluminescence of Acyl Hydrazides 231
  Emil H. White and Robert B. Brundrett

Comparison of Potassium Iodide Quenching of
        3-Amino-Phthalate Fluorescence and
        Luminol Chemiluminescence in
        Aqueous Solution 245
  John Lee and I. B. C. Matheson

The Chemiluminescent Autoxidation of Reduced
        Biisoquinolinium Dications 249
  Carl A. Heller, Ronald A. Henry, and
  John M. Fritsch

Oxygen in Chemiluminescence. A Competitive Pathway
        of Dioxetane Decomposition Catalyzed
        by Electron Donors 265
  Daniel Chia-Sen Lee and Thérèse Wilson

CONTENTS

CHEMICAL MECHANISM IN BIOLUMINESCENCE

Chemical and Enzymatic Mechanisms of Firefly Luminescence 285
   W. D. McElroy and Marlene DeLuca

Model Compounds in the Study of Bioluminescence 313
   Frank McCapra, M. Roth, D. Hysert, and K. A. Zaklika

Aspects of the Mechanism of Bioluminescence 325
   T. Goto, I. Kubota, N. Suzuki, and Y. Kishi

Mechanism of the Luminescent Oxidation of Cypridina Luciferin 337
   Osamu Shimomura and Frank H. Johnson

Mechanism of Bioluminescence and Chemiluminescence Elucidated by Use of Oxygen-18 345
   Marlene DeLuca and Mary E. Dempsey

Structure and Synthesis of a Luciferin Active in the Bioluminescent Systems of the Sea Pansy (Renilla) and Certain Other Bioluminescent Coelenterates 361
   Kazuo Hori and Milton J. Cormier

Bacterial Bioluminescence. Mechanistic Implications of Active Center Chemistry of Luciferase . . 369
   J. W. Hastings, A. Eberhard, T. O. Baldwin, M. Z. Nicoli, T. W. Cline, and K. N. Nealson

Effect of Aldehyde Carbon Chain Length and Type of Luciferase on the Quantum Yields of Bacterial Bioluminescence 381
   John Lee and Charles L. Murphy

Lumisomes: A Bioluminescent Particle Isolated from the Sea Pansy Renilla Reniformis 387
   James M. Anderson and Milton J. Cormier

APPLICATIONS OF CHEMILUMINESCENCE

A Review of Experimental Measurement Methods Based on Gas-Phase Chemiluminescence 393
   Arthur Fontijn, Dan Golomb, and Jimmie A. Hodgeson

Chemiluminescence and Analysis for Trace Elements 427
  W. Rudolf Seitz and David M. Hercules

Chemical Light Product Research and Development 451
  Michael M. Rauhut

Applications of Bioluminescence and Chemiluminescence . . . 461
  H. H. Seliger

Abstracts of Short Contributions 479

Epilogue 499

Author Index 501

Subject Index 505

List of Participants

Name	Affiliation
W. Adam	Univ. of Puerto Rico
J. R. Airey	U.S. Naval Research Laboratory
J. M. Albizo	Bel Air, Maryland
R. C. Allen	Tulane University
J. Anderson	Univ. of Georgia
R. Atkinson	Univ. of California, Riverside
T. Baldwin	Harvard University
A. J. Bard	Univ. of Texas
R. Bellisario	State Univ. of New York
A. Bergendahl	IBM
W. H. Biggley	Johns Hopkins University
C. C. Black	Univ. of Georgia
P. Bolton	Univ. of California, Riverside
D. E. Brabham	Florida State University
J. T. Brownrigg	Bedford, Mass.
R. V. C. Carr	Univ. of Chicago
T. Carrington	York University
E. A. Chandross	Bell Laboratories
M. A. A. Clyne	Queen Mary College
M. J. Cormier	Univ. of Georgia
L. P. Cupitt	Redstone Arsenal
N. Curry	Univ. of Georgia
J. A. Davidson	Univ. of British Columbia
M. DeLuca	Univ. of California, San Diego
M. E. Dempsey	Univ. of Minnesota
G. D. Duda	U.S. Atomic Energy Commission
O. Dunn	York University
C. J. Duthler	Univ. of California
M. H. Eley	Univ. of Texas
M. M. Elstein	Univ. of Georgia
J. J. Ewing	Everett Research Laboratories
G. J. Faini	Univ. of Georgia
L. R. Faulkner	Harvard University
A. Fontijn	Aerochem Research Laboratories
H. Fuhr	Univ. of California, Riverside
J. F. Garst	Univ. of Georgia
B. J. Gates	Univ. of Georgia
W. L. Gamble	Redstone Arsenal
C. W. Gilliam	Univ. of Indiana
M. F. Golde	Univ. of Pittsburgh
D. Golomb	Air Force Cambridge Research Lab.
L. H. Goodson	Kansas City, Missouri
T. Goto	Nagoya University
K. D. Gundermann	Technische Universität Clausthal

T. N. Hall	Naval Ordnance Laboratory
W. M. Hardy	Univ. of Georgia
J. W. Hastings	Harvard University
R. Hautala	Univ. of Georgia
H. Held	West Germany
C. Heller	Naval Weapons Center
D. M. Hercules	Univ. of Georgia
D. Herschbach	Harvard University
J. A. Hodgeson	Environmental Protection Agency
E. K. Hodgson	Chapel Hill, North Carolina
K. Honda	Univ. of Tokyo
K. Hori	Univ. of Georgia
G. R. Husk	Army Research Office
S. E. Johnson	Aerochem. Research Laboratories
F. J. Johnston	Univ. of Georgia
F. Kaufman	Univ. of Pittsburgh
M. Kaufman	Emory University
D. Kearns	Univ. of California, Riverside
C. P. Keszthelyi	Univ. of Texas
A. King	Univ. of Georgia
R. F. Knisely	Frederick, Maryland
D. C-S. Lee	Harvard University
J. Lee	Univ. of Georgia
E. K. C. Lee	Univ. of California, Irvine
K. Legg	California State College
R. Li	Univ. of Georgia
Y. H. Li	Univ. of Georgia
M. C. Lin	Naval Research Laboratory
I. G. Lopp	Univ. of Georgia
P. A. Mallory	Univ. of Georgia
I. B. C. Matheson	Univ. of Georgia
J. C. Matthews	Univ. of Georgia
D. J. McCaa	St. Joseph, Michigan
F. McCapra	Univ. of Sussex
W. D. McElroy	Univ. of California, San Diego
R. J. Marcus	Office of Naval Research
J. W. Meduski	Univ. of California, Los Angeles
J. Modderman	Univ. of Georgia
J. G. Morin	Univ. of California, Los Angeles
C. Murphy	Univ. of Georgia
R. A. Nathan	Battelle-Columbus Laboratories
M. P. Neary	Univ. of Georgia
G. A. Neece	Office of Naval Research
J. C. Newton	Lawrence Livermore Laboratories
M. Nicoli	Harvard University
R. H. Obenauf, Jr.	Pennsylvania State University
D. A. Ogryzlo	Univ. of British Columbia
R. N. Pannell	Univ. of Georgia
H. D. Peck	Univ. of Georgia

PARTICIPANTS

G. E. Philbrook	Univ. of Georgia
E. D. Pierron	Monsanto Chemical Company
A. Pighin	Ottawa, Canada
P. E. Platt	Mobile College, Alabama
J. C. Polanyi	Univ. of Toronto
M. M. Rauhut	American Cyanamid Company
G. T. Reynolds	Princeton University
L. P. Rigdon	Lawrence Livermore Laboratories
H. Roberts	U.S. Army Research Office
R. S. Rogowski	Langley Research Center
D. F. Roswell	Loyola College
P. B. Sackett	Air Force Cambridge Research Lab.
N. G. Sansing	Univ. of Georgia
L. Sawyer	Becton-Dickinson Research Center
A. P. Schaap	Wayne State University
Z. Schelly	Univ. of Georgia
W. R. Seitz	Univ. of Georgia
H. H. Seliger	Johns Hopkins University
R. J. Seltzer	Chemical & Engineering News
T. L. Sheehan	Univ. of Georgia
O. Shimomura	Princeton University
A. Shoaf	Tulane University
D. H. Stedman	Univ. of Michigan
I. W. M. Smith	Univ. of California, Berkeley
G. P. Sollott	Frankford Arsenal
P. E. Stanley	Univ. of Adelaide
J. Stauff	Univ. of Frankfurt
R. H. Steele	Tulane University
J. I. Stevens	U.S. Army Research & Development
R. Strecker	Frankford Arsenal
D. Slawinska	Institute of Technology, Szczecin
J. Slawinski	Agricultural University, Szczecin
H. Tachikawa	Univ. of Texas
V. Tarkkanen	The Netherlands
L. Thorington	Duro-Test Corporation
B. A. Thrush	Cambridge University
A. Timnick	Michigan State University
J. R. Totter	U.S. Atomic Energy Commission
J. R. Travis	Univ. of Georgia
F. Tsuji	Univ. Southern California
J. E. Wampler	Univ. of Georgia
W. Ward	Johns Hopkins University
R. P. Wayne	Oxford University
H. H. Webster	Univ. of Indiana
A. Weller	Max-Planck Institut, Gottingen
T. C. Werner	Union College, New York
E. H. White	Johns Hopkins University
D. C. Williams	Univ. of Georgia
T. Wilson	Harvard University

N. C. Yang   Univ. of Georgia
R. A. Young   York University
K. A. Zachariasse   Max-Planck Institut, Gottingen

MECHANISMS OF CHEMILUMINESCENCE AND BIOLUMINESCENCE

John Lee, David M. Hercules, and Milton J. Cormier

University of Georgia

It would be surprising if there was a scientist active today who did not often find himself in a position of having to describe to a lay public, nay even to justify, the nature of his scholarly activity. Although the phenomenon which is the subject of this volume may be described as "Cold Light", to the layman an explanation with reference to the flash of the firefly is usually more enlightening. A gas phase kineticist might hesitate however to use this example to represent his work since he might have in mind the difficulty of having to untangle collisions between these beetles at hyperthermal velocities.

It is the purpose of this book to explain to the physicist that the light flash from the firefly is not the result of such a simple collisional mechanism and to help the more biologically oriented to understand how such mechanisms and others do lead to efficient excitation processes. It was the purpose of the Conference and of these proceedings to trace this thread of similarity from the mechanisms which give rise to chemiluminescence on reaction of atomic oxygen with nitric oxide in the gas phase for instance, all the way across to the enzyme catalyzed reactions of molecular oxygen with luciferins from marine bioluminescence organisms.

Apart from the fact that most of the reactions under consideration here do result in the emission of visible light, it is perhaps the feature that the light is "cold" that is preserved from the gas phase collision process to the enzyme reaction. This feature is the adiabatic nature of the reaction, wherein the reaction free energy is not rapidly dissipated as heat. For many exothermic gas phase reactions, the free energy is deposited in the products, which find themselves initially in energy states far removed from thermal equilibrium. Usually this is a vibrationally excited state of the product molecule and if the pressure is sufficiently low the hot molecule may have time to radiate before it is deactivated by collisions. Chemiluminescence in the infra-red then results.

Reactions that deposit their energy into electronic excited states of
the product are less common. The adiabatic condition is of course a matter
of the time scale on which the processes occur. For electronic excitation
this is not as restrictive as the vibrational case since the radiating electronic
states are much shorter lived and much less susceptible to deactivating
collisions. As a result electronic chemi-excitation is observed both in the
gas and condensed phases and until recently, these reactions received a good
deal more attention than those giving infra-red emission. This neglect also
arose from the historical fact that infra-red detectors that even approach
the sensitivity of those that have long been available for visible light
detection, are of only recent development. Science is very anthropocentric -
the direction of this field was largely guided by the sensitivity range of the
human eye.

It is the intention of this introductory chapter to trace some of the
common ideas that flow from the gas phase systems to the organic and
biochemical. This will be done mostly with reference to the contributions
that are available in this particular volume and we will therefore not
specifically cite these, except to state that the ideas brought out in this
survey largely arise from material to be found in the papers that follow[1].
We will not attempt to include every paper since some are more peripheral to
this main theme, though no less interesting in their own right.

Theoretical

The hope of what can be understood in each system depends largely
on its level of complexity. At each level there is a theoretical treatment
that is appropriate and practical. What can be expected from an analysis
of collisions between He atoms will certainly exceed that from the gas
reaction of O atoms and NO, which will be far more than for reactions of
large organic molecules in the liquid phase.

However it is valuable to compare systems of increasing complexity
to see what features survive each level of approximation. Like the adiabatic
requirement already mentioned, these features should be truly fundamental
to the chemiluminescence phenomenon.

Two concepts in particular can be given attention as the number of
particles participating in the reaction increases. They are the concept of
the electronic state and the dynamics of the interaction. For two particles
the electronic state is well defined at all separations. Almost classical
dynamics can be used to treat the collision. For the liquid phase, collision
dynamics is meaningless and needs to be replaced by the notion of an
encounter.

In the simplest case two particles recombine along a well defined
potential curve and may find themselves to be in a potential minimum which
corresponds to an excited electronic state of the complex. Radiation to the
ground state then follows. The next most complex case introduces the notion
of curve crossing. At certain separations the electronic potential becomes
"fuzzy", i.e. it is not well defined at all separations. Electronic adiabatic
correlation rules make their appearance at this stage to specify the good
quantum numbers and predict which excited states of the product may be
populated. The ground state could be completely excluded by these rules.

Nitrogen atom recombination is a well studied example of this case.

When the number of atoms is increased to three we find ourselves already at the limit of possible quantitative treatment. There is much more possibility of mixing of electronic states and their symmetry now plays an important role in determining crossing probabilities. In the even more complex cases from an a priori point of view, the formation of an excited state over the ground state would have a vanishingly small probability were it not for the influence of certain correlations, e.g. symmetry or spin, that constrain some of the reactions along a path that leads to the excited state. Correlation is one of the ideas which link together the diverse studies described in this book.

Another fundamental idea that is common to systems of increasing complexity on going from gas to liquid phase, is the adiabatic character of the electron transfer process. In addition, the spin quantum number should remain a strong correlation factor, provided in the liquid phase, spin relaxation rates do not approach those of the electron transfer process.

Ionic intermediate states appear to play a key role in many of these processes. This is evident in quenching of excited states by many good donors or acceptors[2], by the nature of the species involved in electrochemiluminescence reactions and above all, by the efficient vapor phase chemiluminescence of the classical reaction of alkali metals and halogens, the alkali metal having a particularly low ionization potential and therefore being in a good position for electron transfer.

Crossed beam experiments have now demonstrated the direct production of excited states like K^* from these reactions: $Na + KBr \longrightarrow NaBr + K^*$ and give a strong indication of the importance of the transient formation of intermediate ion pair states, e.g. $A^+ + X^- \longrightarrow (A^+ X^-) \longrightarrow A^* + X$. At close approach the coulombic energy of attraction may reduce the energy of the ion pair state below that of the excited electronic state, and thus provides a convenient mechanism of crossing between upper and lower states. More important, ion pair mixing can also enhance singlet-triplet conversion, and in this regard it deserves mention that ions are involved in the electrochemiluminescent annihilations, as well as in most of the examples of efficient organic chemiluminescence and bioluminescence reactions.

Ion pair mixing may be invoked to allow violation of the 3:1, triplet: singlet product ratio expected on pure spin statistical reasons in an electron transfer, and thus provides a means whereby the very high efficiencies of bioluminescence could be achieved. The firefly in the ointment has limited the generalization of most other theoretical excitation processes before this time.

Results

The most active research on a particular chemiluminescence reaction will be generally found at any one time in one of four traditional areas. Briefly these are the identification of reactants, products, intermediates and measurement of quantum efficiencies. Since these reactions often may be highly exothermic, many reaction paths and species may need to be considered before the particular reaction that gives rise to excitation can be sorted out.

Often this reaction might be quite minor and the overall chemistry is then irrelevant. Sometimes even in the gas phase, it is hard to be sure of the nature of the reacting species, particularly when they have been generated by electric discharge.

In addition to the traditional categories, two more recent approaches offer much promise for future work. These are the molecular beam techniques, where the energies of the reacting species can be closely controlled, and infra-red chemiluminescence, where the states of initial energy deposition in the product can be identified. Little promise of application of these techniques to the liquid phase is apparent however.

There is an enormous (traditional) literature on the gas phase chemiluminescence of N atom and NO + O reactions. At first sight the N atom recombination should be a two body process but experimentally both two body and three body processes are observed. Almost all of the possible electronic states are populated and a number of mechanisms such as the inverse of predissociation, play equally important roles. Spin conservation is also an important factor. If hot ground states are formed they do not seem to be able to populate other electronic states.

Although reaction of NO + O would at first sight appear to be a three-body system, the majority of the recombination events occur through the more complex NO + O + M reaction. As in any chemiluminescence, one always hopes that someone else can provide a thorough characterization of the product fluorescence. Unfortunately the NO_2 fluorescence "defies analysis" even at $1.5°K$. There is in addition, a fluorescence lifetime anomaly - it is much too long. It is suggested that there exists considerable interelectronic state mixing with ground states in highly excited vibrational levels, leading to both these difficulties.

A considerable amount of kinetic and spectroscopic data allows the choice of the NO + O reaction mechanism to come down to two possibilities: one is the above three body recombination and the other is a two body formation of a NO_2^* collision complex followed by vibrational, electronic and radiative processes. These models, as usual, allow a large number of adjustable parameters and therefore it is difficult to choose between them. Even though the reaction is complex, most of the events lead to the product emitting state.

In solution direct electron transfer is obtained in the annihilation of cation and anion radicals, which are generated electrochemically in a non-polar solvent. Recombination produces light

$$A^- + D^+ \longrightarrow \longrightarrow {}^1A^* + {}^0D \ .$$

The situation can become quite involved. Depending on the energetics of the process the state $^1A^*$ may be produced directly, via dissociation of an intermediate hetero-eximer $^1(AD)^*$, or from triplet-triplet (^3A) annihilation. Further experimental complications arise from the instability of the radical ions and their strong tendency to quench excited states.

In the case that A and D are the same, some of the details of the annihilation process can be examined. The ion pair is considered as a starting point in the encounter complex, where they are separated by two solvent layers, a distance of $6 - 10 Å$. The energy of the complex can be estimated from the oxidation-reduction potentials measured polarographically in that solvent. To this is then added the energy that is necessary for solvent rearrangement and the coulombic term, which come into effect as the pair are brought into contact. From consideration of correlations between initial and final states, qualitative prediction about relative degrees of population of the final states can be made.

A purely thermodynamic approach yields useful information where A and D are different and various energy relations can be produced by varying their nature. For the energy sufficient condition, where there is more than enough released energy to form the product singlet state directly or from dissociation of the hetero-eximer, $^1(AD)^*$, an interesting correlation is observed between the excitation quantum yield and a term representing the excess energy of reaction. A "resonance" behavior is found where the maximum quantum yield occurs when the energy is just sufficient.

A relationship between the energy of emission and the quantum yield of excitation is also seen in the chemiluminescent oxidation of a series of cyclic hydrazides. There is a hint of a resonance in this data which by analogy would suggest an electron transfer type excitation process here also[3].

In electrochemiluminescence a most important recent development has been the measurement of absolute quantum yields. These are difficult measurements both for reason of the awkward optical configuration and the preponderance of quenching and electrochemical side reactions. Nevertheless general agreement now exists that some of these more efficient reactions have excitation quantum yields around 0.1. It is interesting to note that estimates for the excitation effiency of some of the more efficent peroxy radical annihilation reactions are within an order of magnitude of this value[3].

Bioluminescence quantum efficiencies for a long time exceeded any others in solution and it was thought that the enzyme might play a special role in favoring a route to the excited state. This is now known not to be true, since several organic reactions are now found to have quite high quantum yields. The high quantum yields and the involvement of molecular oxygen in most bioluminescence reactions led to the proposal that a dioxetane, a four membered cyclic peroxide was a key intermediate. This proposal was preferred over an electron transfer, since no efficient one electron processes were realized until quite recently and it was thought that spin statistics would limit a singlet excitation efficiency to a maximum of 0.25. This was a particularly important consideration in the case of the firefly bioluminescence, where the singlet excited state is populated with unit efficiency.

If the dioxetane intermediate were to decompose in a concerted fashion, then application of the Woodward-Hoffmann rules of conservation of orbital symmetry by analogy to the decomposition of cyclobutanes, might predict a tendency to favor an excited state route.

Two lines of evidence appear not to favor this proposal. A number of simple dioxetane structures have now been synthetized and are found to decompose to yield triplets, but little or no singlet states, even though they have more than sufficient exothermicity. If the dioxetane is an intermediate it should contain oxygen incorporated from molecular oxygen

$$O_2 + >C{=}C< \longrightarrow \longrightarrow >\underset{O-O}{\overset{|\;\;|}{C-C}}<$$

In the case of the firefly mechanism, decomposition of an intermediate dioxetanone

$$>\underset{O-O}{\overset{|\;\;|}{C-C}}=O \longrightarrow >>C{=}O + CO_2$$

should yield CO_2 containing one oxygen atom from molecular oxygen, detectable by labelling with oxygen-18.

The oxygen label comes from water and not from O_2. The sea-pansy (Renilla) bioluminescence reaction gives similar results except that both the oxygens in CO_2 are labelled from H_2O. To further complicate matters partial molecular oxygen labelling does occur in yet another type of reaction, the Cypridina bioluminescence.

The high light yield reactions of bioluminescence thus appear ripe for new theoretical proposals.

---

1. This Volume.

2. D. Rehm and A. Weller, Israel J. Chem. 8 259 (1970).

3. R.B. Brundrett, D.F. Roswell and E.H. White, J. Amer. Chem. Soc. 94 7536 (1972).

4. V.A. Belyakov and R.F. Vasilev, Photochem. Photobid. 11 179 (1970).

Chemiluminescence in Gases

Tucker Carrington

York University

Toronto, Ontario, Canada

Chemiluminescence in the gas phase has been observed in systems spanning a wide range of complexity, and theoretical approaches to chemiluminescence are equally varied (Carrington & Garvin 1969, Carrington & Polanyi 1972). The simplest process is the radiative recombination of two atoms on a single potential curve. This can be treated with near-ultimate rigour, though these processes have not attracted wide interest. At the other extreme in the range of complexity, are processes involving polyatomic reactants. Here only qualitative treatments are possible at present, and it seems unlikely that anything quantitative will be possible, or even interesting, in the forseeable future.

I will discuss a sequence of four types of chemiluminescent systems, of increasing complexity. For each type, the theoretical approach that is most appropriate and practical will be outlined, but the main interest is in the relationship among these approaches. As one moves toward increasing complexity, certain basic concepts or good quantum numbers lose their validity, and the number of well-defined observables decreases. Apart from this loss of detail which must occur in principle, there is of course an additional loss due to practical difficulties in carrying out calculations or experiments on polyatomic systems. The main purpose of the paper is to trace out certain "lines of descent" showing the loss of rigour and detail, and the rise of qualitative concepts, in treatments of successively more complex systems. Any common elements which survive during this descent should be truly fundamental to chemiluminescence. A study of a variety of approaches to chemiluminescence should make it possible to examine a process at several different levels of rigour. A comparison of treatments from different points of view is nearly always more fruitful than a treatment locked to a single approximation.

The discussion of each type of process takes the following form. First the valid concepts and good quantum numbers are presented, as illustrated by a representative reaction. Then the theoretical methods are outlined, followed by a brief discussion of the kinds of results which are obtainable theoretically and observable experimentally.

I. Radiative Recombination of Atoms on a Single Potential Curve

In this simplest system, a great many concepts and quantities are well-defined. The electronic state of the emitter is well-defined over a large range of internuclear separation, since there are no nearby states of the same parity with which it can interact. This means that the Born-Oppenheimer separation will be an excellent approximation, and the system can be described accurately using continuum radial wave functions and an r-dependent transition moment. As well as being well-defined in principle, parts of the potential curves involved are often known experimentally from analysis of band spectra. Orbital angular momentum is a good quantum number throughout the collision, since we have two-body central forces. The angular momentum, the initial relative kinetic energy, and the potential function determine the collision completely. To get an observable result, one simply averages over a distribution of initial conditions.

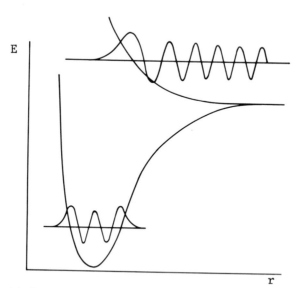

Fig. 1. Potential curves illustrating radiative recombination on a single repulsive curve.

Radiative recombination on a single potential curve can be treated classically or semiclassically from two different points of view. The simplest case is illustrated in Fig. 1. Regarding the recombination as a collision process (Bates 1951), the probability of radiation in a collision with specified initial relative velocity g and impact parameter b is

$$P(g,b) = \int A(r) dt. \qquad (1)$$

The quantity $A(r)$ is the rate coefficient for spontaneous radiation for a pair of atoms with separation r. Integration is carried out along the collision trajectory, determined by the upper potential curve. The probability of radiation in a specified frequency range has the same form, but integration is carried out only over those parts of the trajectory from which radiation in the given frequency range can be emitted, assuming a vertical transition between the upper and lower potential curves. The expression (1) is to be averaged over a Maxwellian distribution of g and b. The emission rate coefficient $A(r)$ is of course the key to the problem. In the crudest approximation one could set $A(r)$ proportional to the product of the electronic transition moment $|R_e(r_e)|^2$ at the equilibrium separation in the lower state and a $\nu^3(r)$ factor determined by the separation of potential curves at r.

An alternative, and simpler classical approach, particularly appropriate for repulsive collisions, is to assume an equilibrium between free atoms and pairs with specified separation (Palmer 1967). Again $A(r)$ is the crucial quantity. It may be approximated as suggested above, calculated theoretically, or in some cases, derived from absorption measurements for the reverse process, photodissociation (Palmer 1967). The rate coefficient for recombination emission is

$$k(\nu) d\nu = A(r) \, K(r) \, \frac{dr}{d\nu} \, d\nu. \qquad (2)$$

Here $K(r) dr$ is the equilibrium constant for pairs with separation in range dr about r,

$$K(r) dr = 4\pi \, r^2 dr \, e^{-U(r)/kT}. \qquad (3)$$

$U(r)$ is the upper state potential function and we have assumed for simplicity unlike atoms with unit statistical weight. This classical equilibrium treatment could be extended to include the effect of orbiting resonances on the equilibrium constant (Jackson & Wyatt 1971).

An example of this simplest case of chemiluminescence is the process (Smith 1967):

$$He(1^1S) + He(2^1S) \rightarrow He_2(A^1\Sigma_u^+) \rightarrow 2He(X^1\Sigma_g^+) + h\nu \; (600 \text{ Å bands}) \qquad (4)$$

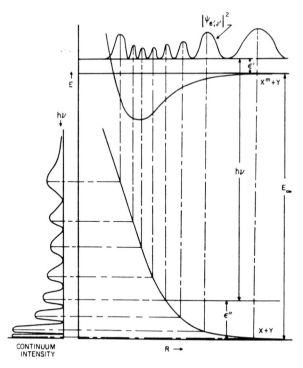

Fig. 2. Radiative recombination in He, Mies & Smith (1966).

The 2^1S state is highly metastable, but it interacts with the ground state atom on the strongly attractive $^1\Sigma_u^+$ potential curve (dissociation energy 58 kcal/mole) for which dipole radiation to the $^1\Sigma_g^+$ ground state of the molecule is allowed, see Fig. 2. The process (4) is properly considered chemiluminescent since the two atoms interact by strong chemical forces.

The classical treatments are reasonably satisfactory when the upper state is repulsive, but they ignore the wave aspects of scattering and cannot predict the periodic structure in the 600 Å band emission from reaction (4) (Smith 1967). Quantum mechanically, one uses Born-Oppenheimer electronic wave functions to calculate the electronic transition moment $|R_e(r)|^2$. Since this is a two-body process, it must occur in the continuum of the upper state potential, and is properly described by continuum radial wave functions, as indicated in Figure 2. The simplest way to predict the structure of the emission spectrum is to use the reflection method, as indicated in the figure (Mies & Smith 1966). This assumes that the continuum intensity is proportional to the square magnitude of the upper state radial wave function. The periodic maxima in the emission continuum correspond to the antinodes in the wave function. The form of this function is very similar to that of the wave function for the highest bound vibrational state, but with nodes somewhat more pushed in toward smaller r values. The spectrum calculated for a particular upper state continuum wave function $\varphi_J(\epsilon)$ must be averaged over a thermal distribution of ϵ and J. For the He_2 A state, with its large vibrational spacing and rotational constant, there is little smearing-out of the structure (Mies & Smith 1966, Mies 1968). The process (4) has also been treated theoretically by Sando (1971), who has emphasized the importance of orbiting resonances, i.e. quasi-bound states with energy above the dissociation limit, but below the barrier in the upper state potential curve (Guberman & Goddard 1972).

The considerable amount of structure observed in the "600 Å band" continuum contains a lot of information about the two potential curves involved. If one curve is known, the other can be determined (Chow & Smith 1971, Sando 1972).

Two-body radiative recombination on a single potential curve also occurs in Ar, where the metastable $^3P_2$ state emits in collision with a ground state atom (Michaelson & Smith 1970). Another example is

$$O(^1S) + Ar(^1S) \rightarrow OAr(^1\Sigma_g^+) \rightarrow O(^1D) + Ar(^1S) + h\nu(5577 \text{ Å}) \quad (5)$$

studied by Corney & Williams (1972), Filseth, Stuhl and Welge (1970) and Hampson & Okabe (1970). Here, at least in the low pressure, two-body limit, the mechanism seems to be that of eqn. (4). Unfortunately, nothing is known of the upper or lower potential curves for OAr, but there is some spectroscopic information for OXe (Cooper, Cobb & Tolnas 1961).

A somewhat different example is (Bates 1951, Bain & Bardsley 1972).

$$C(^3P) + H(^2S) \rightarrow CH(B^2\Sigma^-) \rightarrow CH(X^2\Pi) + h\nu \quad (6)$$

which is of some astrophysical interest. The calculated rate coefficient is of the order 10^{-18} cm^3 sec^{-1} at 100°K.

This is not quite as straightforward as reaction (4) because the atomic states correlate with four molecular states, two quartets as well as the doublets of eqn. (6). The radiative recombination of halogen atoms is a two-body process under some conditions, and has been interpreted in terms of the potential curves involved (Palmer & Carabetta 1968).

II. Radiative Recombination of Atoms via "Curve Crossing"

The previous discussion dealt with recombination occurring on a single potential curve for a well-defined electronic state.

In the present case this simple situation is replaced by one in which two potential curves "cross". This crossing, and the consequent local ambiguity in the concepts of potential curve and electronic state, are the crucial factors which must dominate any treatment of this type of chemiluminescence. Most of the discussion will be rather general, but it may help to have a specific example in mind. The radiative recombination

$$N(^4S) + O(^3P) \rightarrow NO(a^4\Pi) \rightarrow NO(C^2\Pi) \rightarrow NO(X^2\Pi) + h\nu(\delta \text{ bands}) \quad (7)$$

and its inverse, predissociation, have been studied by fluorescence techniques (Callear & Pilling 1970), by direct observation of the recombination emission (Groth, Kley & Schurath 1971, Ackerman & Miescher 1969) and by using the recombination emission as a light source for the photodecomposition of NO (Mandelman, Carrington & Young 1972).

In the theoretical description of a process such as (7), the Born-Oppenheimer separation of electronic and nuclear motion is valid almost everywhere, but fails in the crucial region of the intersection. The simplest thing to do is to assume classical motion of the nuclei, and try to calculate the probability of a transition from one curve to another as the system passes through the crossing point. Again, we have two-body central forces, and the orbital angular momentum of the collision is a good quantum number. The classical path and related semiclassical approximations have recently been discussed by Delos & Thorson (1972), Miller & George (1972), Nikitin (1968), Eu (1971, 1972) and Child (1972). Because the interaction of two electronic states is involved, it is necessary to be very careful in defining the two electronic basis states in terms of which the process is to be discussed. The choice of basis states becomes clearer if one first formulates the collision problem using arbitrary states, φ_1 and φ_2, depending on the internuclear separation r as a parameter. It will then appear that there are several natural ways to choose these basis states so as to simplify the solution of the problem in limiting cases.

Along the classical path, the internuclear separation is a known function of time, and the electronic states can be thought of as depending on the parameter t rather than r. The initial relative velocity and impact parameter distinguish one trajectory from another. For a given classical trajectory we represent the time dependent state of the system by

$$\psi(t) = b_1(t)\,\varphi_1(t)\,e^{-i\int^t \epsilon_1(t')dt'} + b_2(t)\,\varphi_2(t)\,e^{-i\int^t \epsilon_2(t')dt'} \quad (8)$$

where ϵ_1 and ϵ_2 are diagonal elements of the electronic Hamiltonian in the arbitrary basis. Substitution of this into the time-dependent wave equation

$$H(t)\psi(t) = i\hbar \partial \psi(t)/\partial t \qquad (9)$$

leads to a pair of coupled equations for the amplitudes b_1 and b_2 for finding the system in one or the other of the basis states:

$$i\hbar db_1/dt = [V_{12} - \dot{T}_{12}] b_2 \, e^{-i\int^t \epsilon(t')dt'}$$
$$i\hbar db_2/dt = [V_{21} - \dot{T}_{21}] b_1 \, e^{i\int^t \epsilon(t')dt'}. \qquad (10)$$

In these equations, V_{ij} are the off-diagonal elements of the electronic Hamiltonian and $\dot{T}_{ij}$ are the matrix elements of the operator $\dot{T} \equiv i\hbar \, d/dt$, in the arbitrary basis. In the phase factors, $\epsilon \equiv \epsilon_2(t) - \epsilon_1(t)$. Equations (10) say that transitions between the basis states are induced by off diagonal elements of the Hamiltonian and the $\dot{T}$ matrix. The operator $\dot{T}$ is equivalent to the momentum $p = -i\hbar \partial/\partial r$ since r is a known function of t. There are now two obvious ways to choose the basis states. We can diagonalize either V or $\dot{T}$, but not both, since they do not commute. Diagonalizing V gives the <u>adiabatic</u> basis states, with potential curves which do not cross. These should be appropriate for slow collisions and large interactions. If instead we choose the electronic basis so as to diagonalize $\dot{T}$, or the equivalent momentum matrix, we have the <u>diabatic</u> basis states, which cross smoothly without a sudden change of character (Smith 1969, Delos & Thorson 1972, Andresen & Nielsen 1971). These should be most appropriate for fast collisions and weak interactions. It may be advantageous in some cases to choose some other basis set, between the two limiting types. The term diabatic is sometimes used rather vaguely to indicate states whose potential curves cross smoothly, without abrupt change of electronic configuration or orbital occupancy, even though they do not exactly diagonalize the $\dot{T}$ matrix. (Evans, Cohen & Lane 1971, Cohen, Evans & Lane 1971). These might better be called crossing states. They correspond to the ordinary potential curves derived from spectroscopic measurements on unperturbed lines or bands, and frequently form the most useful basis states. For the adiabatic states, on the other hand, the potential curves do not cross and there is a change of electron configuration or orbital occupancy, sometimes quite abrupt, as the system goes through the region of avoided crossing.

It is helpful to think about radiative recombination via curve crossing in terms of its inverse, predissociation. Here it is convenient to choose electronic states with smoothly crossing potential curves. One of these, like $NO(C^2\Pi)$, supports bound vibration-rotation states while the other is either repulsive, or, as with $NO(a^4\Pi)$, the crossing occurs above its dissociation limit.

From this point of view we have the interaction of a bound state, i.e. v, J, in $C^2\Pi$, with a continuum state of the same J on the outer potential curve. Because of this interaction, the bound state is not truly stationary and does not have a well-defined energy. Instead, it has a width Γ, proportional to the strength of the coupling with the continuum and inversely proportioned to its lifetime. (Fano 1961, Harris 1963, Kovacs 1969).

For a kinetic or phenomenological description of predissociation or radiative recombination, it is sufficient to know the level widths, corresponding to total decay rates, and the branching ratios or partial widths for the radiative and dissociative modes of decay. The level widths may depend strongly on energy, and also on the rotational quantum number (Ramsay & Child 1971, Julienne 1971, Comes & Schumpe 1971). When the broadened levels do not overlap, i.e. when rotational structure is resolvable in the absorption or emission spectrum, the rate of recombination into specific angular momentum states is observable. The radiative recombination can be treated simply as a resonance scattering process, using theoretical, experimental or empirical level widths. The simplest procedure is to use the Breit-Wigner form for the cross section (Goldberger & Watson 1964) and average this over collisions with an equilibrium distribution of relative kinetic energy and angular momentum (Carrington 1972b). For the common case in which radiating pairs are in equilibrium with separated atoms, the results agree with treatments which start from this assumption (Golde & Thrush 1972).

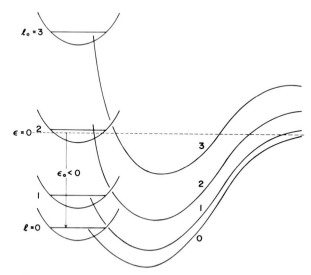

Fig. 3. Crossing potential curves for several values of the angular momentum. The four levels shown are rotational levels in the same vibrational state.

In recombination on a single repulsive potential curve, the angular momentum of the collision frequently does not play a dominant role. In the curve crossing case, the situation can be quite different, as indicated in Fig. 3. There are three types of levels in the inner potential curve. First, the levels below the dissociation limit of the outer curve, having negative energy, are inaccessible to two-body recombination, and have zero width. (Levels 0 and 1 in the figure). Second, the levels above the dissociation limit (positive energy) but below the rotational barrier in the corresponding outer curve will also be inaccessible in most cases (Level 2). Levels of the third type are above the rotational barrier in their outer potential and account for virtually all of the two-body recombination (Levels 3,. . . .). If the lowest level above the dissociative limit has a high J, many levels can be effectively excluded. For example, in the process

$$N + N \rightarrow N_2(^5\Sigma_g^+) \rightarrow N_2(B^3\Pi_g), \tag{11}$$

the level nearest the dissociation limit is $v = 12$, $J = 24$ (Dieke & Heath 1959), but this is inaccessible to two-body radiative recombination, presumably because of the large centrifugal repulsion term in the effective potential for the outer curve. Predissociation is observed to set in only for $J \geq 33$. On the other hand, all the rotational levels in $v = 13$ lie above the dissociation limit and are accessible to two-body recombination (Becker, et al. 1972, Golde & Thrush 1972).

The reactions (7) and (11) are probably the best known cases of two-body radiative recombination via curve crossing. A recent calculation (Julienne & Krauss 1971) for

$$H + O \rightarrow OH(X^2\Pi) \rightarrow OH(A^2\Sigma^+) \tag{12}$$

makes the point that an intersection of potential curves is not really necessary for a transition from one electronic state to another. They simply have to come close enough to be mixed by some appropriate interaction.

A process closely related to the subject of this section is

$$Na^+ + O^- \rightarrow NaO \rightarrow Na^*(^2P) + O \tag{13}$$

which involves a crossing at very long range between the ionic curve on which the collision begins and the nearly flat curve correlating with $Na^* + O$ (Weiner, Peatman & Berry 1971). This is the gas phase analog of the radiative recombination of radical ions in solution, discussed elsewhere in this volume.

III. Three-atom systems

With this step in the sequence, we increase the number of atoms from two to three, which increases the complexity of the system by at least a factor of ten. This is the last step in the sequence which we can hope to understand completely, and that will be well in the future. The potential energy in each electronic state is now a function of three configuration coordinates, rather than one, a surface $V = f(r_1, r_2, r_3)$ in the four dimensional space V_1, r_1, r_2, r_3. It is this increase in the dimension of the configuration space which marks this step in the sequence. In contrast with previous cases, there is virtually never any experimental information on excited state potential surfaces, except perhaps in the immediate neighbourhood of the equilibrium conformation, and some spectroscopic information about energies of excited states in the conformation of the ground state. Theoretical calculations of adiabatic surfaces are available in many cases of interest (Krauss 1970) including some extensive configuration interaction calculations. Some recent results are for NO_2 (Gangi & Burnelle 1971a), O_3 (Hay & Goddard 1972, Peyerimhoff & Buenker 1967) N_2O (Peyerimhoff & Buenker 1968), H_2O (Flouquet & Horsley 1972, Gangi & Bader 1971), and N_2O^+ (Pipano & Kaufman 1972). These calculations are usually for a limited number of conformations of the nuclei, and it is often hard to judge their accuracy.

Experimental or theoretical information about the dependence of the transition moment on conformation of the nuclei is completely lacking. Furthermore, the orbital angular momentum of the collision is no longer a good quantum number, but merely serves as an initial condition. One of the most dramatic differences between diatomic and triatomic systems is the effect of displacement of the nuclei on the symmetry of the system. The vibrations of a diatomic molecule do not alter its axial symmetry. In the triatomic case, on the other hand, vibrations or displacements which distort the symmetry are possible, in fact inevitable. Two adiabatic potential surfaces can intersect if the electronic states have different symmetry, for example A_1 and B_2 in C_2v (isocelles) conformations. When the molecule is distorted to lower symmetry, the distinction between A_1 and B_2 disappears, since both these states are symmetric with respect to the plane of the molecule, the only remaining symmetry element. The two electronic states can now interact and repel one another, avoiding the intersection. Two triatomic surfaces can intersect in a variety of ways, i.e. conical, glancing, asymptotic (Herzberg & Longuet-Higgins 1963, Carrington 1972a). It is in the neighbourhood of these intersections that vibronic interactions can allow transitions from one electronic state to another. The density of vibronic states will be far higher for three atoms than for two, and perturbations linking two electronic states will be correspondingly more important.

The usefulness of classical trajectory studies for reactions on a single potential surface (Carrington & Polanyi 1972) invites extension of this method to situations in which two surfaces interact. One first identifies regions in the V, r_1, r_2, r_3 space where the interaction is strong. A classical trajectory which enters such a region is then considered to hop from one surface to the other with a probability determined primarily by its velocity (Preston & Tully 1971, Tully & Preston 1971). Results are averaged over an equilibrium distribution of collisions. This procedure, which applies quantum mechanics only in certain crucial regions of space, has been examined and extended from a uniformly semi-classical point of view (Miller & George 1972) in which the classical action is used to calculate elements of the scattering matrix. Again, attention focuses on the neighbourhood of surface intersections.

Completely quantum mechanical treatments of reactions on a single potential surface have been carried out (Light 1971), but it seems unlikely that these methods will be useful in the near future for problems in which interacting surfaces are involved.

The theoretical contributions to understanding chemiluminescence in three-atom systems are almost exclusively qualitative at present. The aim is to identify the emitting state, and the mechanism by which it is populated. This often involves locating or postulating an accessible intersection which can connect a precursor state with the state from which emission is observed. On the other hand, when the collision occurs directly on the surface of the emitting state, various competing radiationless transitions will be possible as well, so that intersections are important in this case also.

Here are a few examples of the qualitative interpretation of chemiluminescence in three-atom systems, in terms of potential surfaces and their intersections. The radiative recombination

$$H(^2S) + NO(^2\Pi) \rightarrow HNO(^1A'') \rightarrow HNO(^1A') + h\nu \ (> 6000 \ \text{Å}) \quad (14)$$

is observed to have a negative temperature coefficient. It is second order in its dependence on reactant concentrations, but almost surely involves a third body (Clyne & Thrush 1962). The $^1A''$ state correlates with reactants over a symmetry imposed barrier, indicating that a precursor state is necessary to provide a low energy path. The $^3A''$ state seems to be the only candidate, but two Hartree-Fock calculations have shown that this state is unsuitable, either because it is repulsive at large distances (Krauss 1969), or because there is no suitable crossing with the emitter, $^1A''$ (Salotto & Burnelle 1969). Remember, however, that the reaction requires a third body, so an interpretation in terms of potential surfaces for an isolated HNO molecule may be inadequate.

The most famous radiative recombination in the three-atom domain is

$$O(^3P) + NO(^2\Pi) \rightarrow NO_2^*(^2B_2, {}^2B_1) \rightarrow NO(^2A_1) + h\nu(>3900 \text{ Å}) \quad (15)$$

Kaufman (1973) has given an excellent appraisal of the state of knowledge of this reaction, so only a few comments will be given here. Only the two-body process (Becker, Groth & Thran 1972), really belongs in this section, since the three-body recombination involves four or more atoms. In the two-body recombination, as with the processes in sections I and II, there is no collisional stabilization; only radiation or redissociation. The collision follows a constant energy path until radiation is emitted. If, as seems likely, the interaction is attractive, relative velocities during the interaction will be considerably higher than thermal. From a classical trajectory point of view, the system will have little tendency to seek out the equilibrium (C_{2v}, isocelles) conformation on the initial surface, and will spend most of its time gyrating in C_s, with no symmetry element but the plane of the molecule. Electronic states such as A_1 and B_2, which have the same symmetry with respect to this plane, can interact whenever they approach one another in energy. Correlations, and assignment of the emitting state in C_{2v} symmetry, will have little value when these interactions are strong. The situation is somewhat like that of section II, the collision occurs in the continuum of the initial electronic state, possibly the ground state. Bound vibronic levels in excited states interact with this continuum, providing a mechanism by which the emitting state can be populated. The radiative transition will be more-or-less vertical, but there will be no strong tendency to populate the equilibrium conformation of the ground state. In this respect, the radiative recombination differs from its nominal inverse, predissociation, which starts from states near the vibronic ground state.

Potential surfaces for excited states of NO_2 have been calculated, including configuration interaction (Gangi & Burnelle 1971a) and their implications for the absorption spectrum, fluorescence, photodissociation, and chemiluminescence have been discussed (Gangi & Burnelle 1971b, Busch & Wilson 1972, Kaufman 1973). A state correlation diagram is shown in fig. 4.

Only a few atom exchange reactions

$$A + BC \rightarrow \begin{cases} AB^* + C & (16) \\ AB + C^* & (17) \end{cases}$$

are known to produce chemiluminescence in three-atom systems. No case has received serious theoretical treatment, but it is often possible to make a plausible identification of the important electronic states of the triatomic intermediate ABC (Herschbach

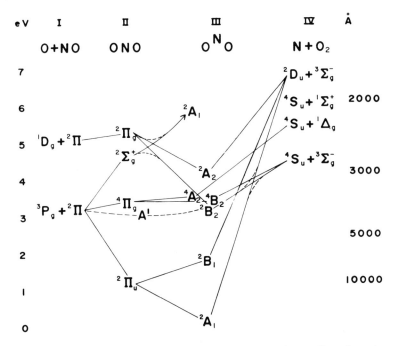

Fig. 4. State correlation diagram for NO_2. See Gangi and Burnelle (1971a).

1973). The reaction (Struve, Kitagawa & Herschbach 1971)

$$Cl(^2P_u) + Na_2(^1\Sigma_g^+) \rightarrow NaCl(^1\Sigma^+) + Na^*(^2P_u) \qquad (18)$$

is enforced by orbital symmetry, and its large cross section (10-100 $Å^2$) has long been explained in terms of an interaction with an ionic surface which is attractive at large distances.

The product of a chemiluminescent reaction will generally be vibrationally and rotationally excited. It is sometimes possible to make a qualitative prediction about internal excitation, based on the shapes of the potential surfaces involved. This has been done for a related process, the photodissociation

$$H_2O(\tilde{X}\ ^1A_1) \xrightarrow{h\nu} H_2O(\tilde{B}\ ^1A_1) \rightarrow H(^2S) + OH^*(^2\Sigma^+) \qquad (19)$$

in which most of the available energy goes into rotation of OH^* (Carrington 1964, Welge, Filseth & Davenport 1970). The potential surface for the $\tilde{B}$ state falls rapidly in energy as the bond angle opens toward $180°$. This is because it must join the ground state surface in a conical intersection in the linear conformation

(Floúquet & Horsley 1972, Gangi & Bader 1971). In this conformation the system will dissociate to ground state OH($^2\Pi$), but some fraction of the system will remain on the upper sheet of the surface, giving OH*($^2\Sigma^+$). The large excitation of the bending vibration associated with the straightening out of the B state is converted to rotation of OH* as the H atom recedes.

IV. Systems of Intermediate Complexity

The next step in the sequence of chemiluminescent systems of increasing complexity brings us to cases where somewhat more than three atoms are involved. A typical example is (Clough & Thrush 1967, Redpath & Menzinger 1971)

$$O_3(^1A_1) + NO(^2\Pi) \rightarrow O_2(^3\Sigma_g^-) + NO_2(^2B_2, {}^2B_1). \qquad (20)$$

An even simpler case, which falls somewhere between types III and IV, is the three-body analog of reaction (15),

$$O(^3P) + NO(^2\Pi) + M \rightarrow NO_2^*(^2B_2, {}^2B_1) + M \qquad (21)$$

If we continue in the spirit of section III, we have, for a five atom system, nine internal degrees of freedom, and the potential function $V(q_1, q_2, \ldots q_9)$ are surfaces in a ten-dimensional space. Furthermore, the density of electronic states in a five-atom system, at energies in the neighbourhood of 4 or 5 eV (100-125 kcal/mole) may be of the order of 10 per eV (4 states in a range of 10 kcal/mole). More important is the density of vibronic states (Forst 1971). At an energy of 2 or 3eV (50-75 kcal/mole) above the ground state, the vibrational level density may well be of the order $10^4/cm^{-1}$ or 10^5/kcal/mole. For a typical triatomic molecule at the same energy, the density of states is only of the order $0.1/cm^{-1}$.

As a result of the high level density in five-atom systems, the concept of electronic states is further degraded, as compared to the triatomic case, by strong vibronic and coriolis interactions, and the corresponding high efficiency of internal conversion (Jortner & Rice 1969, Henry & Kasha 1968, Evans & Rice 1972). In many cases, the only reasonably good quantum number is spin. In discussing reactions such as (20), it is hardly practical to work with specific vibrational or radial eigenfunctions, or to average properly over an initial distribution of internal states and collision parameters. While it was just possible to envision a full treatment of the three-atom problem such a goal is practically unattainable, and in fact of little interest for bigger systems.

We must look for approximations which somehow reduce the number of degrees of freedom to be considered, and are physically appropriate to the process at hand. Reaction (21), with N_2 as third body, is strictly speaking a five-atom system, but the N_2 is probably not electronically or chemically involved. It is sufficient to treat the process in the same way as (15), but with the important difference that energy transfer to N_2 makes it possible to produce NO_2 in states with less than the initial energy of the $O + NO$ collision. These are negative energy states in the language of Section III. We no longer have a constant energy process occurring in the continuum of the initial potential surface. Instead, a range of energies is involved, with different vibronic interactions at every energy. From this point of view, the process (21) is considerably more complex than (15), since we are no longer dealing with a single energy. On the other hand, there may now be some tendency to form NO_2^* in low energy, near-equilibrium conformations where vibronic interactions will be weaker, and symmetry arguments, i.e. correlations in C_{2v}, will be of greater usefulness.

In reaction (20) an oxygen atom is transferred in the collision, having brought its third body with it, so to speak. The interaction is clearly much stronger than in (21). Nevertheless, the emitter in (20) is triatomic. Once it is formed, it must behave much like the product in (21), except that a smaller amount of energy is available. The molecule emits light because of or in spite of various competing radiationless processes. The dynamics of the full five-atom system determines the distribution function for total energy of the emitting molecule, and the distribution of states initially populated. What happens next is a triatomic problem.

Orbital symmetry plays an important role in reaction (20). Since the Π state of NO is doubly degenerate, the reactants can correlate with two electronic states of the intermediate, one of which may lead directly to electronically excited NO_2. This is not possible for the non-degenerate reactants $O_3 + SO$ for example (Thrush, Halstead & Mckenzie 1968).

The role of symmetry is probably quite different in reactions, like (20), with an activation energy considerably greater than kT, as compared with those such as (15) having zero or negative activation energy. In a reaction with high activation energy, a large fraction of the reactive collisions will have energy just above threshold. In a classical path picture, they will pass very close to the lowest saddle point connecting reactants and products. If this point corresponds to a symmetrical conformation, orbital or state correlations in the appropriate point group will be useful, and many degrees of freedom of the nuclei can be ignored. On the other hand, a reaction with zero activation energy will have relatively little tendency to favour symmetrical low energy conformations, and constraints derived from symmetry will be less useful.

It seems appropriate to choose, as the last step in this sequence of increasing complexity, a reaction which is, at the same time, the first step in another sequence of chemiluminescent reactions. The thermal decomposition of dioxetanes in the gas phase (White 1972) is a natural stopping point for our discussion. The reaction in solution (Kopecky & Mumford 1969, White, Wiecko & Roswell 1969) is equally appropriate as a starting point for discussion of organic and bioluminescence, since it seems to be a prototype of chemiluminescent processes in many larger molecules (McCapra 1970, 1973, Wilson 1973).

What one expects of a theoretical discussion of the decomposition is a prediction (or rationalization) of the production of an electronically excited product. One might hope, in additon, for some indication about the activation energy and the order of magnitude of the quantum yield. The simplest theoretical thing to do is to use the approach of Woodward and Hoffman (Woodward and Hoffman 1968, McCapra 1970). Dioxetane is isoelectronic with cyclobutane, and one might expect that the orbital correlation diagram for the decomposition should be qualitatively similar to that for the decomposition of cyclobutane to two ethylene molecules. The now-standard argument indicates that dissociation of cyclobutane to ground state products through D_{2h} (rectangular) or C_{2v} (trapezoidal) paths is symmetry forbidden. State and orbital correlation diagrams predicting a concerted reaction leading to excited carbonyl products have been given by Kearns (1969) and McCapra (1970), see Fig. 5.

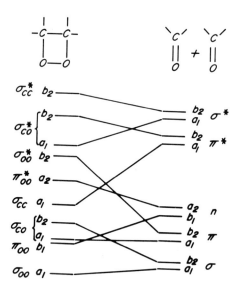

Fig. 5. Orbital correlation diagram for the concerted decomposition of dioxetane, adapted from Kearns (1969).

The meaning and usefulness of these qualitative diagrams is best assessed in the light of other, much more quantitative treatments (Wilson & Wang 1972) which have been carried out for related systems, in particular for the dissociation of cyclobutane (Salem & Wright 1969, Wright & Salem 1972) and the conversion of cyclobutane to cis-butadiene (Hsu, Buenker & Peyerimhoff 1971, Buenker, Peyerimhoff & Hsu 1971, McIver & Komornicki 1972). These calculation use semiempirical or _ab initio_ methods to derive energies as a function of the displacements which are most important for the reaction, i.e. certain sections of the adiabatic potential surface. A systematic search is made for the lowest energy path connecting reactants and products (McIver & Komornicki 1972). In contrast to Woodward-Hoffman approach, these calculations consider several degrees of freedom of the nuclei, and energies of electronic states, including correlation energies of tens of kcal/mole, rather than qualitative energies of one or two molecular orbitals. The results are in general consistant with the simpler approach. Some interesting difficulties arise, however, in locating the lowest saddle point on a multidimensional surface, and apparent discontinuities in the reaction path can occur (Hsu, Buenker & Peyerimhoff 1971, McIver & Komornicki 1972, Gregory & Paddon-Row 1972). All these calculations are done within the Born-Oppenheimer framework, i.e. for clamped nuclei, and no consideration is given to kinetic energy effects.

O'Neal and Richardson (1970) have used thermochemical methods to calculate activation energies and pre-exponential factors for the decomposition of methyl-substituted dioxetanes on the assumption of a biradical mechanism. However, it is not clear that this mechanism "ought" to give excited products. In general, electronically excited products will not be produced unless there is some constraint, often due to orbital symmetry, which makes the statistically more probable ground state inaccessible.

V. Conclusion

We have discussed a sequence of systems of increasing complexity. The evolution of certain central ideas can be traced out as a line of descent through the sequence.

One of these central ideas is the concept of _electronic state_. In I, the electronic state is well-defined for all values of the internuclear distance which are of interest. In II, it is well-defined almost everywhere, but the ambiguity in the concept of electronic state near a "crossing" of potential curves plays the crucial role in the chemiluminescent process. In III there is more mixing of electronic states, due to the lower symmetry of the electronic wave functions and the much higher density of vibronic states. The major feature introduced here is the change of symmetry of the electronic states with displacement of the nuclei. In IV, electronic states are often poorly defined. There is no necessary symmetry element, although reactions with a large activation energy will tend to go through symmetrical transition states. Vibronic level densities are much higher than in III, and mixing is often very strong. Spin may be the only reasonably good quantum number.

In a second line of descent, we can trace the treatment of the _dynamics_ of the interaction. In I, classical or semiclassical methods are usually quite appropriate. One treats collisions with specified relative kinetic energy and orbital angular momentum, and then averages over these initial conditions. In II, we have an inherently quantum mechanical phenomenon, but again one treats individual collisions separately, and then averages the results. In III, the most useful approach is probably the classical trajectory with quantum mechanical surface hopping. In IV, the dynamics often disappears. For a process like (20), a detailed treatment for every set of initial conditions is of no practical interest, but one can try to think about average collisions. For a process with a substantial activation energy, one tries to determine the symmetry and energy of the transition state, and the rest is thermodynamics.

In the various processes we have discussed, _symmetry_, or correlation rules, play several roles. In a decomposition reaction, section IV, the formation of an electronically excited state has a vanishingly small _a priori_ probability, due to the small statistical weight associated with electronic excitation. It is generally symmetry-imposed correlation which constrains the reaction to the statistically unlikely path leading to radiation. In II, on the other hand, symmetry plays a negative role, since it tends to reduce or prevent interaction of the electronic states involved in the curve "crossing". More generally, in II, III, and processes like (20) in IV, symmetry or correlation rules ensure that a more-or-less statistical fraction of collisions will occur on an excited state potential surface.

This work has been supported in part by the National Research Council of Canada.

References

Ackerman, F., and Miescher, E., 1969, J. Mol. Spec., 31, 400.
Andresen, B. and Nielsen, S.E., 1971, Mol. Phys., 21, 523-33.
Bain, R. A., and Bardsley, J. N., 1972, J. Phys. B 5, 277-85.
Bates, D., 1951, M.N.R.A.S. 111, 303-14.
Becker, et.al. 1973, Farad. Disc. Chem. Soc. 53.
Becker, K. H., Groth, W., and Thran, D., 1972, Chem. Phys. Lett. 15, 215.
Buenker, R. J., Peyerimhoff, and Kang Hsu, 1971, J. Am. Chem. Soc. 93, 5005-13.

Busch, G.E. and Wilson, K.R. 1972 J. Chem. Phys. $\underline{56}$ 3638.
Callear, A. B. and Pilling, M.J. 1970 Trans. Farad. Soc. $\underline{66}$ 1618.
Carrington, T. 1964 J. Chem. Phys. $\underline{41}$ 2012.
Carrington, T. 1972 Faraday Disc. Chem. Soc. $\underline{53}$ 27.
Carrington, T. 1972 J. Chem. Phys. $\underline{57}$, 2033
Carrington, T. and Garvin, D. 1969 Chap. 3 Vol.3 <u>Comprehensive Chemical Kinetics</u> Eds. C.H.Bamford & C.F.H.Tipper,Elsevier, N.Y.
Carrington, T. and Polanyi, J.C. 1972 Chap.9 in <u>Reaction Kinetics</u> MTP Biennial Review of Chemistry, Butterworths, University Press Baltimore.
Child, M.S. 1972 Faraday Disc. Chem. Soc. $\underline{53}$.
Chow, K-W and Smith, A.L. 1971 J. Chem. Phys. $\underline{54}$ 1556.
Clough, P.N. and Thrush, B.A. 1967 Trans. Farad. Soc. $\underline{63}$ 915.
Clyne, M.A.A. and Thrush, B.A. 1962 Disc. Farad. Soc. $\underline{33}$ 139.
Cohen, J.S., Evans, S.A. and Lane, N.F. 1971 Phys. Rev. A $\underline{4}$ 2248.
Comes, F.J. & Schumpe, G. 1971 Z. Naturforsch. $\underline{26a}$ 538.
Cooper, C.D., Cobb, G.C. and Tolnas, E.L. 1961 J. Mol. Spec. 7 223.
Corney, A. and Williams, O.M. 1972 J. Phys. B $\underline{5}$ 686.
Delos, J.B. and Thorson, W.R. 1972 Phys. Rev. A $\underline{6}$ 728.
Dieke, G.H. and Heath, D.F. 1959 Johns Hopkins Spec. Report No. 17.
Eu, B.C. 1971 J. Chem. Phys. $\underline{55}$ 5600.
Eu, B.C. 1972 J. Chem. Phys. $\underline{56}$, 2507.
Evans, K. and Rice, S.A. 1972 Chem. Phys. Lett. $\underline{14}$ 8.
Evans, S.A., Cohen, J.S. and Lane, N.F. 1971 Phys. Rev. A $\underline{4}$ 2235.
Fano, U. 1961 Phys. Rev. $\underline{124}$ 1866.
Filseth, S.V., Stuhl, F. and Welge, K.H. 1970 J. Chem. Phys. $\underline{52}$ 239.
Floquet, F. and Horsley, J.A. 1972, Faraday Disc. Chem. Soc. $\underline{53}$.
Forst, W. 1971 Chem. Rev. $\underline{71}$ 339.
Gangi, R.A. and Bader, R.F.W. 1971 J. Chem. Phys. $\underline{55}$ 5369.
Gangi, R.A. and Burnelle, L. 1971a J. Chem. Phys. $\underline{55}$ 843.
Gangi, R.A. and Burnelle, L. 1971b J. Chem. Phys. $\underline{55}$ 851
Goldberger, M.L. and Watson, K.M. 1964 <u>Collision Theory</u>, Wiley.
Golde, M.F. and Thrush, B.A. 1972 Faraday Disc. Chem. Soc. $\underline{53}$.
Gregory, A.R. and Paddon-Row, M.N. 1972 Chem. Phys. Lett. $\underline{12}$ 552.
Groth, W. Kley, D. and Schurath, U. 1971 J.Quant. Spec. Rad. Transf. $\underline{11}$ 1475.
Guberman, S.L. and Goddard, W.A. III 1972 Chem. Phys. Lett. $\underline{14}$ 460.
Hampson, R.F. Jr. and Okabe, H. 1970 J. Chem. Phys. $\underline{52}$ 1930.
Harris, R.A. 1963 J. Chem. Phys. $\underline{39}$ 978.
Hay, P.J. and Goddard, W.A. III 1972 Chem. Phys. Lett. $\underline{14}$ 46.
Henry, B.R. and Kasha, M. 1968 Ann. Rev. Phys. Chem. $\underline{19}$ 161.
Herschbach, D.R. 1973 this volume.
Herzberg, G. and Longuet-Higgins, H.C. 1963 Disc. Farad. Soc. $\underline{35}$ 77.
Hsu, K., Buenker, R.J. and Peyerimhoff, S.D. 1971 J. Am. Chem. Soc. $\underline{93}$ 2117.
Jackson, J.L. and Wyatt, R.E. 1971 J. Chem. Phys. $\underline{54}$ 5271.
Jortner, J., Rice, S.A. and Hochstrasser, R.H. 1969 Adv. Photochem. $\underline{7}$ 149.
Julienne, P.S. 1971 Chem. Phys. Lett. $\underline{8}$ 27.

Julienne, P.S., Krauss, M., and Donn, B. 1971 Astrophys. J. 170 65.
Kaufman, F. 1973 this volume.
Kearns, D.R. 1969 J. Am. Chem. Soc. 91 6554.
Kopecky, K.R. and Mumford, C. 1969 Can. J. Chem. 47 709.
Kovacs, I. 1969 Rotational Structure in the Spectra of Diatomic Molecules American Elsevier, New York.
Krauss, M. 1969 J. Res. Nat. Bur. Stand. 73A 191.
Krauss, M. 1970 Ann. Rev. Phys. Chem. 21 39.
Light, J.C. 1971 Adv. Chem. Phys. 19.
Mandelman, M., Carrington, T. and Young, R.A. 1972 J. Chem. Phys. in press.
McCapra, F. 1970 Pure and Appl. Chem. 24 611.
McCapra, F. 1973 this volume.
McIver, J.W. Jr., and Kormornicki, A. 1972 J. Am.Chem.Soc. 94 2625.
Michaelson, R.C. and A.L. Smith 1970 Chem. Phys. Lett. 6 1.
Miess, F.H. 1968 J. Chem. Phys. 48 482.
Miess, F.H. and Smith A.L. 1966 J. Chem. Phys. 45 994.
Miller, W.H. and George, T.F. 1972 J. Chem. Phys. 56 5637.
Nikitin, E.E. 1968 in Chemische Elementarprozesse. Ed. H.Hartman Springer, Berlin.
O'Neal, H.E. and Richardson, W.H. 1970 J. Am. Chem. Soc. 92 6553.
Palmer, H.B. 1967 J. Chem. Phys. 47, 2116.
Palmer, H.B. and Carabetta, R.A. 1968 J. Chem. Phys. 49 2466.
Peyerimhoff, S.D. and Buenker, R.J. 1967 J. Chem. Phys. 47 1953.
Peyerimhoff, S.D. and Buenker, R.J. 1968 J. Chem. Phys. 49 2473.
Pipano, A. and Kaufman, J.J. 1972 J. Chem. Phys. 56 5258.
Preston, R.K. and Tully, J.C. 1971 J. Chem. Phys. 54 4297.
Ramsey, D.S. and Child, M.S. 1971 Mol. Phys. 22 263.
Redpath, A.E. and Menzinger, M. 1971 Can. J. Chem. 49 3063.
Salem, L. and Wright, J.S. 1969 J. Am. Chem. Soc. 91 5947.
Salotto, A.W. and Burnelle, L. 1969 Chem. Phys. Lett. 3 80.
Sando, K.M. 1971 Mol. Phys. 21 439.
Sando, K.M. 1972 Mol. Phys. 23 413.
Smith, A.L. 1967 J. Chem. Phys. 47 1561.
Smith, F.T. 1969 Phys. Rev. 179 111.
Struve, W.S. Kitagawa, T. and Herschbach, D.R. 1971 J.Chem.Phys. 54 2759.
Thrush, B.A., Halstead, C.J. and McKenzie, A. 1968 J.Phys.Chem. 72 3711.
Tully, J.C. and Preston, J.K. 1971 J. Chem. Phys. 55 562.
Weiner, J., Peatman, W.B. and Berry, R.S. 1971 Phys. Rev.A 4 1824.
Welge, K.H., Filseth, S.V. and Davenport, J. 1970 J.Chem.Phys.53 502.
White, E.H. 1972 unpublished results
White, E.H.,Wiecko, J. and Roswell, D.F. 1969 J.Am.Chem.Soc. 91 5194.
Wilson, E.B. and Wang, P.S.C. 1972 Chem. Phys. Lett. 15 400.
Wilson, T. 1973 this volume.
Woodward, R.B. and Hoffman, R. 1968 Accts. Chem. Res. 1 17.
Wright, J.S. and Salem, L. 1972 J. Am. Chem. Soc. 94 322.

THRUSH: What Dr. Carrington just said reminds me of similar problems in spectroscopy. One can understand diatomic molecules and one can hope to understand the molecules around the size of benzene; between these, things are very complex, and our friends NO_2 and SO_2 are exceptionally hard to understand from the spectroscopic point of view. It is very interesting that at this level you are beginning to see the effect of local symmetry which the organic chemists talk about in the Woodward-Hoffman rules. In the $NO + O_3$ reaction you have in fact two paths governed essentially by local symmetry, one giving the ground state NO_2 and one giving the excited state of NO_2; this is governed by the orientation of the singly occupied π orbital as you go from NO to NO_2. This is one of the smallest systems showing the effect of local symmetry. You can still get the effects of local symmetry very nicely even when you have all sorts of groups tagged on to the basic unit because they have little effect on the reaction path under most conditions.

CARRINGTON: I think this is an important point, that there are intermediate cases which are not simple enough to be really simple but are not complex enough to enjoy the simple behavior often shown by larger molecules.

YOUNG: Is it true that this symmetry is applicable when you know something about the restricted regions which you are attempting to approximate?

CARRINGTON: Some of these processes have been treated much more elaborately by large machine calculations, including configuration interactions, giving potential surfaces which are presumably reasonably accurate. The results of those calculations do seem to substantiate the simple Woodward-Hoffman approach. This Woodward-Hoffman principle of conservation of orbital symmetry or maximum bonding is simply the principal of minimum potential energy during the course of the reaction.

F. KAUFMAN: Whenever one invokes a process such as $NO + O + M \longrightarrow NO_2^* + M$ one should try to indicate whether it is meant to be a direct 3-body process or a succession of 2-body recombination and energy transfer steps. I don't think the 3-body process really exists in this particular system because when you analyze fluorescence and chemiluminescence data of the type which I will discuss on Thursday morning you find that the lifetime of the unstabilized intermediate formed in the 2-body process from O and NO is very much longer than a collision time, i.e., on the order of 10^{-11} seconds and that of course is vastly longer than the lifetime of a "direct" 3-body collision. There is therefore no point in including the direct $NO + O + M$ process.

CARRINGTON: Right, but my point was that in this case you get NO_2^* with distribution of internal energies, and the NO_2 which radiates has less than the total energy of the two separated fragments.

KAUFMAN: That is certainly true, but it is best described by the sequence of collisional energy transfer processes undergone by NO_2^* as a result of which it loses some of its vibrational energy.

CARRINGTON: It is just a vibrational relaxation in the excited NO_2. But that's a 5 atom process if it is colliding with the N_2.

YOUNG: If the lifetime of the 2-body complex is long, there will be many collisions during that lifetime, and so it is not fair to call it a 3-body collision. If the lifetime of a 2-body complex is comparable to the collision time then it is natural to consider 3-body processes.

TRANSVIBRONIC REACTIONS IN MOLECULAR BEAMS

D. R. Herschbach

Department of Chemistry, Harvard University

Collisional processes involving interconversion of translational, vibrational, and electronic energy are referred to as "transvibronic reactions," in analogy to spectroscopic nomenclature. Many such processes can now be studied in molecular beams, thanks to methods for generating collision energies in the previously almost inaccessible hyperthermal range (a few tenths to a few tens of eV) and methods for producing beams of vibrationally excited molecules. The single-collision conditions obtained in beam experiments can be exploited to isolate and identify elementary steps in a chemical mechanism and to determine dynamical properties. These include the dependence of the rate on collision energy and vibrational excitation of the reactants, the distribution in translational and vibrational energy and scattering angle of the products, and further details related to the electronic coupling of the collision partners.

Table I lists examples of transvibronic reactions studied in molecular beams. For these processes the energy transfer is very efficient, often corresponding to "reaction on every collision" or nearly so. Most involve alkali atoms because of historical precedents and experimental convenience. This is a fortunate circumstance from a theoretical viewpoint. The ionization potential of an alkali atom is exceptionally low. Consequently, transvibronic interactions of an alkali atom A with another species X are usually governed by the strongly attractive ion-pair configuration, A^+X^-. The alkali reactions thus provide simple prototypes for a large class of chemiluminescent processes that involve at least partial charge-transfer or electron donor-acceptor mechanisms.

Table I. Transvibronic Reactions Studied in Molecular Beams.[a]

Typical Reactions	Reference	Other Examples
Collisional Excitation: $A + BC \rightarrow A^* + BC$ or $A + BC^*$		
$\underline{K} + N_2 \rightarrow K^* + N_2$	[1,2,3]	H_2, HCl, Cl_2, O_2, CO_2, etc.
$\rightarrow K + N_2^*$	[4]	NO_2, SO_2
$K + \underline{N_2} \rightarrow K^* + N_2$	[5]	CO, CO_2, N_2O, C_2H_4
$Na + N_2^\dagger \rightarrow Na^* + N_2$	[6]	H_2, D_2
Collisional Quenching: $A^* + BC \rightarrow A + BC$		
$Hg^*(^3P_2) + N_2 \rightarrow Hg^*(^3P_1) + N_2$	[7]	H_2, D_2, CO, CH_4, CD_4
Chemiluminescence: $A + BC \rightarrow AB + C^*$ or $AB^* + C$		
$Na + KCl^\dagger \rightarrow NaCl + K^*$	[8]	$KBr^\dagger$
$Cl + Na_2 \rightarrow NaCl + Na^*$	[9]	K_2
$Ba + NO_2 \rightarrow BaO^* + O$	[10]	N_2O, Cl_2
$NO + O_3 \rightarrow NO_2^* + O_2$	[11]	
$Al + O_3 \rightarrow AlO^* + O_2$	[12]	

[a]Underline denotes hyperthermal translational energy, dagger denotes vibrational excitation, asterisk denotes electronic excitation.

This paper deals with two families of alkali reactions. Interpretations or speculations related to the electron-transfer process are emphasized rather than experimental aspects. One family involves collisional excitation without chemical reaction,[1-6]

$$A + X \rightarrow A^+X^- \rightarrow A^* + X \quad \text{(R1)}$$
$$\text{or} \rightarrow A + X^*$$

where the asterisk denotes electronic excitation and $X = H_2$, N_2, O_2, CO_2, SO_2, C_2H_4, C_6H_6, etc. Here the alkali valence electron executes a "double-jump", first switching to the target molecule and then returning to the alkali. In particular, the energy dependence of the cross sections for these processes shows striking features attributable to the electronic structure of the intermediate X^- negative molecule-ion.

The other family comprises chemiluminescent atom exchange reactions involving two alkali atoms, such as[8,9]

$$A + B^+X^- \rightarrow (AB)^+X^- \rightarrow A^+X^- + B^* \tag{R2}$$

and

$$X + AB \rightarrow (AB)^+X^- \rightarrow A^+X^- + B^* \tag{R3}$$

where X is a halogen atom. As indicated by our notation, the bonding is essentially ionic for both the reactant or product alkali halide and the intermediate dialkali halide complex. Here the excitation mechanism involves a single-jump, followed by irreversible dissociation of the $(AB)^+$ ion in the field of the X^- ion.

COLLISIONAL EXCITATION

Reactions (R1) are strongly endoergic. The lowest excited state of $A^*(^2P)$ lies at 1.6 eV for potassium atoms and 2.1 eV for sodium atoms. In the beam experiments, the requisite translational energy (denoted by E_T) or vibrational excitation (denoted by E_V) of the reactants has been supplied in three ways:

	Experimental Method	$E_T(A)$	$E_T(X)$	$E_V(X)$
I	"Beam through a box"	high	low	nil
II	"High temperature oven"	low	high	high
III	"Seeded-jet expansion"	low	high	nil

In each case, the cross section for electronic excitation as a function of E_T or E_V is determined by measuring the fluorescence from A^* or X^*.

The most extensive studies thus far have used Method I, which sends a fast alkali beam through a chamber containing the target gas at a low temperature. The fast beam is generated either by accelerating an ion beam and then neutralizing the ions by resonant charge-exchange or by "sputtering" alkali atoms from a solid surface by ion bombardment. These techniques provide collision energies up to ~20 eV and beyond, although the resolution is not high (~20% at threshold). Cross sections have been measured for a wide variety of target gases (Table I). The examples of Figure 1 bring out several results of interest:

(1) The threshold energy for the A → A* excitation in collisions with noble gas atoms is far higher (>30 eV) than the endoergicity, much higher even than the ionization potentials of both the alkali atom and the noble gas.[1] This implies that the transition from the potential curve for the A + X ground state to the curves for A* + X (which involve both Σ and Π interactions, since A* is a ^2P atom) takes place far up on the repulsive wall. Qualitative arguments,[1] consistent with recent electronic structure calculations, suggest that in fact the A + X and A* + X curves do <u>not</u> intersect. At small distances, the dominant interaction is repulsion of the closed-shell alkali ion core and the noble gas atom. In the excited states, the alkali valence electron is less likely to contribute bonding at small distances than in the ground state (where only a weak van der Waals attraction occurs), since it is less constrained to the region between the atoms. The transition between the A + X and A* + X states is probably due to intersections with A + X* states, which have an open-shell structure and hence foster a less repulsive interaction. This situation is depicted in Figure 2(a).

(2) In contrast, the thresholds for A → A* in collisions with many molecules are at or not much above the endoergicity.[2,3] Again, the potential surfaces for A + X and A* + X almost certainly do not intersect. The link between them is due to the strongly attractive ion-pair configuration A^+X^-, not available with the noble gases. In many cases, the ion-pair potential surface is expected to cross the A + X and A* + X surfaces at separations outside or near the edge of the repulsive region. This accounts for the low thresholds, as indicated in Figure 2(b).

(3) For the molecular targets, the magnitude and qualitative form of the A → A* cross sections versus collision energy are similar for molecules such as (i) H_2, HCl, Cl_2 and saturated hydrocarbons and for (ii), N_2, NO, CO, O_2, olefinic and aromatic hydrocarbons, but differ markedly between the two sets. This difference correlates with the character of the lowest-lying vacant orbital in the target molecule, which receives the itinerant electron in the A^+X^- ion-pair state. For (i), this orbital is a strongly antibonding σ-type orbital. For (ii), it is a weakly antibonding π-type orbital. The increase in bond length on going from the molecule X to the intermediate X^- ion is much larger for (i) than for (ii); it is estimated to be ~0.5 Å for H_2 and Cl_2, but only ~0.08 Å for N_2 and O_2, for example. The data suggest that the large change in bond length required for (i) inhibits the return transition ($A^+X^- \rightarrow A^* + X$) required for excitation.

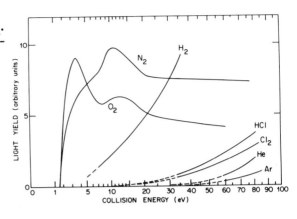

Fig. 1. Cross section vs. collision energy for electronic excitation of K atoms. The ordinate gives photon counts divided by the intensity of the fast beam and the density of the target gas; the units correspond roughly to $Å^2$. The abscissa scale is linear in relative velocity (proportional to $E^{1/2}$), in order to expand the low energy region.

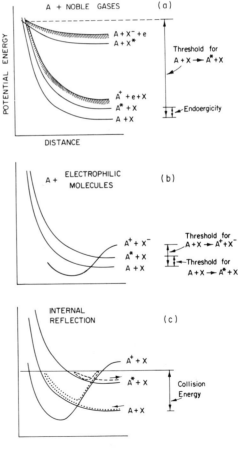

Fig. 2. Schematic potential energy curves for interaction of alkali atoms with (a) noble gas atoms and (b) diatomic molecules, including (c) reaction path illustrating "internal reflection".

(4) Further instructive correlations are found[3] in comparing the excitation cross sections with those for formation of free ion-pairs, via $A + X \to A^+X^- \to A^+ + X^-$. For many molecules, the free ion-pair $A^+ + X^-$ appears at collision energies only a few eV above the $A^* + X$ threshold. The $A^+ + X^-$ yield proves to be much larger for molecules of set (i) than set (ii), whereas the opposite holds for the $A^* + X$ yield. Certain cases offer still more striking evidence[3,4] for competition between decay of the close-coupled A^+X^- configuration to form $A^+ + X^-$ or $A^* + X$. For example, as seen in Figure 3, with $K + SO_2$ there is intense emission of K^* fluorescence at collision energies above 1.6 eV. This emission increases strongly up to about 3 eV and then rapidly decreases. At and above 3 eV there appear ion-pairs, $K^+ + SO_2^-$. This suggests that "internal reflection" occurs for collision energies above the endoergicity for $A^* + X$ (1.6 eV) but below that for formation of $A^+ + X^-$ (3 eV). As illustrated in Figure 2(c), at such energies trajectories which make the $A + X \to A^+X^-$ crossing and intend to exit via the ion-pair channel find it closed. This flux must be reflected back and redistributed at the $A^+X^- \to A^* + X$ and $A^+X^- \to A + X$ crossings until ultimately it escapes via either the excitation channel or the original entrance channel.

(5) Fluorescence from excited molecular states corresponding to $A + X^*$ has also been observed for $K + N_2$ and several other cases.[4] The N_2^* emission occurs in the first positive band, $B^3\Pi_g \to A^3\Sigma_u$, and includes transitions involving several vibrational levels of the A and B states. At collision energies above $\sim$6 eV the N_2^* emission in this triplet system is quite strong. The mechanism suggested for the molecular excitation involves radiationless transitions induced by mixing of X^- orbitals via coulombic interaction with A^+ during the collision.[4] The high efficiency of the singlet $\to$ triplet spin change can be attributed to electron-exchange in the exit stage of the collision, $X^-A^+ \to X^* + A$. This is illustrated in Figure 4. As in analogous electron impact processes ($e + X \to X^- \to X^* + e$), triplet states of X^* are readily excited when the incident electron is captured and another electron emitted.

In the studies discussed above, the target molecules are initially almost solely in the ground vibrational state and the experiments give no information about the final vibrational states. An elegant crossed-beams study[6] dealing with this has been carried out for $Na + N_2$ using Method II, which employs a tungsten furnace operated at temperatures up to $\sim$3000°K. For N_2 such temperatures produce only slight dissociation but populate high vibrational states sufficiently to allow observation of reaction (R1). The translational energy is also enhanced, but is not high enough to give appreciable reaction without vibrational excitation. A mechanical velocity selector was used to separate these contributions. The results show:

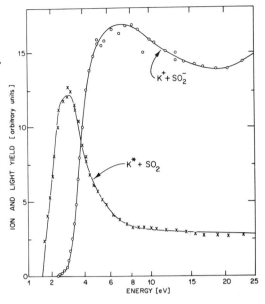

Fig. 3. Cross sections for electronic excitation and ion-pair formation on impact of K atoms with SO_2 molecules.

Fig. 4. Molecular orbital configurations involved in collisional excitation of nitrogen. The alkali valence electron is exchanged for a molecular electron of opposite spin, thereby leaving the molecule in an excited state with unpaired spins.

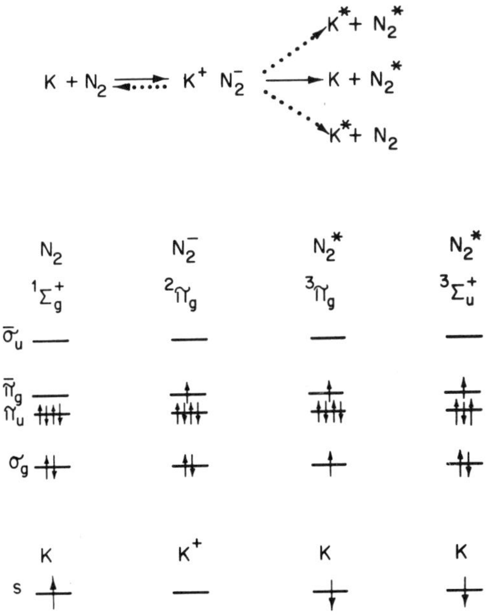

(6) Energy transfer from vibration is much more effective than from translation. For the most probable transitions 7 vibrational quanta are transferred and translation contributes the equivalent of about one quantum. The partial cross sections for some transitions are very large; e.g., of the order of $\sim$300 Å for the v = 10 → v' = 3 transition. These findings are qualitatively consistent with theoretical calculations based on a transient Na^+N_2 collision complex.[13,14] The calculations employ electron impact data, Franck-Condon factors for comparable isoelectronic systems, and semi-empirical correlations to estimate the curve-crossing parameters and transition matrix elements.

The most recent work[5] has brought out another property attributable to the electronic structure of the intermediate X^- molecule-ion. These experiments use Method III, crossing a slow, thermal beam of alkali atoms with a fast beam of molecules. The fast beam is generated by a "seeded-jet" technique, in which a high pressure (800 torr) gas mixture containing $\sim$1% of the species of interest and 99% hydrogen is expanded into vacuum through a small diameter (0.1 mm) nozzle. Collisions in the nozzle bring the heavier molecules to the same exit velocity as the hydrogen. This increases the translational energy by approximately the ratio of the molecular mass to the H_2 mass. A marked collisional "cooling" of vibration and rotation (typically to $\lesssim$100°K) also occurs. The technique gives very high intensity ($\sim 10^5$ higher than Method I!) at translational energies below a few eV and also good energy resolution (typically $\sim$5%). This makes feasible accurate measurements of the "threshold-law" for reaction (R1), i.e., the energy dependence of the cross section as it drops to zero at threshold. Figure 5 shows data for the two cases studied thus far.

(7) A threshold-law of the form $(E - E_o)^n$ fits the data satisfactorily, with n = 5/2 for N_2 and n = 3/2 for CO. The threshold energy E_o = 1.6 eV for both, which corresponds to excitation of K* with N_2 or CO remaining in the v = 0 ground vibrational state. The range of collision energy covered extends $\sim$1 eV above the v = 0 → 0 threshold, thus includes those for the 0 → 1 and 0 → 2 processes, but no vibrational structure is seen.

The exponents n (experimental uncertainty about ±1/2 unit) show an intriguing qualitative correlation with the N_2^- and CO^- "shape resonances" found in electron impact spectroscopy.[15] The theory of threshold laws suggests n = ℓ + 1/2, where ℓ is the orbital angular momentum (in units of $\hbar$) of the added electron.[16,17] The "shape resonances" arise from tunneling through the centrifugal barrier associated with this angular momentum. From the molecular polarizability, the centrifugal barriers are estimated to be a few eV high and a few Å wide. The tunnelling rate for an electron is very rapid, however, as evidenced by the observed autoionization lifetimes of $\sim 10^{-14}$ sec.[15] In forming N_2^-, the added electron

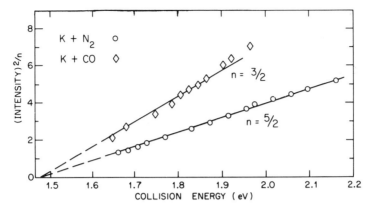

Fig. 5. Energy dependence of cross sections for electronic excitation of K atoms by collision with nitrogen and carbon monoxide. Ordinate gives the 2/n-power of the cross section, with n = 5/2 for N_2, n = 3/2 for CO. Lines extrapolate below the endoergic threshold at 1.6 eV because of imperfect velocity definition in the parent beams.

enters a π_g molecular orbital. The angular momentum ℓ therefore has a projection of unity on the molecular axis but must be even to provide inversion symmetry. Hence, $\ell = 2$ is the lowest allowable value. In CO^- the added electron likewise enters a π orbital but the inversion symmetry is absent. Hence, $\ell = 1$ is the lowest value.

The observed threshold laws thus suggest that transfer of the alkali valence electron between the atom and the molecule involves tunneling, as in electron impact. This interpretation is still speculative, of course; it must be tested with other molecules e.g., $C_6H_6^-$ involves an e_{2u} orbital, $\ell = 3$ predicted) and should provoke theoretical analysis.

CHEMILUMINESCENT REACTIONS

The intense atomic sodium D-line emission observed in Polanyi's classic studies of (Na, Cl_2) diffusion flames had an important role in developing early ideas about transvibronic reactions.[18-20] The early experiments clearly linked the chemiluminescence to the very exoergic (∼3.5 eV) reaction of Cl atoms (released in the primary Na + Cl_2 step) with Na_2 dimer molecules (present as ∼1% mole fraction at thermal equilibrium). However, a major question remained unresolved: whether the excited Na* atoms come (1) <u>directly</u> from Cl + Na_2 → NaCl + Na* or (2) <u>indirectly</u> from Cl + Na_2 → NaCl† + Na, followed by energy transfer from the vibrationally excited NaCl† in subsequent collisions.

Reaction (R2) corresponds to the vibrational-to-electronic energy transfer process. This was studied by means of a "triple-beam" experiment.[8] Vibrationally excited $KBr^\dagger$ was formed at the intersection of crossed beams of K and Br_2. Previous work had determined the vibrational energy distribution, which peaks at ~ 2 eV. This $KBr^\dagger$ entered a second scattering chamber containing an Na cross beam, where fluorescence from K* was observed. The cross section found for this "reactive transfer" process,

$$Na + KBr^\dagger \rightarrow NaBr + K^*$$

is large, ~ 10-100 $Å^2$. The reaction is much more efficient than "nonreactive transfer" via

$$K + NaBr^\dagger \rightarrow K^* + NaBr,$$

which was studied by interchanging the K and Na beams. The latter process gave no detectable fluorescence signal, indicating the cross section was at least tenfold smaller.

This result again indicates the effectiveness of electron transfer as an excitation mechanism. The initial energy distributions for the $Na + KBr^\dagger$ and the $K + NaBr^\dagger$ experiments are nearly the same, as shown in Figure 6, and the same set of potential surfaces is accessible in both cases. Therefore, on energetic or statistical grounds, one would expect both processes to form K^* + NaBr with the same probability. However, if in the intermediate $(AB)^+X^-$ complex the $(AB)^+$ ion decomposes rapidly and irreversibly in the field of the X^- ion, configurations in which A and B are symmetrical with respect to both charge-sharing and interaction with X^- may seldom be traversed. The chemical exchange process $(A + B^+X^- \rightarrow A^+X^- + B^*)$ is then more favorable for electronic excitation because it involves charge-transfer $(B^+ + e \rightarrow B^*)$ whereas the nonreactive process $(A + B^+X^- \rightarrow A^* + B^+X^-)$ does not involve charge-transfer.

A quite detailed picture of this energy transfer mechanism has been obtained recently from calculations of the potential energy surfaces for dialkali halide systems.[21,22] A "pseudopotential" method was employed which greatly simplifies the problem, reducing it to that of a single valence electron moving in the field of the closed-shell A^+, B^+, and X^- ion cores. For brevity, only one instructive aspect of the results is noted here. The potential surfaces for the ground and excited states prove to be widely separated except for configurations in which the X^- ion is interposed between A^+ and B^+. For those configurations, the valence electron density switches abruptly between A^+ and B^+ according to which is more distant from the X^- ion.

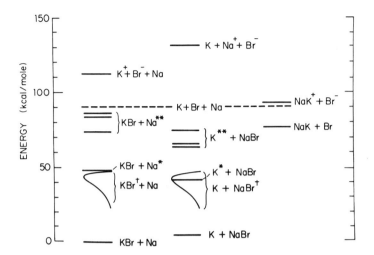

Fig. 6. Energetics of reactions involving Na + KBr† and K + NaBr† systems. The initial distribution of vibrational energy is indicated in each case and the location of various excited or ionized states, with reactants or products at infinite separation.

Reaction (R3) was studied[9] using a beam of Cl atoms (obtained by 90% dissociation of Cl_2 in a graphite tube at ~1600°K) and an alkali beam containing up to ~30% mole fraction Na_2 (obtained by a nozzle expansion at source pressures of ~200 torr). The D-line emission was visible to the naked eye. It was shown to vary linearly with Cl and with Na_2 but not with Na by changing the source conditions and by deflecting away the Na atoms with an inhomogeneous magnetic field. Variation of the Cl beam temperature (up to 2000°K) showed no discernable activation energy. Polarization measurements showed the emission to be spatially isotropic. The cross section for forming NaCl + Na* was estimated as $\gtrsim 10$ Å^2, comparable to or perhaps larger than that for forming NaCl + Na without electronic excitation.

This demonstration that direct electronic excitation occurs in the Cl + Na_2 reaction is consistent with theoretical predictions obtained from estimates of the potential surface crossings[19,22] due to the electron-jump configuration, Cl^- + Na_2^+, and from a statistical model for energy partitioning.[23] There is a simple qualitative aspect of general interest. The reactant halogen atom has three-fold orbital degeneracy, corresponding to location of its valence-shell "hole" in the p_x, p_y, or p_z orbital, whereas the reactant and product molecules are in nondegenerate electronic states. Thus the reactants give rise to three distinct potential energy surfaces but only one of these can lead to the nondegenerate ground state of the product alkali atom and the other two surfaces must lead to electronically excited states. Such "degeneracy-induced" excitation might be inhibited by an activation energy. This seems unlikely for $Cl + Na_2$, however, in view of the extremely large reaction exoergicity and the ready formation of $Cl^- + Na_2^+$, which results from the high electron affinity of Cl and low ionization potential of Na_2^+. The potential surface calculations indeed indicate no activation energy.[22] The possible importance of degeneracy-induced chemiluminescence was recognized long ago,[19] but was ignored in all subsequent discussions of the Polanyi flame system until revived by the molecular beam experiments.

CONCLUDING REMARKS

Since chemiluminescence in liquid solutions is often interpreted in terms of electron transfer, some aspects of the processes discussed here may perhaps offer useful analogies. In particular, the analogy seems quite direct for electrogenerated chemiluminescence[24,25] involving reaction of ion-pairs, $A^+ + X^- \rightarrow A^* + X$. We should emphasize, however, that the transient formation or rearrangement of ion-pairs in reactions (R1), (R2), (R3) occurs only at close distances of approach. At large distances, the ion-pair state $A^+ + X^-$ typically lies well above both the ground $A + X$ and excited $A^* + X$ states; only the energy gained by coulombic attraction allows the ion-pair state to become lowest at close approach. As illustrated in Figure 2, the essential role of the ion-pair state in energy transfer is to act as an "elevator" which runs up and down between the ground and excited potential surfaces as the reactants approach and the products separate.

Ion-pair states may play this role in many systems for which it has not been customary to invoke ionic intermediates. The energy required for electron transfer with the reactants at infinite separation, $A + X \rightarrow A^+ + X^-$, is given by the difference in the ionization potential of the donor and the electron affinity of the acceptor, $\Delta = IP(A) - EA(X)$. The energy required with the reactants at a finite distance R is approximately $\Delta - (e^2/R)$, as long as R is large enough to make repulsive overlap forces small. Thus, if $\Delta \sim 5$ eV, the electron transfer is endoergic for $R > 3$ Å and exoergic for $R < 3$ Å. The parameter $\Delta \sim 5$ eV for $Na + N_2$ and a comparable value probably occurs for many organic systems involving interactions of large hydrocarbons (IP ~ 7 eV) with molecules containing electrophilic groups (EA ~ 2 eV).

Another aspect that deserves emphasis is the ease of singlet-triplet conversion via electron exchange. As illustrated in Figure 4, the process

doublet + singlet → doublet + triplet

can be readily induced by the mixing of molecular orbitals caused by coulombic interaction of the transient ion-pair at close distances. This mechanism should be particularly likely when the molecule undergoing the singlet → triplet conversion has an unfilled π-orbital not greatly higher in energy than the lowest filled π-orbital.

Acknowledgements

The Harvard work reviewed in this article has been supported by the National Science Foundation.

References

1. R. W. Anderson, V. Aquilanti, and D. R. Herschbach, Chem. Phys. Letters 4, 5 (1969).

2. V. Kempter, W. Mecklenbrauck, M. Menzinger, G. Schuller, D. R. Herschbach, and Ch. Schlier, Chem. Phys. Letters 6, 97 (1970); 11, 353 (1971).

3. K. Lacmann and D. R. Herschbach, Chem. Phys. Letters 6, 106 (1970).

4. P. R. LeBreton, W. Mecklenbrauck, A. Schultz, and Ch. Schlier, J. Chem. Phys. 54, 1752 (1971); V. Kempter and P. R. LeBreton (to be published).

5. H. J. Loesch, R. A. Larsen, J. R. Krenos, and D. R. Herschbach, J. Chem. Phys. (to be published).

6. J. E. Mentall, H. F. Krause, J. Fricke, and W. L. Fite, Disc. Faraday Soc. 44, 157 (1967); J. Chem. Phys. 56, 4593 (1972).

7. L. J. Doemeny, F. J. Van Itallie, and R. M. Martin, Chem. Phys. Letters 4, 302 (1969).

8. M. C. Moulton and D. R. Herschbach, J. Chem. Phys. 44, 3010 (1966).

9. W. S. Struve, T. Kitagawa, and D. R. Herschbach, J. Chem. Phys. 54, 2759 (1971).

10. C. D. Jonah, Ch. Ottinger, and R. N. Zare, Chem. Phys. Letters 5, 243 (1970); 9, 65 (1971); J. Chem. Phys. 56, 271 (1972).

11. A. E. Redpath and M. Menzinger, Can. J. Chem. 49, 3063 (1971).

12. J. L. Gole and R. N. Zare, J. Chem. Phys. (to be published).

13. A. Bjerre and E. E. Nikitin, Chem. Phys. Letters 1, 179 (1964).

14. E. Bauer, E. R. Fischer, and F. R. Gilmore, J. Chem. Phys. 51, 4173 (1969).

15. J. N. Bardsley and F. Mandl, Rept. Prog. Phys. 31, 471 (1968).

16. E. Wigner, Phys. Rev. 73, 1002 (1948).

17. T. F. O'Malley, Phys. Rev. 137, A116 (1965).

18. M. G. Evans and M. Polanyi, Trans. Faraday Soc. 35, 178, 192, 195 (1939).

19. J. L. Magee, J. Chem. Phys. 7, 652 (1939); 8, 687 (1940).

20. K. J. Laidler, *The Chemical Kinetics of Excited States* (Clarendon Press, Oxford, 1955), p. 85.

21. A. C. Roach and M. S. Child, Mol. Phys. 14, 1 (1968).

22. W. S. Struve, Mol. Phys. (to be published).

23. P. Pechukas, J. C. Light, and C. Rankin, J. Chem. Phys. 44, 794 (1966).

24. D. M. Hercules, Accts. Chem. Res. 2, 301 (1969).

25. A. J. Bard and L. R. Faulkner, in *Creation and Detection of the Excited State*, W. R. Ware, Ed., Vol. 3 (Marcel Dekker, New York, to be published).

STEDMAN: Dr. Carrington pointed out that a plane of symmetry is preserved in a 3-atom collision. The effect of this plane is to force the reaction $Ar\ ^3P_{2,0} + N_2 \longrightarrow Ar's + N_2\ (^3\pi_g)$ to produce only the π^+ state. The reaction $K + N_2 \longrightarrow N_2\ (C\ or\ B^2\pi)\ ^g+ K$ has the same symmetry property and would thus be predicted to produce only the π^+ states at threshold.

STUDIES OF VIBRATIONALLY EXCITED MOLECULES BY INFRARED CHEMI-
LUMINESCENCE

IAN W.M. SMITH

Department of Physical Chemistry, University Chemical
Laboratories, Lensfield Road, Cambridge, England

I. INTRODUCTION

As the proceedings of this conference demonstrate, the subject of chemiluminescence is primarily concerned with the observation of radiation from electronically excited species in the visible region of the spectrum or at shorter wavelengths. Chemiluminescence in the infrared is generally associated with the formation of reaction products which are in their ground electronic state but excited vibrationally. Largely due to the later development of infrared detectors which even approach the efficiency of those used for shorter wavelengths, experimental studies of infrared chemiluminescence are all of quite recent origin, the first laboratory observation of spontaneous emission in the instrumental infrared from a chemically formed, vibrationally excited molecule being reported by Cashion and Polanyi in 1958.[1]

Since that time Polanyi and his coworkers have developed their experimental technique to a high level of sophistication. They[2], and other workers[3,4] have concentrated almost entirely on discovering the initial quantum state distributions of diatomic hydride molecules produced in simple atom-exchange reactions, such as
$$H + Cl_2 \rightarrow HCl + Cl, \quad \Delta H^o_0 = -45.1 \text{ kcal mole}^{-1}.$$

Their more recent experiments have been carried out at very low total pressures ($\sim 10^{-4}$ Torr), thus effectively 'arresting' gas-phase relaxation and making it possible to observe the product distribution from the reaction directly. This technique has proved so powerful that rotational, as well as vibrational, level distributions have been determined for several reactions.[2a]

The detailed state distributions which are obtained in experiments of this kind are particularly important in two ways. First, together with the results from crossed molecular beam experiments and from Monte Carlo trajectory calculations on reactive collisions, they have greatly increased our knowledge of the factors which affect how energy released in a chemical reaction is shared among the degrees of freedom of chemical reaction products. Secondly, reaction product distributions indicate which chemical reactions may be capable of sustaining laser action on the vibration-rotation transitions of the excited product and are essential information if these lasers are to be understood and systematically improved.[5] An interest in both reaction dynamics and in the processes occurring in molecular gas lasers has stimulated the programme of research in infrared chemiluminescence which we have followed in Cambridge over the last four years. In this paper this work is reviewed in three sections: (i) the determination of the relative rates at which the related reactions

$$O(^3P) + CS \rightarrow CO + S, \Delta H_0^o = -85 \text{ kcal mole}^{-1}$$

and $O(^3P) + CSe \rightarrow CO + Se, \Delta H_0^o \sim -117 \text{ kcal mole}^{-1}$,

populate the energetically accessible vibrational levels of CO[6,7]; (ii) the measurement of vibrational deactivation rates of CO[8], HCl[9], and HF and DF[10] in high states of excitation; (iii) attempts to find a satisfactory system for producing a relatively simple OH chemical laser. Before discussing the results of these investigations, a brief description is given of the experimental apparatus and general method of interpretation.

II. METHOD

Observations of spontaneous infrared chemiluminescence have nearly all been carried out in discharge-flow systems of the type also commonly used to study electronic chemiluminescence from simple, gas-phase reactions. In our own experiments, which have been described fully elsewhere,[6b,8b] atoms are generated by passing a diatomic gas, such as O_2, H_2 or Cl_2, highly diluted in Ar through a microwave discharge cavity. The diatomic parent species is partially dissociated and the atoms are subsequently mixed with the second reagent someway downstream from the discharge. This mixing occurs in a 1 liter spherical vessel which is internally coated in gold to collect as much as possible of the infrared emission. The radiation is viewed through a window on the side of the vessel with a spectrometer which can be fitted with a variety of infrared detectors--most usually cooled or uncooled PbS photoconductive cells.

In our work, the total pressure is in the range 0.5-20 Torr, but the concentrations of the reagent species entering the reaction vessel are kept as low as is consistent with obtaining good S/N ratios in the observed spectra. Typically, the reagents form $\lesssim$ 1 part in 10^4 of the gas mixture. The range of pumping speeds employed is $\sim$0.5-6.0 liter/sec.

In order to interpret the results of chemiluminescent experiments, it is first necessary to relate the observed spectral intensities to the relative steady state concentrations of excited species. In the case of infrared chemiluminescence from diatomic hydrides this part of the analysis presents no problems since individual vibrational-rotational lines can be resolved and their intensities are related to the excited state concentrations by well-known formulae.[11] Some allowance must be made for the way in which the detector sensitivity changes with wavelength but this can be determined by recording the continuous spectrum emitted from a regulated black-body source. With non-hydride molecules, the structure in the infrared bands cannot be resolved and deducing the relative vibrational state populations (hereafter denoted by N_V) from the overlapped spectrum is more difficult. Measurements on $CO^\dagger$ (the $\dagger$ is used here to indicate vibrational excitation) have been made almost entirely on the first overtone (i.e., $\Delta v=2$) bands. These are at wavelengths accessible to PbS photoconductive detectors and, as the spectra in figure 1 show, the bands are partially resolved in that they show minima at the wavelengths corresponding to the band origins, the maxima in between arising largely through the constructive overlap of P and R branches in neighbouring bands. A computer technique, based on that used by Karl, Kruus and Polanyi,[12] has been developed to find the N_V from these spectra, and spectra simulated subsequently with the derived N_V have been shown[8b] to give an excellent match with those observed experimentally.

The main difficulty in interpreting infrared chemiluminescence experiments arises because the N_V deduced from an observed spectrum generally do not correspond to the relative rates at which the vibrational levels are populated (which are here termed R_V) by the reaction under investigation, due to the effects of relaxation. The situation may be contrasted with that for electronic chemiluminescence, when the mean radiative lifetimes are typically $\sim 10^{-7}$ sec and the radiative process completely removes molecules from the range of excited levels. Because of the short radiative lifetime, it is frequently possible to perform experiments at low enough pressures for collisions not to occur during the excited state lifetime and then the distribution of species among the vibration-rotation levels of the chemiluminescent

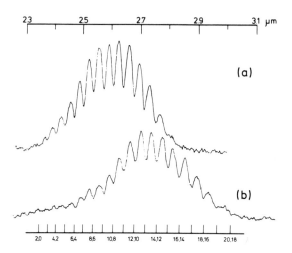

Figure 1. CO overtone spectra. Flow rates (μmole/sec): (a) 0.018 CS_2, 0.2 partially dissociated O_2, 650 Ar; (b) 0.008 CSe_2, 0.1 partially dissociated O_2, 650 Ar. In both cases the total pressure was 11.7 Torr, the pumping speed 1 liter/sec. (from ref. 7b, reprinted by permission of the Faraday Society).

state will be that actually produced in the step forming electronically excited molecules. In the case of vibrational excitation, however, spontaneous radiative lifetimes are typically $\sim 10^{-2}$ sec, and consequently much lower total pressures must be reached if collisions are to be totally avoided within the radiative lifetime. A second complicating feature is that spontaneous radiative decay and collisional relaxation do not remove molecules in a single step from the manifold of excited levels but transfer molecules in mainly single quantum transitions down the 'ladder' of vibrational states.

Relaxation can be allowed for using a steady state analysis originally proposed by Charters and Polanyi,[13] and developed by Hancock and Smith[8b] to include the effects of a quenching additive on the luminescence. The total rate of formation of CO in each vibrational level by (a) direct chemical production (at a rate = R_V), (b) spontaneous radiative transitions from higher levels $\left\{ \sum_{n=1}^{n_{max}} A_{v+n,v} N_{v+n} \right\}$, and (c) collisional deexcitation of higher levels $\left\{ \sum_{\text{all } q} \sum_{n=1}^{n_{max}} k^q_{v+n,v} [q] N_{v+n} \right\}$ is equated to the total rate of loss by (d) spontaneous emission $\left\{ \sum_{n=1}^{v} A_{v,v-n} N_v \right\}$, (e) collisional deactivation $\left\{ \sum_{\text{all } q} \sum_{n=1}^{v} k_{v,v-n}^{q} [q] N_v \right\}$, and (f) surface removal and pumping, which is allowed for by a first order term $(k_w + k_p) N_v$. This yields

$$R_v = \left\{ \sum_{n=1}^{v} A_{v,v-n} N_v - \sum_{n=1}^{n_{max}} A_{v+n,v} N_{v+n} \right\}$$

$$+ \left\{ \sum_{\text{all } q} \sum_{n=1}^{v} k_{v,v-n}^{q} [q] N_v - \sum_{\text{all } q} \sum_{n=1}^{n_{max}} k_{v+n,v}^{q} [q] N_{v+n} \right\}$$

$$+ (k_w + k_p) N_v. \tag{1}$$

Under the conditions of arrested relaxation used in Polanyi's laboratory, excited molecules strike the walls before significant radiative relaxation, and having suffered very few gas-phase collisions. If they are then removed or completely deactivated, equation (1) holds with the $(k_w+k_p)N_V$ term dominant on the right-hand side so that R_V and N_V are directly proportional to one another. Unfortunately, it is difficult to apply this technique to the O + CS and O + CSe reactions. One reason is that CS and CSe are not stable species but have to be produced in the primary reaction of atomic oxygen with CS_2 or CSe_2; any reduction in the reagent concentrations and increase in pumping speed to reach very low total pressures would rapidly decrease the intensity of emission from $CO^\dagger$. Secondly, there is the difficulty that with CO it is necessary to observe the overtone bands, rather than the intrinsically stronger fundamental bands as in the case of diatomic hydrides. Finally, it seems that $CO^\dagger$ is less likely to be completely deactivated in single wall collisions than the hydrides.

If equation (1) is to be used to allow for relaxation it is important to choose the experimental conditions carefully. Since the spontaneous emission coefficients are accurately known, we try to minimise the magnitude of all the non-radiative terms on the right of the steady state equation. To achieve this, the reagent flows are kept as low as possible and Ar, which is extremely inefficient at relaxing CO, HF and HCl, is always added in great excess to slow down diffusion and hence reduce wall deactivation. One considerable advantage in the systems producing $CO^\dagger$ is that this molecule is, broadly speaking, much less liable to collisional deactivation than diatomic hydrides. Only vibrational-vibrational (V-V) energy exchange with $CO(v=0)$ causes significant relaxation and then only for the lowest levels ($v \leq 6$) for which results are quoted.[7a,8b] The total concentration of CO product is always $\lesssim$

5×10^{-4} Torr; that of <u>excited</u> molecules is about 2 orders of magnitude less, because the mean lifetime of excited states is much less than the mean residence time of molecules in the reaction vessel. Hancock and Smith give a detailed description and justification of the steady state model including how k_w is estimated.[8b]

The case where a non-reactive, quenching gas is added to the system is covered by simplifying the equations like (1) by including only $\Delta v=1$ and 2 radiative transitions and $\Delta v=1$ collisional processes, and then summing the resultant equations from v to v_{max} yields

$$\sum_{n=0}^{n_{max}} R_{v+n} = \cdot (A_{v,v-1} + A_{v,v-2} + \sum_q k^q_{v,v-1} [q]) N_v$$
$$+ A_{v+1,v-1} N_{v+1} + (k_w + k_p) \sum_{n=0}^{n_{max}} N_{v+n}. \qquad (2)$$

If a distinction is made between collisional relaxation by species (q) normally present and by added species (Q), equation (2) represents the system in the absence of Q, whilst an extra term, $k^Q_{v,v-1} [Q] N_v$ must be added to its right-hand side when Q is present. If Q is inert $\sum_{n=0}^{n_{max}} R_{v+n}$ is unchanged and may be eliminated between the two equations. This leads to an equation which is analogous to the Stern-Volmer relationship frequently used to describe the quenching of electronic emission and with which the experiments reviewed in section IV are analysed:

$$\frac{N_v^o}{N_v} + \frac{A_{v+1,v-1}}{k'_v} \left\{ \frac{N_{v+1}^o - N_{v+1}}{N_v} \right\} + \frac{(k_w + k_p)}{k'_v} \left\{ \sum_{n=0}^{n_{max}} \frac{N_{v+n}^o - N_{v+n}}{N_v} \right\}$$
$$= 1 + \frac{k^Q_{v+1,v-1}}{k'_v} [Q]; \qquad (3)$$

N_v^o denotes the populations when Q is absent and $k'_v = (A_{v,v-1} + A_{v,v-2} + A_{v,v-2} + \sum_q k^q_{v,v-1} [q])$.

III. THE FORMATION OF CO† IN THE REACTIONS OF O ATOMS WITH CS AND CSe

The reactions occurring in mixtures of CS_2 with oxygen have attracted considerable interest in the last few years since it is possible to sustain chemical laser action on lines in the CO fundamental bands under a variety of conditions. Pollack[14a] discovered this laser in 1966, using flash photolysis to initiate reaction. In 1969, as a result of our early observations of spontaneous emission from CO† when O atoms were mixed with CS_2, we proposed[6a] that the reaction producing CO† was that between O atoms and CS with at least 50% of the reaction energy being channelled into the CO vibration, and that in the flash photolysis lasers[14] this was preceeded and accompanied by the reactions

$CS_2 + h\nu \rightarrow CS + S$,
$S + O_2 \rightarrow SO + O$,
$O + CS_2 \rightarrow SO + CS$.

After some controversy, it now seems to be generally accepted[15-17] that the O + CS reaction is that forming CO† in all CS_2-oxygen chemical lasers.

In more recent experiments, we have concentrated on determining the R_V for the O + CS reaction using the steady state analysis outlined in section II. Because errors increase to lower v levels as collisional relaxation and wall removal become more significant relative to radiative decay, the R_V distribution is most accurate at the highest v levels and can be meaningfully extended only down to v=7. In Table 1, the distribution deduced from infrared chemiluminescence measurements is compared with those found with two other techniques,[17,18] and with that derived for the O + CSe reaction.[7b] The O + CS distributions are in broad agreement, although that derived by Foster and Kimbell[17] falls off less rapidly to low v levels. It seems that this may be due to very rapid spreading of the initial distribution by V-V energy exchange between pairs of highly excited CO molecules in Foster's laser. Certainly an effect of this kind was observed in the laser amplification experiments[18] when the excited state concentrations were apparently at least two orders of magnitude less than in Foster's system and the time resolution was better. The way in which the vibrational levels decay in such a system is undoubtedly complex and not characterised by a single decay time. Consequently it may not be sufficient to test whether V-V exchange is occurring by observing the change in the CO† distribution over a limited range of time or concentrations.

In order to compare the distributions from the O + CS and O + CSe reactions, the R_V have been plotted in figure 2 as a

Table 1

Relative Rates of Excitation of CO(v) in the Reactions: $O + CS \rightarrow CO + S$ and $O + CSe \rightarrow CO + Se$

	$O + CS \rightarrow CO + S$			$O + CSe \rightarrow CO + Se$
	(i)	(ii)	(iii)	(iv)
v = 5		0		
6		0.05	0.37	
7	0.1	0.17	0.44	
8	0.21	0.32	0.69	
9	0.49	0.41	0.82	
10	0.61	0.55	0.91	0.25
11	0.73	0.65 ~0.6	1.0	0.4
12	0.91	0.85 0.87	1.0	0.44
13	1.0	1.0 1.0	1.0	0.47
14	0.92	0.90 0.72		0.63
15	0.3	0.58 0.28		0.74
16	0	0.32 0		0.94
17		0.18		0.94
18		0		1.0
19				0.66
20				0.2
21				0

(i) from the infrared chemiluminescence experiments described in this paper;[7a] (ii) from time resolved measurements using a CO cw laser to probe the CO vibrational distributions.[18] CS is formed vibrationally 'hot' by photolysis of CS_2[19] and this causes higher CO levels to be excited in the O + CS reaction. Values of R_v in the right of this column were derived from experiments where N_2O was added to deexcite CS† before it reacted; (iii) R_v deduced by Foster and Kimbell[17] from measurements on the output of a CS_2-oxygen chemical laser; (iv) from infrared chemilumincescence.[7b]

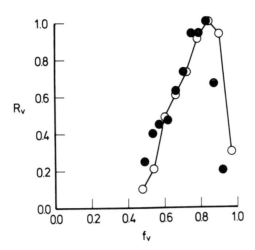

Figure 2. The R_v distributions for (○) $O + CS \to CO + S$, (●) $O + CSe \to CO + Se$, plotted against f_v, the vibrational energy expressed as a fraction of the total energy available (from ref. 7b, reprinted by permission of the Faraday Society).

function of f_v, which is the energy of the vth level expressed as a function of the total energy Q available for distribution among the reaction products. Q is approximately equal to $-\Delta H_0^o + 5/2RT$.[20] The excitation of CO(v=15) in the O + CS reaction indicates that the value until recently accepted for $D_0^o(CS)$ is too large, since its direct formation would then not be energetically possible. Recently, Okabe[21] has redetermined $D_0^o(CS)$. His value removes this anomaly and leads to the values of ΔH_0^o for the O + CS reaction which was quoted earlier.

The similarity of the R_v distributions for the CS and CSe reactions indicates that the reaction dynamics and therefore the potential hypersurfaces for the two reactions are probably very alike. Almost certainly the activation energy for both reactions is less than 1 kcal/mole, and it seems likely that most of the reaction energy is released as attraction[22] as the reagent species approach one another. Reaction through the OCS singlet ground state would lead to a more random sharing of the reaction energy than is found, and the results are more consistent with reaction occurring via a spin-allowed path over a triplet hypersurface which relative to $CO + S(^3P)$ is only weakly anti-bonding. We have reviewed elsewhere[7a] some fragmentary evidence which supports this hypothesis.

The possibility of chemical laser action from the O + CSe reaction is indicated by the data in Table 1. Wittig and Smith[23] have recently operated such a laser in a pulsed discharge system. Stimulated emission has been observed on P-branch transitions in CO bands as high as (18,17).

IV. VIBRATIONAL DEEXCITATION OF $CO(v \leq 13)$, $HCl(v \leq 3)$ $HF(v \leq 5)$ AND $DF(v \leq 3)$

One of the main requirements in attempting to model infrared molecular lasers is a set of rate coefficients for the major processes which transfer molecules between the range of vibrational states which are populated in the laser medium. Unfortunately, most relaxation measurements only provide data on transfer between the lowest and first excited vibrational levels and using these results to estimate the rates of processes involving more highly excited molecules has its dangers. Quenching measurements of steady state infrared chemiluminescence, under carefully selected experimental conditions, can provide this valuable data directly. A detailed description of the method has been given by Hancock and Smith[8b] who used it to derive rate coefficients for transfer of energy from individual vibrational levels of CO ($4 \leq v \leq 13$) to a number of collision partners. Subsequently, measurements have been made on $HCl(v \leq 3)$[9] and $HF(v \leq 5)$.[10] Because of the interest in molecular lasers, vibrational relaxation is a field of intense research activity at the present time. In the present paper, we largely confine ourselves to a comparison of some very recent experimental and theoretical data with the rates of vibrational deexcitation which have been obtained by the infrared quenching technique.

For CO, Hancock and Smith determined probabilities (P) of V-V exchange with $CO(v=0)$, NO, N_2, O_2, OCS, N_2O and CO_2. One important aspect of being able to obtain data for several v states is that rates can be compared for processes which differ only in the levels of excitation of one collision partner and the discrepancy, Δv, between the two vibrational transition energies. According to harmonic oscillator selection rules, for a process like $AB(v=m) + CD(v=n) \rightarrow AB(v=m-1) + CD(v=n+1)$,

$P_{m,m-1}^{n,n+1} = m(n+1) P_{1,0}^{0,1}$, so the dependence on excitation level can be allowed for by calculating a reduced probability, $P_{m,m-1}^{n,n+1}/m(n+1)$.

These reduced probabilities, plotted for V-V exchange in the exothermic direction, fall off with increasing energy discrepancy once $\Delta v \geq 60$ cm^{-1}. From this plot (see Fig. 4 of ref. 8b), some particularly important comparisons with other data can be made, since estimates of probabilities can be made for the processes

Table 2

Rate Coefficients ($Torr^{-1} sec^{-1}$) for Collisional Deexcitation of $HCl(v \leq 3)$, $HF(v \leq 5)$ and $DF(v \leq 3)$

	(a)	(b)		ref.
$HCl(v=1) + HBr$	2.7(4)	3.4(4)		28(a)
$HCl(v=1) + CO_2$	6.7(4)	9.4(4)		28(b)
$HCl(v=2) + CO_2$	2.2(5)			
$HCl(v=3) + CO_2$	5.0(5)			
$HCl(v=2) + HCl$	4.5(4)	8.4(4)		29(a)
		9(4)		29(b)
		10(4)		29(c)
$HCl(v=3) + HCl$	5.8(4)			
$HCl(v=1) + Cl$	3.1(4)	3.5(5)		30
$HCl(v=2) + Cl$	1.0(5)			
$HCl(v=3) + Cl$	2.6(5)			
$HF(v=2) + HF$	6.0(5)	6.6(5)		31(a)
		5.0(5)		31(b)
$HF(v=3) + HF$	6.4(5)			
$HF(v=4) + HF$	>1.7(6)			
$HF(v=5) + HF$	>2.6(6)			
$HF(v=1) + CO_2$	4.0(4)	3.7(4)	350°K	32
		5.9(4)		33
$HF(v=2) + CO_2$	1.6(5)			
$HF(v=3) + CO_2$	2.4(5)			
$HF(v=4) + CO_2$	4.8(5)			
$HF(v=5) + CO_2$	>8(5)			
$DF(v=1) + HF$	7.0(4)	3.4(4)		34
$DF(v=2) + HF$	9.5(4)			
$DF(v=1) + CO_2$	1.3(5)	1.7(5)	350°K	32
		1.8(5)		35
$DF(v=2) + CO_2$	2.2(5)			
$DF(v=3) + CO_2$	5.2(5)			

(a) from infrared quenching experiments, $T = 300(\pm 5)°K$; 2.7(4) corresponds to 2.7×10^4 $Torr^{-1} sec^{-1}$; (b) other measurements, all using laser excited fluorescence.

$$CO(v=1) + NO(v=0) \rightleftarrows CO(v=0) + NO(v=1),^{24}$$
$$CO(v=1) + N_2(v=0) \rightleftarrows CO(v=0) + N_2(v=1),^{25,26}$$
$$CO(v=1) + OCS(000) \rightleftarrows CO(v=0) + OCS(001),^{27}$$

and compared with the values obtained by other methods. Hancock and Smith suggested that earlier measurements on $CO-N_2$ exchange might be in error and this proposal now appears to be confirmed by Zittel and Moore's,[25] and Green and Hancock's,[26] laser excited fluorescence experiments. All of these measurements are in good agreement with estimates from the infrared quenching experiments and increase our confidence in the steady state interpretation of our measurements. Current experiments in our own laboratories are also designed partly to test the analysis further. Here, quenching measurements are being made on the chemiluminescence from the O + CSe reaction. Rates derived for processes involving $CO(v \leq 13)$ should coincide with those obtained already if the steady state model adequately describes the relaxation behaviour. These experiments will also extend data on relaxation of $CO^{\dagger}$ up to $v = 20$.

Table 2 presents a summary of data on HCl and HF relaxation compared to results obtained by various other groups using laser excited fluorescence. This comparison reveals one major discrepancy: between the deactivation rates of $HCl(v=1)$ by Cl atoms determined by Ridley and Smith[9a] and by Craig and Moore.[30] Experiments are under way to redetermine the rate of this process using a technique which combines some of the better features of the two previous methods. Otherwise the level of agreement between our data and results from laser excited fluorescence measurements is reasonably good. Moreover, new theoretical calculations[36] are able to predict the high probabilities found for many V-V exchange processes of importance in chemical lasers even where Δv is several hundreds of cm^{-1}. It also seems likely that these calculations will account for the initially surprising discovery that exchange probabilities for $CO(v \leq 9) + CO(v=1) \rightarrow CO(v+1) + CO(v=0)$ are not significantly different at 100°K from those at 300°K.[37]

V. INFRARED CHEMILUMINESCENCE FROM $OH^{\dagger}$

One of the first reactions known to be extremely efficient at exciting a product vibrationally is

$$H + O_3 \rightarrow OH^{\dagger} + O_2, \quad \Delta H_0^o = -77 \text{ kcal/mole.}$$

Experiments by Polanyi and his coworkers[38] have shown that this reaction proceeds fastest into $OH(v=9)$ and that $\sim 90\%$ of the total energy is channelled into vibrational excitation of OH. Despite this, attempts to construct chemical lasers based on this reaction have not been very successful.[39] Undoubtedly there are two major reasons for this: the low transition probabilities for the OH vibration-rotation bands,[40] and the susceptibility of OH to reaction or fast vibrational relaxation. We have explored two systems which it was hoped might be capable of sustaining laser action on OH. In both, an attempt was made to use secondary reactions to remove OH formed in the primary process; the hope being that the secondary reaction might occur (a) more rapidly than vibrational relaxation, and (b) at a rate approximately independent of the vibrational level of OH, so that the OH distribution formed initially in the first reaction might be maintained in a steady state system.

In the first series of experiments, O atoms were mixed with HBr or HI. The reactions,

$$O + HBr \rightarrow OH + Br, \quad \Delta H_0^o = -13.7 \text{ kcal/mole}$$

and

$$O + HI \rightarrow OH + I, \quad \Delta H_0^o = -29.8 \text{ kcal/mole},$$

are of similar exothermicity to those of Cl atoms with HBr and HI, which form the basis of HCl lasers.[41] It was expected that the $O + HX$ (X = Br or I) reaction would be followed by

$$OH + HX \rightarrow H_2O + X,$$

which might act in the manner suggested in the previous paragraph. However, no OH chemiluminescence could be detected from either O + HBr or O + HI. Later rate measurements[42] indicate that a major cause is that the O-atom reactions are much slower (at least 2 orders of magnitude) than either the corresponding Cl-atom reactions or the OH + HX reactions. In view of the indicated activation barrier, it seems unlikely that the O + HX reactions form OH^{+} efficiently and, in any case, the rapid secondary reaction is unlikely to permit the OH concentration to reach the level required for laser action even if the OH vibrational distribution were favorable.

In a second set of experiments, a preliminary study has been made of the effects on the OH vibrational chemiluminescence of adding CO to a reaction mixture containing H-atoms and O_3. In the absence of added CO, the OH vibrational distribution was the heavily relaxed one observed previously in experiments similar to ours.[43] CO was added to initiate the reaction,

$$OH + CO \rightarrow CO_2 + H,$$

which regenerates H-atoms, and could maintain the inverted distribution of OH from the $H + O_3$ reaction if it successfully competes with vibrational relaxation. Although the effect of adding CO was less dramatic than had been hoped, population inversions were obtained, in this slow-flow, high pressure system, between OH(v=4) and OH(v=3) and between (v=3) and (v=2). The greatest inversion, $N_2:N_3:N_4 \sim 1:4.8:6.2$, was found under the following conditions: relative flows-discharged $H_2:O_3:CO:Ar = 1:4:20:3500$, total pressure = 3.5 Torr, pumping speed 1 liter/sec. It was possible to estimate the absolute vibrational level concentrations by comparing the intensity of the OH overtone bands with the intensity of the NO $C^2\pi-A^2\Sigma^+$ electronic chemiluminescence, which is emitted at 1.22 μm at a known rate during the preassociation of N and O atoms. The concentrations which were determined ($\sim 10^{-4}$ Torr) are about two orders of magnitude below those likely to be needed in a laser medium. No scaled-up experiments have yet been performed but it does seem possible that this method might form the basis of an OH cw laser system.

It is a pleasure to record my gratitude to my collaborators in the research summarised in this paper: J.R. Airey, M. Braithwaite, G. Hancock, C. Morley, B.A. Ridley, and C. Wittig. I should also like to thank S.R.C. for support and Drs. J.F. Bott, H.-L. Chen, T.A. Dillon, K.D. Foster, W.H. Green, J.J. Hinchen, C.B. Moore, and A. Szoke for sending me details of their work prior to publication.

REFERENCES

1. J.K. Cashion and J.C. Polanyi, J. Chem. Phys. 29, 455 (1958); 30, 1047 (1959).

2. (a) K.G. Anlauf, P.E. Charters, D.S. Horne, R.G. Macdonald, D.H. Maylotte, J.C. Polanyi, W.J. Skrlac, D.C. Tardy, and K.B. Woodall, J. Chem. Phys. 53, 4091 (1970); (b) K.G. Anlauf, P.J. Kuntz, D.H. Maylotte, P.D. Pacey, and J.C. Polanyi, Disc. Farad. Soc. 44, 183 (1967); D.H. Maylotte, J.C. Polanyi, and K.B. Woodall, J. Chem. Phys. 57, 1547 (1972); K.G. Anlauf, D.S. Horne, R.G. Macdonald, J.C. Polanyi, and K.B. Woodall, J. Chem. Phys. 57, 1561 (1972); J.C. Polanyi and K.B. Woodall, J. Chem. Phys. 57, 1574 (1972).

3. R.L Johnson, M.J. Perona, and D.W. Setser, J. Chem. Phys. 52, 6372, 6384 (1970).

4. N. Jonathan, C.M. Melliar-Smith, and D.H. Slater, Mol. Phys. 20, 93 (1971); N. Jonathan, C.M. Melliar-Smith, D. Timlin, and D.H. Slater, Appl. Optics 10, 1821 (1971).

5. C.B. Moore, Ann. Rev. Phys. Chem. 22, 387 (1971).

6. (a) G. Hancock and I.W.M. Smith, Chem. Phys. Lett. 3, 469 (1969); (b) G. Hancock and I.W.M. Smith, Trans. Farad. Soc. 67, 2856 (1971).

7. (a) G. Hancock, B.A. Ridley, and I.W.M. Smith, J.C.S., Farad. Trans. II, in press; (b) C. Morley, B.A. Ridley, and I.W.M. Smith, J.C.S., Farad. Trans. II, in press.

8. (a) G. Hancock and I.W.M. Smith, Chem. Phys. Lett. $\underline{8}$, 41 (1971); (b) G. Hancock and I.W.M. Smith, App. Optics $\underline{10}$, 1827 (1971).

9. (a) B.A. Ridley and I.W.M. Smith, Chem. Phys. Lett. $\underline{9}$, 457 (1971); (b) B.A. Ridley and I.W.M. Smith, J.C.S., Farad. Trans. II, $\underline{68}$, 1231 (1972).

10. J.R. Airey and I.W.M. Smith, J. Chem. Phys. $\underline{57}$, 1669 (1972).

11. J.K. Cashion and J.C. Polanyi, Proc. Roy. Soc. A (London) $\underline{258}$, 529 (1960).

12. G. Karl, P. Kruus, and J.C. Polanyi, J. Chem. Phys. $\underline{46}$, 224 (1967).

13. P.E. Charters and J.C. Polanyi, Disc. Farad. Soc.

14. (a) M.A. Pollack, Appl. Phys. Lett. $\underline{8}$, 237 (1966); (b) D.W. Gregg and S.J. Thomas, J. Appl. Phys. $\underline{39}$, 4399 (1968).

15. C. Wittig, J.C. Hassler, and P.D. Coleman, J. Chem. Phys. $\underline{55}$, 5523 (1971).

16. H.S. Pilloff, S.K. Searles, and N. Djeu, Appl. Phys. Lett. $\underline{19}$, 9 (1971).

17. K.D. Foster and G.H. Kimbell, 14th Intl. Symp. Comb., in press; K.D. Foster, J. Chem. Phys. $\underline{57}$, 2451 (1972).

18. G. Hancock, C. Morley, and I.W.M. Smith, Chem. Phys. Lett. $\underline{12}$, 193 (1971).

19. A.B. Callear, Proc. Roy. Soc. A (London) $\underline{276}$, 401 (1963).

20. K.G. Anlauf, D.H. Maylotte, J.C. Polanyi, and R.B. Bernstein, J. Chem. Phys. $\underline{51}$, 5716 (1969); J.C. Polanyi and D.C. Tardy, J. Chem. Phys. $\underline{51}$, 5717 (1969).

21. H. Okabe, J. Chem. Phys. $\underline{56}$, 4381 (1972) gives $D_0^o(CS)$ = 170.4(±0.7) kcal/mole. This value is about 10 kcal/mole higher than that derived from JANAF tables but in agreement with an earlier estimate of $\Delta H_0^o(CS)$ by V.H. Dibeler and J.A. Walker, J. Opt. Soc. Am. $\underline{57}$, 1007 (1967) which seems to have been generally overlooked.

22. P.J. Kuntz, E.M. Nemeth, J.C. Polanyi, S.D. Rosner, and C.E. Young, J. Chem. Phys. 44, 1168 (1966); J.C. Polanyi, Acc. Chem. Res. 5, 161 (1972).

23. C. Wittig and I.W.M. Smith, Appl. Phys. Lett., in press.

24. N. Basco, A.B. Callear, and R.G.W. Norrish, Proc. Roy. Soc. A (London) 269, 180 (1962).

25. P.F. Zittel and C.B. Moore, Appl. Phys. Lett. 21, 81 (1972).

26. W.H. Green and J.K. Hancock, Proc. of 3rd Conf. on Chem. and Mol. Lasers to be published in I.E.E.E., J. Quart. Electron.

27. M. Lev-On, D.C. Richman, D.J. Miller, and R.C. Millikan, J. Chem. Phys., in press.

28. (a) H.-L. Chen and C.B. Moore, J. Chem. Phys. 54, 4072, 4080 (1971); (b) H.-L. Chen, J.C. Stephenson, and C.B. Moore, Chem. Phys. Lett. 2, 593 (1968).

29. (a) B.M. Hopkins and H.-L. Chen, to be published; (b) I. Burak, Y. Noter, A.M. Ronn, and A. Szoke, to be published; (c) S. Leone and C.B. Moore, to be published.

30. N.C. Craig and C.B. Moore, J. Phys. Chem. 75, 1622 (1971)

31. (a) R.M. Osgood, Jr., A. Javan, and P.B. Sackett, Appl. Phys. Lett. 20, 469 (1972); (b) J.F. Bott, J. Chem. Phys. 57, 96 (1972).

32. R.R. Stephens and T.A. Cool, J. Chem. Phys. 56, 5863 (1972).

33. J.K. Hancock and W.H. Green, J. Chem. Phys., in press.

34. J.J. Hinchen, J. Chem. Phys., in press.

35. J.F. Bott and N. Cohen, to be published.

36. T.A. Dillon and J.C. Stephenson, Phys. Rev. A, in press.

37. C. Wittig and I.W.M. Smith, Chem. Phys. Lett., in press.

38. K.G. Analuf, R.G. Macdonald, and J.C. Polanyi, Chem. Phys. Lett. 1, 619 (1968); P.E. Charters, R.G. Macdonald, and J.C. Polanyi, Appl. Optics 10, 1747 (1971).

39. The suggestion for an OH chemical laser was made by J.H. Kiefer in 1962. Pulsed lasers have been reported by A.B.

Callear and H. van den Berg, Chem. Phys. Lett. $\underline{8}$, 17 (1971); E. Vietzke, H.I. Schiff, and K.H. Welge, Chem. Phys. Lett. $\underline{12}$, 429 (1971). No report of cw operation has appeared.

40. R.E. Murphy, J. Chem. Phys. $\underline{54}$, 4852 (1971).

41. J.R. Airey, I.E.E.E., J. Quant. Electron. $\underline{3}$, 208 (1967); J.R. Airey, J. Chem. Phys. $\underline{52}$, 156 (1970).

42. C. Morley, Ph.D. thesis, University of Cambridge, 1971.

43. D. Garvin, H.P. Broida, and H.J. Kostkowski, J. Chem. Phys. $\underline{32}$, 880 (1960).

44. R.A. Young and R.L Sharpless, Disc. Farad. Soc. $\underline{33}$, 228 (1962).

STEDMAN: Are you able to observe in your OH excitation the rotational lines, or are they quenched out at the pressure you are using?

SMITH: In none of these experiments were we anywhere near getting initial rotational populations. In all our work we were rotationally equilibrated: after all we worked at 10 torr, so there's no doubt.

LIN: I have a comment about the OCS system. We flashed photolysed OCS and naturally we got CO laser emission. In this case the emission we got was from the singlet state. We are planning to do laser problems in which we can determine initial populations and to see how we can compare the results of these systems.

CHEMICAL LASERS PRODUCED FROM O(^{3}P) ATOM REACTIONS. II. A MECHANISTIC STUDY OF 5-μm CO LASER EMISSION FROM THE O + C_2H_2 REACTION

M. C. Lin

Naval Research Laboratory

Washington, D. C.

I. INTRODUCTION

The reaction of O(^{3}P) atoms with C_2H_2 has been the subject of many investigations. At low temperatures, the mechanism of CO formation, studied by a variety of methods,[1-6] is now generally accepted as[7]

$$O + C_2H_2 \rightarrow CO\dagger + CH_2 + 47 \text{ kcal/mole} \quad (1)$$

$$O + CH_2 \rightarrow CO\dagger + 2H + 75 \text{ kcal/mole} \quad (2)$$

Other minor reactions yielding HC_2O and C_2O are also believed to occur.[5] IR chemiluminescence experiments have been performed independently by Creek et al.[8] and by Thrush and co-workers[9] employing low-pressure, fast-flow methods. At low C_2H_2 concentrations, the distribution of CO extends to v'=14, which is attributed to reactions (1) and (2).[8,9] Additional distribution extending up to v'≈33 was detected at high C_2H_2 flows and in the presence of H_2. This is believed to be due to reaction (3),[9]

$$O + CH \rightarrow CO\dagger + H + 176 \text{ kcal/mole} \quad (3)$$

Other processes such as:

$$O + C_2O \rightarrow CO^* + CO\dagger + 65 \text{ kcal/mole} \quad (4)$$

$$2O + C_2H_2 \rightarrow O + [C_2H_2O]^* \rightarrow 2CO\dagger + 2H + 122 \text{ kcal/mole} \quad (5)$$

have also been proposed to account for the production of CO in high vibrational levels.[5,8]

We have investigated the mechanism of CO† formation in the reactions of $O(^3P)$ with C_2H_2 and CH_2 through 5-μm CO laser emission measurements. The present results, combined with those obtained from the CO laser emission study of the $O(^3P)$ + CH reaction,[10] indicate that reaction (2) is the primary laser pumping process and that reaction (3) is unimportant in the $O(^3P)$-C_2H_2 system under our experimental conditions.

II. EXPERIMENTAL

The vacuum U.V. flash apparatus used in these experiments has been described elsewhere.[11] Briefly, a Suprasil laser tube (1m long, 2.2 cm in I.D.) fitted with BaF_2 windows at the Brewster angle was positioned in an optical cavity formed by two 2.5-cm diameter, 3-m radius gold-coated mirrors at a separation of about 1.2m; one of the mirrors had a 1-mm coupling hole at its center. Two quartz flash lamps, each 40 cm long and 3.4 cm in I.D., were sealed concentrically to the Suprasil laser tube. The space between the two flash lamps, about 2 cm, was used as a sample inlet, directly connected to the vacuum system which could be readily pumped down to 10^{-5} torr.

The flash output had a rise time of ∼ 15 μsec and a half-width of ∼ 45 μsec when 10 torr of 1% Xe/Ar mixture was used as the flash source. The width lengthened slightly when a higher pressure was employed. Laser emissions were analyzed by passing the beam through a 0.5-m 305M13 Minuteman grating monochromator, and were observed by a Ge:Au detector in conjunction with an oscilloscope.

All gases used in this work, except CH_2Br_2 (Aldrich), were obtained from the Matheson Gas Products. The condensable gases were subjected to trap-to-trap distillation at 77°K, while the noncondensable were delivered under high pressure and were used without further purification. All experiments were conducted at room temperature (24 ± 2°C).

III. RESULTS AND DISCUSSION

CO laser emission at 5μm was detected when SO_2 and C_2H_2 (or CH_2Br_2) were flash photolyzed in vacuum U.V. (λ ≥ 165nm) in the presence of large amounts of SF_6. The preliminary results of these experiments have been reported together with several other laser systems involving $O(^3P)$ atoms.[11] This paper presents the results obtained from the parametric study of CO laser emissions from both the O + C_2H_2 and the O + CH_2 reactions.

In the present work, CH_2Br_2 was chosen as the CH_2 radical source instead of the commonly used chemicals, such as CH_2CO and

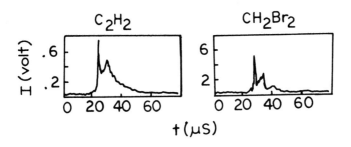

Fig. 1. The Total Laser Emission Traces.

CH_2N_2, to avoid the complication that may arise from the presence of $CO^{\dagger}$ and $N_2^{\dagger}$ formed in the initial photodissociation reactions. CO laser emission at 5 μm was actually detected when CH_2CO was flash-photolyzed under similar experimental conditions.[12] The advantage of using SO_2 as the $O(^3P)$ atom source lies in its high extinction coefficient below 200 nm, the relatively high stability of both SO_2 and SO, and the insignificance of $O(^1D)$ atom production above 165nm.[11]

Fig. 1 shows the total laser emission pulses observed from 20 torr of $1:1:20/SO_2:C_2H_2(\text{or } CH_2Br_2):SF_6$ mixtures flashed with 1.6 kj energy. The emissions begin at ~ 15-20 μsec, and continue until approximately 35-40 μsec; this width of ~ 15-20 μsec is about the same as that observed from the $O(^1D)$ + CN reaction,[13] and it is only 10% of that detected from the $O(^1D)$ + C_3O_2 reaction.[14]

1. DEPENDENCE ON FLASH ENERGY

The peak power of laser emission from both C_2H_2 and the CH_2Br_2 systems varies with flash energy (E_λ). The intensity of the $SO_2 - C_2H_2$ system increases linearly with E_λ, whereas that of the $SO_2-CH_2Br_2$ system varies linearly with $(E_\lambda)^2$; the results are shown in Fig. 2. These observations lead directly to the following conclusions. (1) The rates of production of O and CH_2, in the range of flash energies applied, are limited by the initial photodissociation processes:

$$SO_2 + h\nu \rightarrow O(^3P) + SO\ (^3\Sigma^-),\ I_a \propto E_\lambda$$

$$CH_2Br_2 + h\nu \rightarrow CH_2 + 2Br,\ I_b \propto E_\lambda$$

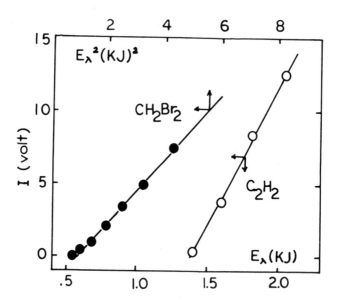

Fig. 2. Dependence of Laser Intensity on Flash Energy.

(2) Under the conditions employed, the rate of reaction (2) is so fast that the side reactions involving O and CH_2 such as:

$$CH_2 + SO \rightarrow products \qquad (6)$$

$$CH_2 + CH_2 \rightarrow C_2H_4 \text{ or } C_2H_2 + H_2 \qquad (7)$$

$$CH_2 + C_2H_2 \rightarrow C_3H_4 \qquad (8)$$

$$O + SO + M \rightarrow SO_2 + M \qquad (9)$$

are not important. The linear relationships given in Fig. 2 are not expected to hold for higher flash energies at which the radical concentrations become greater. This was found to be the case in the SO_2-$CHBr_3$ CO laser system.[10]

2. DEPENDENCE ON PRESSURE AND RELATIVE PROPORTIONS

The effect of pressure on total laser intensities is shown in Fig. 3a for 2:1:20/SO_2:C_2H_2:SF_6 and 1:1:20/SO_2:CH_2Br_2:SF_6 mixtures flashed with 1.8 kj energy. For C_2H_2, the optimum pressure occurs at 10 torr; beyond that, the intensity drops rapidly, and laser oscillation ceases to take place at P > 40 torr. The emission from the CH_2Br_2 mixture is stronger; its

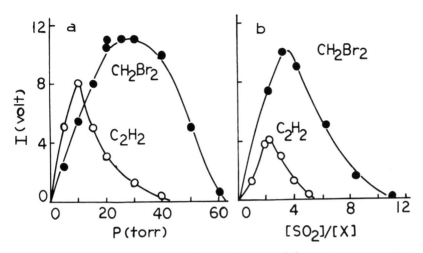

Fig. 3. Dependence on Laser Intensity on (a) Total Pressure and (b) Relative Proportions, X = C_2H_2, CH_2Br_2.

optimum pressure range extends from 20 to 40 torr. Laser action terminates at P > 60 torr.

The observed total pressure effects in both systems may result from many contributions: The increase in CO↑ relaxation rate, the reduction in rate of CO↑ production by third-body effect and the enhancement of termination reactions (6)-(9) as the total pressure (reactants + diluent) increases. Some of these individual effects will be examined later.

Fig. 3b shows the dependence of total emission intensities on the $[SO_2]:[X]$ ratio, where X = C_2H_2, CH_2Br_2. The experiments were carried out by adding various amounts of SO_2 to 10 torr of 1:20/X:SF_6 mixtures; E_λ=1.6 kj for all flashes. For C_2H_2, the mixture with $[SO_2]:[C_2H_2]$=2 produces the highest power; it drops drastically as the $[SO_2]:[C_2H_2]$ ratio exceeds 2. The CH_2Br_2 system generates comparatively higher output; its maximum occurs at $[SO_2]:[CH_2Br_2]$ = 3.5. A similar rapid decrease in intensity is also noted when the ratio passes 3.5. Since $[SO_2]$ is the only variable in these experiments, the drop in power at higher $[SO_2]$ in both systems is attributed to the increasing importance of reaction (6). The analogous reaction $CH_2 + O_2$, as will be shown in Section 4, is very effective in quenching the laser emissions.

The more pronounced effect of excess SO_2 on the laser intensity in the C_2H_2 system is probably due to its lower gain, resulting from the inherently slower rate of reaction (1). On the other hand, if reaction (1) is the main pumping step, one does not expect such a drastic quenching by SO_2 at $[SO_2]:[C_2H_2] > 2$. This indirectly leads us to the tentative conclusion that, in both systems, the primary pumping process is reaction (2). The conclusion will be further substantiated by other more direct means.

3. INERT GAS EFFECTS

In a chemical laser system, an inert diluent usually plays two important but mutually nullifying roles. It prevents the rise of rotational-translational temperature of the laser system and thus enhances its output. On the other hand, the presence of large excess of a buffer gas increases the rate of relaxation of the lasing species and, accordingly, lowers the power. Because of these two cancelling effects, a bell-shaped curve such as that shown in Fig. 4 is obtained. The data presented in the figure were taken from mixing various amounts of SF_6 with 2 torr of $2:1/SO_2:C_2H_2$ and $1:1/SO_2:CH_2Br_2$ mixtures flashed with 1.8 and 1.6 kj of energies, respectively.

It is interesting to note the difference in the effects of SF_6 on these two systems. The emission intensity of the C_2H_2 mixture reaches its maximum when 10 torr of SF_6 is added; it decreases steeply at higher SF_6 pressures and disappears totally at $[SF_6] > 20$ torr. The CH_2Br_2 mixture, however, has a broader plateau, between 20-60 torr, with maximum outputs. The emission declines as the pressure increases but still persists quite strongly at $[SF_6] > 110$ torr. Apart from the fact that the

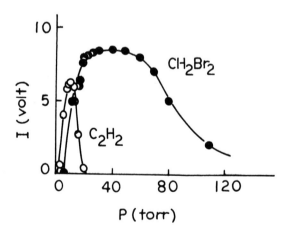

Fig. 4. Inert Gas (SF_6) Effect on Laser Output.

CH_2Br_2 system has a higher gain due to a greater $CO\dagger$ concentration, (based on our gas analysis results given in Section 5), the clear difference between the two systems seems to indicate the possibility of a deactivation effect other than the vibrational relaxation of $CO\dagger$.

It has been widely assumed that reaction (1) takes place via the CH_2CO molecule, formed by a hydrogen atom migration:

$$O(^3P) + C_2H_2 \rightarrow CH_2CO^* \rightarrow CH_2\,(^3\Sigma^-) + CO\dagger$$

In a solid matrix, the presence of CH_2CO was confirmed by Haller and Pimentel[15] in their study of the $O + C_2H_2$ reaction. If the spin-conservation rule holds for this reaction, CH_2CO must be in its triplet electronic state, which is about 61 kcal/mole above the singlet ground state.[16] The newly formed CH_2CO^* molecule therefore possesses about 68 kcal/mole of internal energy, or about 47 kcal/mole above the $CH_2(^3\Sigma^-) + CO(^1\Sigma^+)$ dissociation limit.

The intermediate of reaction (2) is likely to be a vibronically excited CH_2O molecule, which may be in its quintet, triplet or singlet state.

$$O(^3P) + CH_2(^3\Sigma-) \rightarrow CH_2O^* \rightarrow CO\dagger + 2H \text{ (or } H_2\dagger)$$

The CH_2O^* molecule thus produced contains as much as 180 kcal/mole of internal energy. The threshold energy for the formation of H atoms is 87 kcal/mole, the value of $D(HCO-H)$;[17] that for the $CO + H_2$ formation is not known. However, it might not differ very much from the above value because the threshold for the elimination of H_2 from C_2H_4 has been estimated to be ~ 80 kcal/mole.[18] Both steps may take place concurrently in reaction (2). On these grounds, the excess energy possessed by CH_2O^* can be as much as ~ 90 kcal/mole. On account of the higher excess energy (cf. 47 kcal/mole for CH_2CO^*) CH_2O^* should experience less stabilization than CH_2CO^* by third-body. Besides, the $O + C_2H_2$ reaction encounters dual deactivations described above because of its two consecutive steps.

Fig. 4 also indicates that SF_6 dilution is indispensable to laser action in both systems; no laser emission was detected when it was absent. Nor was emission observed when He, Ar, CF_4 or CH_4 was used instead. The high efficiency of SF_6 as a diluent for many laser systems including hydrogen halides can be ascribed to its large heat capacity as well as its small deactivation cross sections for these molecules.

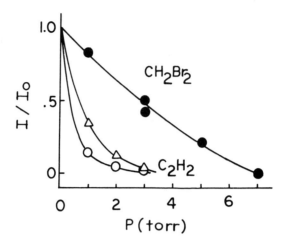

Fig. 5. Effects of O_2 and C_2H_2 on Emission Intensity. Circles, O_2; Triangles, C_2H_2.

4. INHIBITION BY O_2 AND C_2H_2

The CH_2 radical is known to be scavenged effectively by oxygen and olefins. We have investigated their effects on the laser outputs in both systems; the results are shown in Fig. 5. The experiments were done by adding small amounts of O_2 or C_2H_2 to 10 torr of $2:1:20/SO_2:C_2H_2:SF_6$ and $1:1:20/SO_2:CH_2Br_2:SF_6$ mixtures. The flash energy was maintained at 1.8 kj. According to Fig. 5, addition of 1 torr O_2 and C_2H_2 each to the C_2H_2 mixture reduces its output by 86 and 66% respectively, and 3 torr of each additive quenches the laser action almost entirely. The effect of O_2 on the emission from the CH_2Br_2 mixture, although less drastic, is still quite significant. For example, 3 torr of O_2 reduces the intensity by as much as 50%.

These effects cannot be due to the vibrational relaxation of CO† by the additives, because O_2 was actually found to enhance CO laser emissions in the photodissociations of CH_2CO and OCS.[12] The effect of C_2H_2 on the emissions from both systems is probably due to reaction (8); gas analysis carried out for these runs indeed shows an increase in the C_3H_4 production. The inhibition effect of O_2 is probably the result of removal of the CH_2 radical via the following mechanism:[19]

$$CH_2 + O_2 \rightarrow CH_2O + O + 62 \text{ kcal/mole} \qquad (10)$$

$$CH_2 + O_2 \rightleftarrows \cdot CH_2OO \cdot \dagger \rightarrow HCOOH\dagger + 184 \text{ kcal/mole} \qquad (11)$$

The excited HCOOH molecule contains ~ 184 kcal/mole of energy. It can readily undergo various decomposition reactions yielding $H_2O + CO$ ($\Delta H=6.3$ kcal/mole), or $HCO + OH$ ($\Delta H=107$ kcal/mole), $HCOO + H$ ($\Delta H=107$ kcal/mole) or $H + COOH$ ($\Delta H=93$ kcal/mole). Gas analysis of the flashed mixtures in the presence of O_2 indicates increases in CO and CO_2 yields.

The drastic quenching of the C_2H_2 output by both O_2 and C_2H_2 strongly suggests that reaction (2) is the major laser pumping step in both systems and also that the population of CO formed in reaction (1) is not inverted. Otherwise, the oscillation should continue to occur at lower vibrational levels even after reaction (2) is completely stopped.

5. PRODUCT ANALYSIS

The product analysis of both flashed and unflashed mixtures, employing a CEC-620 mass spectrometer, shows that a single flash of a 10 torr $2:1:20/SO_2:C_2H_2:SF_6$ mixture with 1.6 kj energy consumed ~ 20% C_2H_2 and ~ 10% SO_2. A trace amount of C_3H_4 (m/e=40) was also detected. The relative yield of CO, measured by the ratio of peak heights of mass 28 to 35 (SF_2^{++}), was [28]/[35]=0.31. A similar flash of 10 torr $1:1:20/SO_2:CH_2Br_2:SF_6$ mixture led to ~ 12% SO_2 conversion and the relative CO yield [28]/[35]=0.47. Thus, the maximum CO† production rate via reaction (2) in the C_2H_2 system is only 0.16, which is only 1/3 as much as that measured in the CH_2Br_2 system. This is in total agreement with a stronger emission observed in the CH_2Br_2 flash experiments.

6. IDENTIFICATION OF TRANSITIONS

Analysis of laser emissions from the flash ($E_\lambda=18$kj) of 10 torr $2:1:20/SO_2:C_2H_2:SF_6$ mixture indicated that $P_{13,12}$ (9) and $P_{12,11}$ (9)-(12) were present. A similar analysis of the emission from 20 torr $3:1:20/SO_2:CH_2Br_2:SF_6$ mixture identified the following 15 transitions: $P_{9,8}$ (10)-(12), $P_{8,7}$ (10)-(13), $P_{7,6}$ (10)-(14), $P_{6,5}$ (12)-(14) and $P_{5,4}$ (12).

The absence of the transitions above $v'=14$ and below $v'=11$ in the $O-C_2H_2$ system is significant. Without the higher transitions, one may conclude that reaction (3) is unimportant, since very strong CO laser emissions with $v' > 14$ were observed in the $O + CH$ reaction carried out under similar experimental conditions.[10]

The same conclusion also holds for the CH_2Br_2 system.

The absence of lower transitions ($v'<11$) in the C_2H_2 system, in contrast to those observed in CH_2Br_2 indicates that the vibrational temperature of CO molecules below $v'=11$ formed by the $O + C_2H_2$ reaction is lower than that of the molecules produced by the $O + CH_2$ reaction. Essentially, this implies that the CO molecules produced by reaction (1) are much colder than those by reaction (2), in accordance with the differences in exothermicities and the sizes of reaction intermediates for the two reactions. The presence of these colder molecules from reaction (1) in the C_2H_2 system pushes the gain created by reaction (2) up to higher vibrational levels (i.e., v'=13, 12) probably by Treanor type pumping.[20] This explains why the C_2H_2 system oscillates at higher v'.

IV. CONCLUSION

CO laser emission at 5 μm was observed from both $O + C_2H_2$ and $O + CH_2$ reactions. Product analysis and detailed study of the effects of pressure and inhibition by O_2 and C_2H_2 lead to the conclusion that the stimulated emission observed in these systems is primarily a consequence of the $O + CH_2$ reaction. Based on the results of frequency analysis, it is concluded that the $O + CH$ reaction is unimportant under the present conditions. Reactions (4) and (5) can also be excluded because of the similarities in all aspects, except observed laser transitions, between the C_2H_2 and the CH_2Br_2 systems.

V. REFERENCES

1. J. O. Sullivan and P. Warneck, J. Phys. Chem., 69 1749 (1965).
2. C. A. Carrington, W. Brennan, G. P. Glass, J. V. Michael and H. Niki, J. Chem. Phys. 43 525 (1965).
3. J. M. Brown and B. A. Thrush, Trans. Faraday Soc. 63 630 (1970).
4. G. S. James and G. P. Glass, J. Chem. Phys. 50 2268 (1969).
5. D. G. Williamson and K. D. Bayes, J. Phys. Chem. 73 1232 (1969).
6. A. A. Westenberg and N. de Haas, J. Phys. Chem. 73 1181 (1969).
7. Throughout the text, "†" will be used for vibrational excitation and "*" for electronic excitation.
8. D. M. Creek, C. M. Melliar - Smith and N. Jonathan, J. Chem. Soc. (A) 646 (1970).
9. P. N. Clough, S. E. Schwartz and B. A. Thrush, Proc. Roy. Soc. (London) A317 575 (1970).
10. M. C. Lin, "5 μm CO Laser Emission from the O + CH Reaction", to be published.
11. M. C. Lin, Int. J. Chem. Kinetics, in press.
12. M. C. Lin, to be published. We have recently observed 5 μm CO laser emissions from the photodissociations of both CH_2CO and SCO in vacuum U.V. above 165 nm.
13. L. E. Brus and M. C. Lin, J. Phys. Chem. 76 1429 (1972).
14. M. C. Lin and L. E. Brus, J. Chem. Phys. 54 5423 (1971).
15. I. Haller and G. C. Pimentel, J. Am. Chem. Soc. 84 2855 (1962).
16. J. S. E. McIntosh and G. B. Porter, Can. J. Chem. 50 2313 (1972).
17. R. Walsh and S. W. Benson, J. Am. Chem. Soc. 88 4570 (1966).

18. A. W. Kirk and E. Tschuikow-Rowx, J. Chem. Phys. $\underline{51}$ 2247(1969).
19. The proposed mechanism for the $CH_2 + O_2$ reactions can account for the results obtained by Noyes and co-workers [A. N. Strachan and W. A. Noyes, Jr., J. Am. Soc. $\underline{76}$ 3258 (1954), R. A. Holroyd and W. A. Noyes, Jr. J. Am. Chem. Soc. $\underline{78}$ 4831 (1956)] in their studies of the effect of O_2 on the kinetics and mechanism of CH_2CO photolysis. When O_2 was added they detected small amounts of CO_2 and H_2CO and the quantum yield of CO was found to be increased. Many other condensable products such as CH_3COOH, $(CH_3CO)_2O$, $HCOOH$ and $HCOOCOCH_3$ were also identified. These compounds are very likely the secondary products of H_2O ($H_2O + H_2CCO \rightarrow CH_3COOH$) and $HCOOH$ ($HCOOH + H_2CCO \rightarrow HCOOCOCH_3$), etc. The small amount of CO_2 detected may be the decomposition product of both HCOO and COOH radicals. H_2CO may result from reaction (10).
20. N. Djeu, Private Communication. The author thanks Dr. Djeu for his critical comments on this work.

THRUSH: How much dissociation of SO_2 do you get in your $SO_2 + C_2H_2$ system?

LIN: Based on our gas analysis there is about 10% destruction in a single shot; for the CH_2 we are losing about 30%, for acetylene about 20%.

THRUSH: The reason I asked this question is because in this case CH_2 will react some three times faster with C_2H_2 than with O. The reaction of CH_2 with C_2H_2 gives C_3H_4 and we have shown that the reaction of oxygen atoms with methyl acetylene gives highly vibrationally excited CO which could well be a source of laser action.

LIN: We have the apparatus to study this system. We have not done that yet, but if you compare the characteristics of both systems, $O + CH_2$ and $O + C_2H_2$, I think that you will be convinced that the pumping step is mainly $O + CH_2$.

THE NITROGEN AFTERGLOW

B. A. Thrush and M. F. Golde

University of Cambridge, Department of
Physical Chemistry, Lensfield Road,
Cambridge, England

Recent progress towards understanding the mechanisms by which the recombination of ground state nitrogen atoms populates excited electronic states of N_2 is reviewed. It is shown that the areas of agreement between various studies now substantially exceed the areas of disagreement.

INTRODUCTION

Afterglows in which electronically excited molecules are formed by the combination of atoms with each other or with molecules are the most common chemiluminescent reactions in the gas phase. The long lived yellow afterglow of active nitrogen is the most familiar of them and has attracted an immense bibliography (Wright and Winkler 1968) since its discovery by Morren in 1865. All the theoretically predicted states of N_2 which could be populated by the recombination of ground state nitrogen atoms have now been identified spectroscopically except for $^5\Sigma_g^+$. All these states have also been observed in the nitrogen afterglow except for $^3\Delta_u$ and $^1\Delta_u$, and failure to detect these states more probably arises because they lack strong transitions and are readily quenched rather than because they are not populated at a significant rate in active nitrogen.

Tremendous progress towards understanding the mechanism of the nitrogen afterglow has been made in the last twenty years, but some areas of (healthy) disagreement still remain. The afterglow shows many features of interest; electronically excited nitrogen molecules are formed by both two body and three body recombination processes and their subsequent relaxation, quenching and energy transfer processes with both molecules and atoms are typical of the reactions of excited molecules in which there is so much current interest. This review attempts to collate the conclusions of recent experimental studies although the references cited are necessarily selective.

ACTIVE SPECIES

Active nitrogen is usually studied in discharge flow systems at pressures of 0.1 to 10 Torr. Some 0.1s after a discharge through nitrogen, the only active species present in significant concentrations are ground state nitrogen atoms, $N(^4S)$, typically 0.1 - 1% plus a few percent of vibrationally excited ground state N_2 molecules mainly in v = 1. Relaxation of the latter can heat the gas, but they play no chemical role in the afterglow. Highly vibrationally excited ground state N_2 molecules have been postulated in active nitrogen, but their concentration is very low except in the complex short-lived pink afterglow of nitrogen which is not discussed here.

EXCITED ELECTRONIC STATES

$A(^3\Sigma_u^+)$

Vegard-Kaplan band emission (N_2, A → X) is very weak in active nitrogen at normal pressures. Noxon's measurements (1962) at pressures up to atmospheric can be interpreted to show that the intensity obeys the relation $I = I_o[N][M]$ (Thrush 1967). Experiments where the mercury resonance line at 253.7 nm is excited by energy transfer from $N_2(A)$ yield similar conclusions (Brennen and Kistiakowsky 1966, Thrush and Wild 1972). The accepted mechanism for population is

$$N + N + M = N_2(A) + M \qquad (1)$$
$$N_2(A) + N = N_2 + N \qquad (2)$$
$$N_2(A) = N_2(X) + h\nu \qquad (3)$$

Since $k_2 = 3 \times 10^{13}$ cm^3 mol^{-1} s^{-1} (Young and St. John 1968) and $k_3 = 0.5$ s^{-1} (Shemansky 1969), the Vegard-

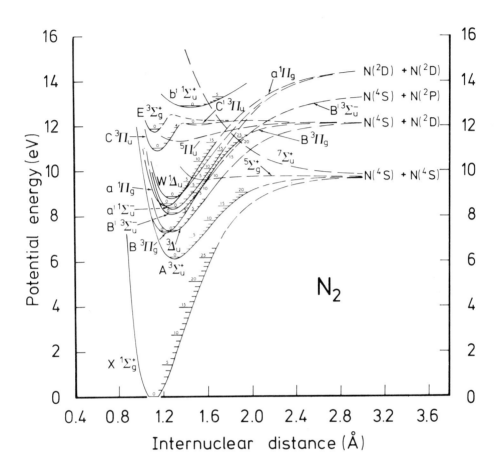

Figure 1 - Potential curves of N_2 after Gilmore (1965) but including data of Benesch and Saum (1971).

Kaplan band emission is very weak. Only v = 0 and 1 are normally detected, apparently because vibrational relaxation with $\Delta v = -2$ is rapid, being nearly resonant with ground state N_2 (Perner 1972). There is experimental (Thrush and Wild 1972) and theoretical evidence (Shui, Appleton and Keck 1970) that the A state is populated in roughly 60% of the three body recombinations of ground state nitrogen atoms, for which the rate coefficient is 1.4×10^{15} cm^6 mol^{-2} s^{-1} with M = N_2 and Ar.

$B(^3\Pi_g)$

The yellow First Positive emission (N_2, B → A) is the most prominent and most controversial feature of the nitrogen afterglow. At normal pressures, the emission is enhanced by partial replacement of the nitrogen carrier with an inert carrier. Jeunehomme's measurements (1966) of the radiative life of $N_2(B)$ clearly demonstrate that this effect is due to quenching by N_2, and the simplified mechanism

$$N + N + M = N_2(B) + M \qquad (4)$$
$$N_2(B) + N_2 = N_2(A) + N_2 \qquad (5)$$
$$N_2(B) = N_2(A) + h\nu \qquad (6)$$

gives the observed emission kinetics $I = I_0[N]^2$. Further $k_4 = 6 \times 10^{14}$ cm^6 mol^{-2} s^{-1} (Campbell and Thrush 1967), that is $k_1 \sim k_4$, which is not surprising since spin-allowed radiation and quenching of the B state would populate the A state, but it does raise a 'chicken and egg' problem.

Bayes and Kistiakowsky (1960) were the first to classify the First Positive bands according to their probable kinetic origin. Here we need distinguish only two regions.

(a) Below v = 9 where the population increases steadily but not smoothly to v = 0 and where some 70% of the emission lies. The changes in population of these levels which Kistiakowsky and Warneck (1957) observed between the ^{14}N and ^{15}N afterglows show that these levels of the B state are being populated from the A state. There is evidence from the excitation of higher triplet levels of mercury (Thrush and Wild 1973) that the A state population at these energies does not greatly exceed the B state population.

(b) Levels v = 9 - 12 immediately below the predissociation limit where the vibrational distribution shows a peak which is shifted to lower energies by inert carriers and to higher energies by ammonia, etc. Recent studies of the nitrogen afterglow at pressures down to 1 m Torr in the large reactor at Bonn (Becker et al 1972) have provided new information about this population distribution. These workers observed two body population of level v = 13 by inverse predissociation from the $^5\Sigma_g^+$ state. Their data show that Benson's suggestion (1968) whereby these levels of the B state are populated by vibrational relaxation from v' = 13 cannot account for the observed intensity.

The Bonn group favour the mechanism of Berkowitz, Chupka and Kistiakowsky (1956), whereby two nitrogen atoms approach each other along the $^5\Sigma_g^+$ curve and make a collision induced transition to high levels of the B state, over the suggestion of Campbell and Thrush (1967) that the A state is the precursor. In the BCK mechanism some molecules (e.g. CO_2, H_2O) would have to induce the spin-forbidden crossing with rate constants of 4×10^{14} cm^3 mol^{-1} s^{-1}, orders of magnitude faster than other spin forbidden processes involving excited states of nitrogen (Golde and Thrush 1972a).

B' $^3\Sigma_u^-$

This state has been studied by the infra-red B' → B emission (Bayes and Kistiakowsky 1960) and by the vacuum ultra-violet B' → X emission (Golde and Thrush 1971). Levels v = 5 - 8 predominate, and apart from their low pressure fall-off their kinetic behaviour follows that of N_2(B, v = 8 - 12) from which they are almost certainly populated by collision induced transitions. Similarly, the emission shows $[N]^2$ kinetics, being quenched by N_2, although not as efficiently as N_2B (Golde and Thrush 1972b). The population of B' is almost certainly less than that of the B state.

a $^1\Pi_g$

The Lyman-Birge-Hopfield emission (a → X) in the vacuum ultra-violet has only recently been studied in detail (Golde and Thrush 1971, Becker et al 1972). Although the emission kinetics at pressures around 1 Torr are formally complicated,

$$I = I_0[N]^2/([N_2] + \gamma) + I_0'[N]^3/[N_2]$$

the mechanisms are of considerable interest. Because the a → X transition is allowed by magnetic dipole but not electric dipole radiation (t = 1.5 x 10^{-4} s), collisional processes predominate in the removal of N_2 a, and in the 1 Torr pressure range absolute rate of population of the a state are obtained from electronic energy transfer to added CO and observation of the corresponding but fully allowed CO Fourth Positive emission ($A^1\Pi \to X^1\Sigma^+$) which is not significantly quenched (Golde and Thrush 1972b). The vibrational levels of CO(A) populated from a given level of N_2a can be identified and do not fit a "vertical transition" model well.

The first term in the above relation gives rates of population which decrease from v = 6 to v = 0. It arises by two body population of N_2a, v = 6, J > 13 by inverse predissociation through the $^5\Sigma_g^+$ state; this is followed by rotational and vibrational relaxation, both of which have high collisional efficiencies

$$N(^4S) + N(^4S) \rightleftharpoons N_2(a, v = 6, J > 13) \quad (5, -5)$$
$$N_2(a, v = 6, J > 13) + N_2 \rightleftharpoons N_2(a, v=6, J \leqslant 13) + N_2 \quad (6, -6)$$
$$N_2(a, v=6, J \leqslant 13) + N_2 = N_2(a, v < 6) + N_2 \quad (7, -7)$$
$$N_2(a) = N_2(X) \text{ or } N_2(a') + h\nu \quad (8)$$

Detailed analysis of this mechanism over a wide pressure range is complex because the predissociation and preassociation rates (k_5, k_{-5}) increase with J and this changes the relative roles as well as the overall importance of rotational and vibrational relaxation. Statistical mechanical calculation of k_5/k_{-5} combined with observed rates of predissociation and relaxation give excellent agreement with the absolute rates of formation and values of γ found both in normal pressure and very low pressure experiments (Golde and Thrush 1972b, 1973).

The $[N]^3$ formation term is observed with levels v ⩽ 3 and increases towards lower energies. In argon carriers, this term shows a similar enhancement to emission by higher levels of N_2B. The kinetics are in quantitative agreement with population by the reaction

$$N(^4S) + N_2(B, v \geqslant 6) = N_2(a, v \leqslant 3) + N(^4S) \quad (9)$$

As there is no theoretical reason why a similar process

involving high levels of N_2A should not populate one component of the $a\,^1\Pi_g$ state, it is possible that the populations of very high levels of N_2A show a similar enhancement in N_2 carriers to those of N_2B.

$a'\,^1\Sigma_u^-$

Only level $v = 0$ of the forbidden $a' \to X$ emission is observed in active nitrogen, apparently because higher levels are removed by vibrational relaxation or quenching into the less metastable a state (Campbell and Thrush 1969). Kinetically, the a' state behaves like an extension of the a state and the dominant $[N]^3$ population term arises mainly from direct N atom induced population from the triplet system, as with the a state (Golde and Thrush 1972b). Quenching by N atoms as well as by N_2 molecules is observed (Campbell and Thrush 1969). Competition between these quenching processes presumably arises because of slow spin allowed quenching of the a' state from which (unlike the a state) only the ground state can be populated.

DISCUSSION

Despite the baffling complexity of the literature on active nitrogen, there is now substantial agreement on the mechanism by which various excited states of N_2 are populated in the recombination of ground state nitrogen atoms

(1) Inverse predissociation along the $^5\Sigma^+$ curve yields $N_2(B\,^3\Pi_g,\ v = 13)$ and $(a\,^1\Pi_g,\ v = 6,\ J > 13)$ from which two body chemiluminescence is observed. This has a positive temperature coefficient, unlike the luminescence associated with three body recombination. Collisional relaxation from these states is only a minor source of three body recombination but is a major source of singlet chemiluminescence. These processes show excellent agreement with the theoretical predictions of statistical mechanics (Thrush 1968) or resonance scattering (Carrington 1972).

(2) Nitrogen atoms produce a net flow of nitrogen molecules from excited triplet to excited singlet states showing that the former are more highly populated in active nitrogen. Diamagnetic species have much lower efficiencies in these processes.

(3) There is no evidence that recombination into high vibrational levels of the ground state leads to the formation of electronically excited nitrogen molecules. This is as expected, since the ground state lies well

below all the other states except near the convergence limit.

(4) 40% of the three body recombinations of ground state N atoms populate the ground state of N_2 when $M = N_2$ or Ar; this proportion may differ for more complex third bodies.

(5) The remaining 60% of recombination proceeds via the A and B states. Of this, 45% is most probably three body recombination into the A state from which levels $v < 9$ of the B state are populated. The remaining 15% of recombination is responsible for the most conspicuous and controversial feature of the nitrogen afterglow, the yellow emission from $N_2 B$, $v = 9 - 12$. This most probably arises from two N atoms approaching on the $^5\Sigma_g^+$ or $A^3\Sigma_u^+$ curve and undergoing a collision induced transition to the $B^3\Pi_g$ state. Since there is strong evidence of the importance of spin conservation in collision processes involving excited N_2 molecules, we favour the latter explanation.

REFERENCES

Bayes, K.D. and Kistiakowsky, G.B. 1960 J. Chem. Phys. 32, 992.

Becker, K.H., Fink, E.H., Groth, W., Jud, W. and Kley, D. 1972 Faraday Disc. Chem. Soc. 53.

Benesch, W.M. and Saum, K.A. 1971 J. Phys. B, 4, 732.

Benson, S.W. 1968 J. Chem. Phys. 48, 1765.

Brennen, W.R. and Kistiakowsky, G.B. 1966 J. Chem. Phys. 44, 2695.

Campbell, I.M. and Thrush, B.A. 1967 Proc. Roy. Soc. A 296, 201.

Campbell, I.M. and Thrush, B.A. 1969 Trans. Faraday Soc. 65, 32.

Carrington, T. 1972 J. Chem. Phys. 57, 2033.

Gilmore, F.R. 1965 J. Quant. Spec. Rad. Transfer, 5, 369.

Golde, M.F. and Thrush, B.A. 1971 Chem. Phys. Letters, 8, 375.

Golde, M.F. and Thrush, B.A. 1972 Faraday Disc. Chem. Soc. 53, 52.

Golde, M.F. and Thrush, B.A. 1972 Proc. Roy. Soc. A 330, 79, 97, 109, 121.

Golde, M.F. and Thrush, B.A. 1973 in course of publication.

Jeunehomme, M. 1966 J. Chem. Phys. 45, 1805.

Kistiakowsky, G.B. and Warneck, P. 1957 J. Chem. Phys. 27, 1417.

Morren, M.A. 1865 Ann. Chem. Phys. 4, 293.

Noxon, J.F. 1962 J. Chem. Phys. $\underline{36}$, 926.
Perner, D. 1972 Faraday Disc. Chem. Soc. $\underline{53}$.
Shemansky, D.E. 1969 J. Chem. Phys. $\underline{51}$, 689.
Shui, V.H., Appleton, J.P. and Keck, J.C. 1970 J. Chem. Phys., $\underline{53}$, 2547.
Thrush, B.A. 1967 J. Chem. Phys., $\underline{47}$, 3691.
Thrush, B.A. 1968 Ann. Rev. Phys. Chem., $\underline{19}$, 371.
Thrush, B.A. and Wild, A.H. 1972 J. Chem. Soc. Faraday II, $\underline{68}$.
Thrush, B.A. and Wild, A.H. 1973 J. Chem. Soc. Faraday II, to be submitted.
Wright, A.N. and Winkler, C.A. 1968 'Active Nitrogen' (Academic Press, New York).
Young, R.A. and St. John, G.A. 1968 J. Chem. Phys., $\underline{48}$, 2572.

STEDMAN: On the question of two electron jumps being forbidden in collisions, there is experimental evidence provided by the excited Ar (^{3}P) + CO data, and the $2N_2(A)$ energy pooling. The only states populated are the ones that are allowed both by spin and one electron jump rules.

WAYNE: Since radiative loss from CO is so efficient, is there any danger, at any stage, of radiation trapping in the system?

THRUSH: In fact there is a bit of self reversal by CO for transitions to the ground vibrational state but all of the measurements were done on levels of CO which did not go down to the ground vibrational level.

THE AIR AFTERGLOW REVISITED*

Frederick Kaufman

Department of Chemistry, University of Pittsburgh

Pittsburgh, Pa. 15213

A. INTRODUCTION

To begin with a little soul-searching: The fact that science and technology are advancing at an ever accelerating rate, while both the number of investigators working in a given field and their level of support are at best levelling off, has put us in a new bind. Blessed with marvellous gadgets of all types, photomultipliers capable of detecting a few light quanta, tunable lasers of great power, narrow spectral width, and ultrashort repetitive flash duration, with ultrahigh vacuum techniques and ultrafast time resolution, with a bulging arsenal of analytical methods for measuring neutral or charged species in specific quantum states, etc., etc., our ability to study physical and chemical processes ever more minutely appears to be without limit. This widens immensely the discrepancy between what is doable technologically and what can actually be undertaken, particularly when we remember how expensive most of our new 'toys' are. The bind is, then, the increasing selectivity which we must exercise in our choice of systems and problems to work on, since neither the 'Everest' justification ('because it is there') nor the 'Part XXIV in a series of continuing studies' justification is quite good enough.

Serendipitously, it turns out that the air afterglow - the O + NO chemiluminescence - is a fine example of a system well worth studying for many reasons ranging from environmental and societal to purely scientific. Very briefly, these are as follows: (1) The phenomenon is observed in the normal and perturbed upper atmosphere[1] where its potential usefulness in measuring the local or column integrated [O] [NO] product has not yet been fully realized; (2) It is an extremely sensitive monitor of either O- or NO- concentrations wherever one or the other is known and has therefore been successfully used in laboratory O-atom kinetic studies[2] for about 15 years, but is equally applicable to NO measurements in combustion and pollution studies, where it may sometimes be preferred over its much more weakly emitting sister-reaction, O_3 + NO; (3) Following the work of Fontijn, Meyer, and Schiff,[3] it has become a secondary actinometric standard for the study of other glows in the visible or near ultraviolet, a function which is now being expanded in our laboratories to the infrared; and (4) As a fundamental chemiluminescent system it is one of the best characterized, most thoroughly studied, and represents our opportunity to analyze and resolve in a small system the increasingly insoluble problems of larger molecules. The present review will first attempt to bring the reader up to date on the recent results of direct experimental studies and then to discuss and interpret these results in the light of supporting work on the spectrum, fluorescence, photodissociation, and quantum theoretical calculations of NO_2.

B. REVIEW OF EXPERIMENTAL RESULTS

B. 1. The Pressure Dependence

The once raging battle whether the O + NO chemiluminescence is a truly bimolecular radiative recombination whose rate constant, $I_o \equiv I/[O][NO]$, is pressure independent or whether it is principally a termolecular process at its high pressure limit (even at as low a pressure as 1 Torr because of the long radiative lifetime of NO_2^*) which requires I_o to fall off with decreasing pressure, this battle is now ended. I_o does decrease, but does not do so indefinitely. It apparently levels off again at very low pressures ($\leq 10^{-4}$ torr) in what appears to be a compromise solution of the problem, albeit one which requires about 90% termolecular and only 10% bimolecular character. This experimental result is much more satisfactorily represented by a model which involves the bimolecular formation of an unstabilized NO_2^* collision complex followed by vibrational energy transfer within NO_2^*, electronic energy transfer to ground-state NO_2, (both of them collisional processes), or spontaneous radiative decay to NO_2, with the exact nature of NO_2^* to be discussed later. The experimental basis for this interpretation is shown in Fig. 1 which shows an approximate summary of all work on pressure-dependence on a logarithmic pressure scale, arbitrarily normalized (and extended) to "high" pressure. Its chronology is not without irony. In 1964, both Harteck's[4] and Jonathan's[5] groups reported pressure independence down to 3 millitorr, but this was challenged by Kelso and myself[6] on the basis of experiments in a 2.5 cm diameter flow tube down to 30 millitorr. A few additional experiments of Harteck's group[7] seemed to confirm the pressure independence, but Jonathan and Petty's[8] measurements in flow tubes of 2.5, 5.0, and 12 cm diameters clearly showed a fall-off in I_0. On the other hand, our early measurements undoubtedly overshot the mark, i.e. showed too large a fall-off due to diffusion and surface recombination effects. All other recent studies have shown good agreement on the decrease of I_0 with decreasing pressure. McKenzie and Thrush[9] studied pressure and M-effects from 10 to 100 millitorr and their points in Fig. 1 were obtained from their Stern-Volmer type expression (they refer to Ar carrier gas whereas all others refer to M=O_2), and the extensive work of Becker et al.,[10,11] first in a 200 liter and then in a 2×10^5 liter sphere, and that of Cody[12] in my laboratory span the much larger pressure range of 0.1 to 100 millitorr. Our data were obtained in a 15 cm diameter flow tube pumped by any suitable combination of forepump (15 lit/sec), Roots blower (130 lit/sec), and 6 inch oil diffusion pump (about 700 lit/sec with cooled baffle). The emission was viewed by a cooled photomultiplier through any one of 12 interference filters from 4050 to 7850 Å, 6 of which could be mounted on a filterwheel. The O-atom concentration was measured by resonance absorption near 1300Å across the flow tube at the same axial position where the chemiluminescence signal was obtained in order to avoid assumptions regarding the catalytic behavior of surfaces over large pressure ranges. Fig. 2 shows a schematic diagram of the apparatus.

The points plotted in Fig. 1 should have variable but reasonable large error bars ($\pm$ 10 to 30%) attached to them, but it is abundantly clear that a substantial fall-off exists and that the two most extensive studies, at Bonn and Pittsburgh, are in excellent agreement. I should also be said that only the points and curve of Becker, Groth, and Thran[11] represent I_0 values properly integrated across the emission spectrum, but since substantially smaller fall-off ratios only occur at the blue end of the spectrum (see B. 3. below) which contributes relatively little to the total emission, the comparison of Fig. 1 is valid.

B. 2. The M-Dependence

This question is complementary to that of the pressure dependence, because, in a truly bimolecular recombination, there is, of course, no M-effect, and in a termolecular one (or in a vibrational energy transfer sequence) it would be astonishing and improbable if there were none. It should be emphasized, however, that large M-effects would be unlikely, since one really observes the ratio of two M-effects, that for the termolecular formation of an NO_2^* (or its vibrational relaxation in the energy transfer mechanism) to that for electronic quenching of NO_2^*, and although it is unlikely that the relative efficiencies of different M should be constant

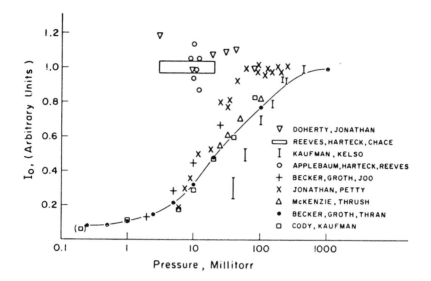

FIGURE 1

Pressure dependence of I_o

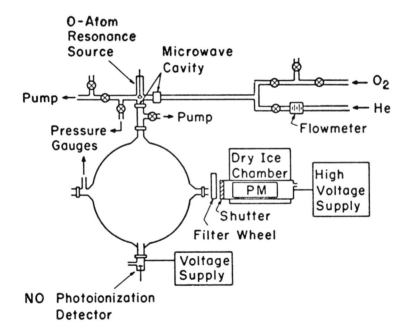

FIGURE 2

Diagram of apparatus. Top view looking down along axis of 15 cm diameter flow tube.

in the two processes (leading to no effect) it is also likely that, for the simple atoms and molecules investigated, there should be some parallelism and partial cancellation leading to small M-effects.

Experimental data are shown in Table I which includes our earlier results and various other entries. Surprisingly good agreement abounds, when one considers the small magnitude of the relative effect, except for the results of ref. 7 which show no M-effect as well as no pressure dependence. The entries for ref. 11 are their normalized d/q ratios for $\lambda=4450$Å and for 5600 or 6320Å where a second value is given. If the truly bimolecular part of I_o and the radiative lifetime of NO_2^* are considered constant in these experiments, d/q equals the ratio of k_D^M, the termolecular rate constant for the formation of NO_2^* to its quenching rate constant, k_q^M. The results of McKenzie and Thrush[9] also lend qualitative support to the findings of an M-effect as, for example, in the ratio I_o (green)/I_o (blue) which, when normalized as $p \to 0$, approaches high pressure values in the order $Ar > O_2 > CO_2$, N_2O, $CH_4 > SF_6$.

TABLE I

Relative Emission Efficiencies, $R^M \equiv I_o^M/I_o^{O_2}$

M	Ref. 6	Ref. 7	Ref. 11 (a)	Other Work
He	1.25	1.03	1.28, 1.30	1.25[13]
Ne	--	--	1.20	
Ar	1.06	1.06	1.08	1.0[13]
N_2	1.15	--	1.10	
H_2	--	--	1.12	
CO_2	0.79	1.03	0.74, 0.79	0.75[8]
N_2O	0.86	1.02	--	
CH_4	0.7	--	--	
CF_4	0.50	1.02	--	
SF_6	0.52	--	0.52, 0.51	
H_2O	0.57	--	--	

(a) First value: $\lambda=4450$Å; second value: $\lambda=6320$Å (He), $\lambda=5600$Å (CO_2, SF_6)

B. 3. The Spectral Dependence on p and M

Recent data of the pressure shift of the emission spectrum and of fairly monochromatic fall-off measurements which provide I_o^λ (p) over a wide pressure range represent the clearest proof for the correctness of the termolecular/energy transfer mechanism. This was first shown qualitatively by Freedman and Kelso[14] in my former laboratory, then by McKenzie and Thrush,[9] and in much greater detail by Becker et al.[11] and by Cody.[12] The overall spectrum shifts towards the blue with decreasing pressure, i.e. although the spectral threshold remains unchanged, the relative intensities increase at the short wavelength end and decrease at long wavelengths. There is a concomitant change in the fall-off characteristics of I_o^λ (p) - which is also a function of M - such that there is increasing fall-off as λ is increased from its threshold near 3970Å, as predicted by the energy transfer mechanism, since NO_2^* molecules which have lost some of their vibrational energy can not emit radiation at the blue end of the spectrum. Conversely, the emission very close to 3970Å must come mainly from unstabilized NO_2^* and should therefore show no fall-off. This is supported by the recent data of Becker et al[11] and of Cody[12] as shown in Table II. Considering the very different experimental techniques, the agreement is very good.

TABLE II

Fall-off Ratios, $I_o^\lambda(p\to\infty)/I_o^\lambda(p\to 0)$

as Function of Wavelength for $M=O_2$

λ, Å											Ref.
3980	4030	4050	4360	4450	4700	5360	5600	5660	6320	7250	
1.56	2.13	--	--	4.52	--	9.1	10.5	--	13.5	--	11
--	--	1.5	4.5	--	5.5	--	--	7.9	--	9.6	12

The data of Becker et al. are probably more accurate, but the discrepancy at the long wavelength end is due at least partly to the manner of extrapolation from the highest pressures used in these experiments, which was near 100 millitorr in both studies, to $I_o^\lambda(p\to\infty)$ which should be reached near 1 torr. This amounted to 25-30% in Becker's and to about 15% in our studies. For the spectrally integrated I_o- ratio Becker et al. report 15.2 which represents a considerable, further increase over a value of about 12.5 at their lowest measured pressure of 0.2 millitorr. This leads them to set $I_o^o = 4.2 \times 10^{-18}$ cm^3 $molecule^{-1}$ sec^{-1} for the rate constant of the purely two-body process by normalization to the high pressure I_o of 6.4×10^{-17} by Fontijn et al[3]. Although all of our data have not yet been fully analyzed, we would favor and integrated I_o- ratio in the 10-12 range.

A spectral M-dependence was first seen qualitatively by Freedman and Kelso[15] at high pressures, and recently in much greater detail by Becker et al[11]. In terms of the energy transfer mechanism, such a dependence of spectrally resolved fall-off curves on M is, of course, absolutely required if a total, integrated M-effect exists as the data in Table I seem to indicate. Since the residual, low pressure, truly bimolecular emission process can not depend on M and must therefore be identical in all gas mixtures, this high pressure M-effect suggests different efficiencies for vibrational energy transfer for different M in such a way that an M for which R in Table I is greater than 1 (such as He) should have larger spectrally resolved fall-off ratios in the red, and an M such as SF_6 should have smaller ones. Thus, ref. 11 reports a $I_o^\lambda(p\to\infty)/I_o^\lambda(p\to 0)$ of 8.45 for CO_2 and 5.8 for SF_6 at 5600Å compared to 10.5 for O_2, but one of 17.2 for He at 6320Å compared with 13.5 for O_2. The mechanistic picture appears to be entirely consistent with the available experimental data.

B. 4. The Temperature Dependence

The temperature dependence has been experimentally studied only at high pressures, first by Clyne and Thrush[13] between 200 and 300°K who reported, for $M=O_2$, $I_o^\infty = 5 \times 10^{-18} \exp[(1500 \pm 400)/RT]$ cm^3 molecule^{-1} sec^{-1} or an analogous T^{-n} expression of $6.4 \times 10^{-17} (T/300)^{-3\pm 0.8}$. Hartunian et al.[16] covered the range 500 - 1200°K in their glow discharge shock tube experiments and reported a somewhat lesser negative temperature dependence of $T^{-1.55}$ in T^{-n} form but a somewhat larger exponential dependence of $\exp(2200/RT)$. Parkes[17] studied the temperature dependence for $M=O_2$, He, Ar, CF_4, and SF_6 at 1.5 torr from 170 to 370°K in a simple, yet accurate flowtube experiment with three matched photomultiplier tubes of which one viewed the chemiluminescence at a position whose temperature was varied, while the other two simultaneously measured the intensity at 300°K up-and downstream of that position. This provided accurate intensity ratios as the temperature of the center section was varied slowly and continuously. Neither a pure T^{-n} nor an $\exp(E/RT)$ gave a fully satisfactory fit, but a two parameter expression $T^{-m} \exp(-300/RT)$ provided a good description with m=2.50 for O_2, 2.65 for He and Ar, and 1.90 for CF_4 and SF_6, which predicts that in the 900 to 1200°K range the emission will have equal intensity for all 5 M-gases. This effect can be qualitatively ascribed to the increasing probability of re-excitation of vibrationally quenched NO_2^* at higher temperatures and to unequal energy increments in the vibrational ladder.

Vanpee et al.[18] recently reported a new measurement of the integrated rate constant I_o^∞ of 6.8×10^{-17} cm^3 molecule^{-1} sec^{-1} ±

35% for M=O_2 and N_2 in a free jet at 367°K at pressures near 1 torr. They also reported a greater contribution of near infrared radiation to the spectrum which is responsible for a 12% increase in the spectrally integrated I_0^∞ under otherwise identical conditions. With this correction and the above temperature dependence measured by Parkes[17], a value of $(9.3 \pm 3.2) \times 10^{-17}$ is obtained for comparison with that of $(6.4 \pm 1.9) \times 10^{-17}$ of Fontijn et al.[3] at 296°K.

B. 5. Infrared Measurements

The absolute intensity measurements of Fontijn et al.[3] extended to 1.4μm where the signal was found to be negligible, but this was primarily a detectability problem. Since then, the infrared chemiluminescence and vibraluminescence has been studied by Stair and Kennealy[19] who measured the emission spectrum to 7μm using Fourier interferometer spectroscopy and found a considerably stronger infrared component of the emission than the earlier workers. I_0^∞ decreased about 20 fold from 1.25 to 3.3μm, but then increased about 3 fold to a peak at 3.7μm and fell sharply to near zero at 4.0μm. Vanpee's[18] spectral measurements to 2.0μm are in qualitative agreement with the more extensive work of Stair-Kennealy, but they indicate a somewhat too great intensity decrease between 1.2 and 2.0μm. Roche and Golde[20] in my laboratory are now doing absolute intensity measurements using liquid nitrogen cooled PbS and In Sb detectors and a circularly variable filter for spectral dispersion. Preliminary results at total pressures from 0.3 to 3 torr are in good agreement with Stair's results on the spectral distribution including the peak at 3.7μm and suggest a larger absolute rate constant in the overlapping 1.0 to 1.5μm range than either Fontijn or Vanpee.

C. DISCUSSION AND INTERPRETATION

C. 1. The NO_2^* Energy Transfer Model

It is clear from the preceding sections that either simultaneous two-body and three-body recombination (Model I) or the two-body formation of an unstabilized NO_2^* collision complex followed by vibrational, electronic, and radiative processes (Model II) can be used in the interpretation of experimental data. Model I can be set down as follows:

$$NO + O \xrightarrow{1} NO_2 + h\nu$$
$$NO + O + M \xrightarrow{2} NO_2^* + M$$
$$NO_2^* + M \xrightarrow{3} NO_2 + M$$
$$NO_2^* \xrightarrow{R} NO_2 + h\nu$$

which, with NO_2^* in steady state, yields

$$I_o^M = k_1 + \frac{k_2 k_R [M]}{k_R + k_3 [M]} \qquad (1)$$

with limiting values $I_o^M(p \to 0) = k_1$ and $I_o^M(p \to \infty) = k_1 + (k_2 k_R/k_3)$ i.e. a fall-off ratio of $1 + (k_2 k_R/k_3 k_1)$. Although [M] has cancelled out in this expression, the fall-off is, of course, M-dependent, since the rate constants k_2 and k_3 will be different for different M, and should perhaps have better been labelled k_2^M and k_3^M. In this, its simplest form, Model I is unsatisfactory as most Stern-Volmer type models, because it recognizes only a single energy level in the excited state and can therefore not explain spectral shifts due to p or M. In this form it may be compared with Stern-Volmer data on NO_2 fluorescence quenching[21], and the second term of eq. 1 can be re-written $k_2[M]/(1+a[M])$ where $a = k_3/k_R$ is the quenching constant, on the (unreasonable) assumption that the same single excited state is reached in both processes. Model I can now be generalized by allowing for a large number of levels in NO_2^*, summing over all of their steady state concentrations and introducing k_{2i}^M and k_{3i}^M as well as k_{Ri}. With that number of adjustable parameters, one could, of course, explain the spectral p and M dependence as well as pressure fall-off plots as complex as the silhouette of a giraffe or the oboe part of the first movement of Beethoven's Seventh. Yet, some physical unreasonableness would remain in this complicated model, since it would require that all levels of NO_2^* have to be formed and quenched in single collisions, no matter how large the energy gap between them and the NO + O continuum or the NO_2 ground state, i.e. it retains the "strong collision" approximation.

Model II relies on a stepladder process of vibrational energy transfer:

$$O + NO \underset{-1}{\overset{1}{\rightleftarrows}} NO_2^{*o}$$

$$NO_2^{*o} + M \begin{array}{c} \overset{Vo}{\to} NO_2^{*1} + M \\ \overset{Eo}{\to} NO_2 + M \\ \overset{Ro}{\to} NO_2 + h\nu \end{array}$$

$$NO_2^{*i} + M \begin{array}{c} \overset{Vi}{\to} NO_2^{*i+1} + M \\ \overset{Ei}{\to} NO_2 + M \\ \overset{Ri}{\to} NO_2 + h\nu \end{array}$$

where the collision complex, NO_2^{*0}, is capable of rapid redissociation, k_{-1}, in addition to three other modes of transformation, V (vibrational energy transfer), E (electronic energy transfer), and R (spontaneous radiation). If now, in the simplest approximation, all V_i are set equal, $V_i = V$, all $E_i = E$, and all $R_i = R$, the steady state approximation is applied to all NO_2^{*i}, and (very reasonably) $k_{-1} \gg (V[M], E[M], R)$, (to avoid having a flood of subscripted k's, the rate constants are labelled V, E, and R), one obtains, setting $k_1/k_{-1} \equiv K$,

$$I_0(p) = KR [1 + \frac{1}{1+S+X} + (\frac{1}{1+S+X})^2 + \cdots (\frac{1}{1+S+X})^n] \quad (2)$$

where $S \equiv E/V$, $X \equiv R/(V[M])$, and n is the number of vibrational levels which are able to radiate at a given emission wavelength, i.e. $n = 1+(\nu_0-\nu_\lambda)/\Delta\nu_V$ with $\nu_0 = 25,160$ cm^{-1}, the full O + NO bond energy, ν_λ the reciprocal of the emitted wavelength, and $\Delta\nu_V$ the average vibrational energy (cm^{-1}) transferred per collision with M. It must be pointed out that, in addition to all other simplifications, eq. 2 neglects the spectral differences of the I_0^i contributions coming from the different NO_2^{*i} and therefore does not explicitly show the spectral pressure shift. As $p \to \infty$, $X \to 0$, so that the model predicts $I_0^\lambda(p \to \infty)/I_0^\lambda(p \to 0)$ ratios to equal $1 + 1/(S+1) + \cdots 1/(S+1)^n = (1-B^{n+1})/(1-B)$ where $B = 1/S+1 = V/(E+V)$, and, for $n \to \infty$ the ratio approaches $1 + (V/E)$.

Model II was first proposed by Keyser, Kaufman, and Zipf[22] on the basis of NO_2 fluorescence studies in which E and V were fitted to data on monochromatic, steady excitation of NO_2 with monochromatic fluorescence intensity measurements, and R was measured by the phase shift method using modulated excitation[23]. As applied to the O + NO chemiluminescence, it predicted surprisingly accurately the now experimentally observed fall-off ratio of about 10 for the spectrally integrated, high pressure emission rate constant, I_0^∞, to its low pressure limit, I_0^0. It was also used by Becker et al.[11] to extract information on the magnitudes of E, V, and $\Delta\nu_V$ from their fall-off data, i.e. $E \simeq 2 \times 10^{-12}$ cm^3 molecule^{-1} sec^{-1}, $V \simeq 2 \times 10^{-10}$ cm^3 molecule^{-1} sec^{-1}, and $\Delta\nu_0 \simeq 800$ cm^{-1}, based on a radiative lifetime of 60µsec for NO_2^*. Similar analysis of Cody's[12] data appears to give $V = 1$ to 2×10^{-10}, $E \sim 1 \times 10^{-11}$ cm^3 molecule^{-1} sec^{-1} and $\Delta\nu_V \sim 500$ to 1000 cm^{-1}. The model does, of course, correctly predict the decreasing values of the spectrally resolved fall-off ratios (see Table II) at lower λ, because as λ is decreased so is n, the number of NO_2^{*i} levels which are able to emit, and thereby the number of contributing terms in eq. 2.

In its above form, Model II assigns equal statistical weights to the vibrational states of NO_2^*, which is unreasonable, because with three oscillators there is a higher density of states at higher energy. Schwartz and Johnston[24] introduced an RRK statistical weight, $g_j = (j+s-1)!/j!(s-1)!$, with j = total number of average vibrational quanta (set equal to 1250 cm^{-1}) above the origin of NO$_2$* (set equal to 12,500 cm^{-1}) and s=3, in their study of NO$_2$ fluorescence. For NO$_2$* states near the O + NO continuum, j therefore equals 10 and g_j=66, whereas the equivalent g_j for ground state NO$_2$ (j=20) equals 231, and by coming down 5 vibrational steps (which is equivalent to going from about 4000 to 5300Å in emission threshold) g_j would decrease to 21 in NO$_2$* and to 105 in the ground state. In future calculations this effect, which will tend to deplete higher energy levels faster than lower ones, should be included and g_j calculated more realistically. The experimentally fitted parameters, V, E, and $\Delta\nu_V$ are in surprisingly good agreement with the results of two fluorescence studies,[23,24] although it is puzzling (a) that for M=O$_2$, $\Delta\nu_V$ should be in the range 500 to 1000 cm^{-1} when ν(O$_2$) = 1580 cm^{-1}, and (b) that the energy transfer rate to O$_2$ would still be nearly gas-kinetic. On the whole, however, the physical picture is remarkably consistent, and has good predictive power with an economical minimum of assumptions and adjustable parameters. It requires only a single electronically excited state which it reasonably assumes to be identical with that reached in fluorescence, and, using its known (average) lifetime, interprets all aspects of the radiative recombination in terms of an energy transfer model which has had many successful applications in simple, non-radiative processes. We may be tempted to stop here and bask in its simple glory, but there are promises to keep, and miles to go before we sleep.

C. 2. Connections with NO$_2$ Spectroscopy, Fluorescence, Photodissociation and Quantum Calculations

Before we launch into this general discussion, a few experimental facts: (1) The absorption spectrum of NO$_2$ in the visible from 4000 to 8000 Å is amazingly complex, has, with the exception of a band progression near 4000Å which Douglas and Huber[25] showed to arise from a $^2B_1 \leftarrow ^2A_1$ transition, resisted analysis, even at 1.5°K.[26] Discrete features of the resonance fluorescence spectrum of NO$_2$ excited at 5145Å with an argon ion laser were analyzed by Abe, Myers, McCubbin, and Polo[27] and ascribed to a $^2B_2 \leftarrow ^2A_1$ transition. Still more recently, Stevens, Swagel, Wallace, and Zare[28] have analyzed the fluorescence spectrum excited at 5934 to 5940 Å by a narrow band, pulsed, tunable dye laser and report features due to both $^2B_1$ and $^2B_2$ states. (2) The extensive quantum calculations of Gangi and Burnelle[29] indicate that the two lowest excited states of NO$_2$, $^2B_1$ and $^2B_2$, lie 1.75 and 3.33 eV above the $^2A_1$ ground state in its equilibrium configuration and that their radiative lifetimes, τ_R, are 1.53 and 0.125μsec, respectively. Of the next three states, $^4B_2$, $^4A_2$, and $^2A_2$, only the $^4A_2$ state has an appreciable transition probability to the ground state corresponding to a τ_R of 8μsec. The $^2B_1$ state has a linear or nearly linear equilibrium configuration and the $^2B_2$ state is more strongly bent than the ground state (~110°).

(3) Photodissociation studies of NO_2 at 3471Å by Busch and Wilson[30] have shown that the rapidly dissociating (or predissociating) state reached in the initial absorption is predominantly of B_2 symmetry. Their upper limit of $\sim 2 \times 10^{-13}$ sec for the dissociation lifetime is in good agreement with the results of high pressure photolysis experiments by Gaedtke, Hippler, and Troe.[31] (4) Finally, we must consider the increasing number of partly contradictory fluorescence lifetime studies and their relevance to the air afterglow. Ever since Neuberger and Duncan's[32] direct measurement of 44µsec for τ_R, the discrepancy between that value and 0.26µsec calculated from the integrated absorption coefficient has been under experimental and theoretical scrutiny. The experimental search for a second, very much faster state gave uniformly negative results[23,24,33] with exception of recent, indications[34] that a very small fraction of the total emitted fluorescence may have a shorter lifetime (0.5 to 3.7 µsec) at a few wavelengths. This observation, even if correct, does, of course, in no way help to bridge the lifetime anomaly. Surprisingly, then, at least six studies[23,24,32,33,35,36] are in broad agreement that the measured lifetime is in the 60 to 70 µsec range when viewing geometry corrections are applied[23,24,35], although a minor controversy remains whether τ_R remains essentially constant[23] over the 4000 to 6000Å excitation range or exhibits fairly discontinuous variations of 10 to 20% when the excitation band width is narrowed to 1 to 5 Å. Only the recent findings of Zare's group[28] of non-exponential decays in the unresolved fluorescence at 5934.5Å, which is then spectrally resolved into stronger features coming from a $^2B_2$ state with $\tau_R = 30 \pm 5$ µsec and weaker features believed to be coming from a $^2B_1$ state with $\tau_R = 115 \pm 10$ µsec, are in some conflict with the other data, but although such experiments are clearly most important and desirable, they will have to be repeated for an extremely large number of lines in the complex NO_2 absorption spectrum before conclusions may be drawn about their importance in the overall process. It is significant to note that four of the above six lifetime studies[23,32,33,35] directly or indirectly looked for non-exponential behavior and found either none or very little.[35]

The extent to which the lifetime anomaly is explainable in terms of well-known physical processes was characterized very clearly by Douglas[33] who suggested four independent causes, the ν^3 frequency effect, transition moment variation with internuclear distance, vibrational level mixing, and interelectronic state mixing. Of these, the first may contribute small factors, e.g. excitation (or recombination) at an effective excitation wavelength of 4000Å and radiation at 6000Å provides for a factor of 3.4; the second is difficult to assess, but is unlikely to be large, at least by analogy with the behavior of diatomic molecules; the third, although surely applicable in NO_2 whose vibration frequencies are low and whose excited states have substantially different equilibrium geometries from the ground state, is more likely to help explain the very complex absorption spectrum, because each band is spread into overlapping weaker bands, than the lifetime anomaly except again through the ν^3 and R_e effects; it seems, therefore, that interelectronic state mixing must supply the bulk of the total factor of nearly 300 which needs to be rationalized, i.e. a factor of 30 to 100. Two prerequisites must then be fulfilled: (a) the states must interact strongly, and (b) their level density ratio must be sufficiently large. The second of these requirements, in conjunction with the known energetics of all possible states,[29] strongly points to the ground state as the cause of the perturbation, i.e. both in fluorescence and chemiluminescence, vibrationally highly excited $NO_2(X^2A_1)$ is strongly mixed with the $^2B_1$ and $^2B_2$ states. For $^2B_1$, the interaction can not be vibronic, as there are no vibronic symmetry species common to both states, but may be due to Coriolis interaction. The latter will increase strongly with increasing rotational energy, but may not be sufficiently strong to provide the extensive mixing which is required. For $^2B_2$, vibronic interactions are possible. In the absence of detailed information on the vibration frequencies of the two excited states no quantitative estimates of level density ratios can be made, but certain limiting approximations provide useful information. At the high levels of vibrational excitation, at least for the ground state, the semiclassical expression of Marcus and Rice[37] $N(\varepsilon)=(\varepsilon+\varepsilon_z)^{s-1}/[\Gamma(s)\prod_s h\nu_i]$ should be applicable. Here ε_z is the total zero-point energy and s, the number of oscillators, equals 3. For $NO_2(X)$, $\varepsilon_z=1850$ cm^{-1} and since ε values of 15,000 to 25,000 cm^{-1} are of interest here, ε_z may be neglected. If it is further assumed that excited state frequencies are roughly equal to those of $NO_2(X)$, the level density ratio is given by $(\varepsilon/\varepsilon')^2$ where ε and ε' are the total vibrational energies of the ground and excited states. If the frequencies are not equal it will be more likely that the excited state frequencies will be smaller than those of the ground state, and in that case, the N/N' ratio will be less than $(\varepsilon/\varepsilon')^2$. Now, for the $^2B_1$ state, both experiment and theory put its energy minimum at or below about 12,000 cm^{-1} (corresponding to an origin of the $^2B_1 \rightarrow ^2A_1$ band system at or beyond 8000Å) which, at excitation frequencies (or recombination energy) of 20,000 to 25,000 cm^{-1}, will make $\varepsilon/\varepsilon' \lesssim 2$ and the level density ratio $\lesssim 4$, i.e. far too small to explain the lifetime anomaly. For the higher lying $^2B_2$ state the situation is somewhat better, since its energy minimum is calculated[29] to be about 2.2 eV above that of the ground state which, depending on excitation frequency, may give ε/ε' in the 3 to 10 range and corresponding density ratios of 10 to 100. The danger and possible fallacy of this argument is, however, that, particularly in the excitation range where N/N' is large, its magnitude is primarily controlled by that of N' which is decreasing rapidly as the origin of the $^2B_2 \leftarrow ^2A_1$ band system is approached and one would therefore predict a strong dependence of observed fluorescence lifetime on excitation frequency, i.e. an increase of τ_R up to about 6000Å, and a sharp decrease beyond that wavelength. Schwartz and Johnston[24] did report a small increase of τ_R in the excitation wavelength range from 4000 to 5600Å and a small decrease from there to 6000Å, but our own work did not support even these minor trends and showed τ_R to be constant. Nevertheless, it would be interesting to measure τ_R in the low pressure limit as far into the red as possible. On the whole, the fluorescence lifetime anomaly seems less resolved than ever even though the great complexity of the absorption and emission spectrum comes as no surprise. The weakness of the level density argument has all but eliminated one of the principal potential explanations, the pseudo-degeneracy of the upper state by mixing with excited ground state.

The implications of this large body of related work for the O + NO afterglow are unclear. The $^2B_1$ is favored as the emitter by its direct correlation with $O(^3P) + NO(^2\Pi)$ whereas the $^2B_2$ state can be reached only through an intermediate state in a pre-association for which Burnelle et al[29] have suggested a high-lying $^2A_1$ ($^2\Sigma_g^+$) state as a likely intermediate. Carrington[38] points out that, since the recombination proceeds by way of highly unsymmetrical configurations of NO_2, i.e. in C_s rather than C_{2v} symmetry, the $^2A'$ state corresponding to $^2B_2$ may possibly be reached directly. The fact that the spectral onset of chemiluminescence corresponds closely to the full O-NO bond energy makes it unlikely that the curve crossing occurs appreciably below the continuum threshold. The smoothness of the emission spectrum, on the other hand, argues (weakly) against invoking both excited states as does the success of the earlier parametric representation of experimental data.

The very large vibrational relaxation rate constant suggests that the role of rotational excitation in the collision complex and subsequent rotational relaxation also be considered. In their recent high pressure photolysis studies of NO_2, Troe and co-workers[39] have been able to measure the limiting high pressure, second-order rate constant for the total O + NO reaction, $k^\infty = 2 \times 10^{-11}$ cm^3 $molecule^{-1}$ sec^{-1}, which includes all initial recombination pathways and is therefore an upper limit to k_1 in Model II. This k^∞ corresponds to a maximum impact parameter of 1Å for collisions of spherical particles and even when allowance is made for the exclusion of wrong end collisions of NO, the impact parameter is unlikely to be larger than 2Å which means that rotationally highly excited NO_2^* will not be an important intermediate.

Lastly, the question to what extent recombination to form NO_2^* is distinct from the known overall three-body recombination process may be examined semi-quantitatively. Assuming R to be constant and equal to 1.6×10^4 sec^{-1}, and KR, the low pressure limit for I_0, to be 5 to 6×10^{-18} cm^3 $molecule^{-1}$ sec^{-1}, $K = k_1/k_{-1}$ equals 3 to 4×10^{-22} $molecule^{-1} cm^3$. If, at sufficiently high pressure ($\gtrsim 0.1$ torr) all NO_2^* are vibrationally relaxed with rate constants in the 1 to 2×10^{-10} cm^3 $molecule^{-1}$ sec^{-1} range as obtained in section C1 above, the effective three-body recombination through NO_2^* is of the order

3 to 6×10^{-32} cm^6 molecule^{-2} sec^{-1} which is in the range of the reported overall rate constants of 6 to 8×10^{-32}, i.e. the process represents a major part of the total recombination. This had been suggested by us[6] earlier in terms of the three body NO_2^* mechanism and was confirmed recently by Becker et al[11]. The unimolecular dissociation rate constant of unstabilized NO_2^{*o}, k_{-1}, is then $\leq$ 5 to 7×10^{10} sec^{-1} which is a reasonable value under thermal conditions at 300°K compared with the much higher 5×10^{12} sec^{-1} for NO_2^* with 10 to 12 kcal of excess energy from photodissociation[30] and high pressure photolysis[31]. The close equality of the radiative and total rate constants further supports the notion of extensive state mixing in the excited state and decreasing meaning of pure state labels. Perhaps, in desperation and exhaustion, we need not worry, then, whether $^2B_1$ or $^2B_2$ labels apply to the air afterglow emitter, because both (or more than two) are variably and complicatedly mixed with the ground state.

REFERENCES

*This work was supported by the Department of the Air Force under Contract No. F19628-70-C-0255 and by the Advanced Research Projects Agency under ARPA Order No. 826.

1. D. J. Baker and R. O. Waddoups, J. Geophys. Res. 72, 4881 (1967); 73, 2546 (1968)
2. F. Kaufman, J. Chem. Phys. 28, 352 (1958); Proc. Roy. Soc. A247, 123 (1958); P. Harteck, R. R. Reeves, and G. Mannella, J. Chem. Phys. 29, 1333 (1958)
3. A. Fontijn, C. B. Meyer, and H. I. Schiff, J. Chem. Phys. 40, 64 (1964)
4. R. R. Reeves, P. Harteck, and W. H. Chace, J. Chem. Phys. 41, 764 (1964)
5. G. Doherty and N. Jonathan, Faraday Discussions 37, 73 (1964)
6. F. Kaufman and J. R. Kelso, Symposium Chemiluminescence, Duke University, 1965
7. D. Applebaum, P. Harteck, and R. R. Reeves, Photochemistry and Photobiology 4, 1003 (1965)
8. N. Jonathan and R. Petty, Trans. Faraday Soc. 64, 1240 (1968)
9. A. McKenzie and B. A. Thrush, Chem. Phys. Lett. 1, 681 (1968)
10. K. Becker, W. Groth, and F. Joo, Ber. Bunsenges. Physik. Chem. 72, 157 (1968)
11. K. H. Becker, W. Groth, and D. Thran, Chem. Phys. Lett. 6, 583 (1970); Ber. Bunsenges. Physik. Chem. 75, 1137 (1971); Forschungsbericht SHA/8, Instit. Physik. Chem. Univ. Bonn (1971); Chem. Phys. Lett. 15, 215 (1972); XIVth Combustion Symposium, The Combustion Institute, Pittsburgh, Pa., to be published
12. R. J. Cody and F. Kaufman, J. Chem. Phys., to be published
13. M. A. A. Clyne and B. A. Thrush, Proc. Roy. Soc. A269, 404 (1962)
14. E. Freedman and J. R. Kelso, Bull. Am. Phys. Soc. 11, 453 (1966)

15. E. Freedman and J. R. Kelso, 153rd Meeting, Am. Chem. Soc., April 1967, Section R, Abstract 29.
16. R. A. Hartunian, W. P. Thompson, and E. W. Hewitt, J. Chem. Phys. $\underline{44}$, 1765 (1966)
17. D. A. Parkes and F. Kaufman, J. Chem. Phys., to be published
18. M. Vanpee, K. D. Hill, and W. R. Kineyko, AIAA Journal $\underline{9}$, 135 (1971)
19. A. T. Stair and J. P. Kennealy, J. de Chimie Physique $\underline{64}$, 124 (1967)
20. A. E. Roche, M. F. Golde, and F. Kaufman, to be published
21. G. H. Myers, D. M. Silver, and F. Kaufman, J. Chem. Phys. $\underline{44}$, 718 (1966)
22. L. F. Keyser, F. Kaufman, and E. C. Zipf, Chem. Phys. Lett. $\underline{2}$, 523 (1968)
23. L. F. Keyser, S. Z. Levine, and F. Kaufman, J. Chem. Phys. $\underline{54}$, 355 (1971)
24. S. E. Schwartz and H. S. Johnston, J. Chem. Phys. $\underline{51}$, 1286 (1969)
25. A. E. Douglas and K. P. Huber, Can. J. Phys. $\underline{43}$, 74 (1965)
26. G. W. Robinson, M. McCarty, Jr., and M. C. Keelty, J. Chem. Phys. $\underline{27}$, 972 (1957)
27. K. Abe, F. Myers, T. K. McCubbin, and S. R. Polo, J. Mol. Spec. $\underline{38}$, 552 (1971)
28. C. G. Stevens, M. W. Swagel, R. Wallace, and R. N. Zare, Phys. Rev. Lett., in press
29. R. A. Gangi and L. Burnelle, J. Chem. Phys. $\underline{55}$, 843, 851 (1971)
30. G. E. Busch and K. R. Wilson, J. Chem. Phys. $\underline{56}$, 3638 (1972)
31. H. Gaedtke, H. Hippler, and J. Troe, Chem. Phys. Lett., in press
32. D. Neuberger and A. B. F. Duncan J. Chem. Phys. $\underline{22}$, 1693 (1954)
33. A. E. Douglas, J. Chem. Phys. $\underline{45}$, 1007 (1966)
34. P. B. Sackett and J. T. Yardley, Chem. Phys. Lett., $\underline{9}$, 612 (1971)
35. P. B. Sackett and J. T. Yardley, J. Chem. Phys., $\underline{57}$, 152 (1972)
36. K. Sakurai and G. Capelle, J. Chem. Phys., $\underline{53}$, 3764 (1970)
37. R. A. Marcus and O. K. Rice, J. Phys. Colloid Chem., $\underline{55}$, 894 (1951)
38. T. Carrington, This Volume, p.
39. H. Gaedtke, K. Glänzer, H. Hippler, and J. Troe, XIVth Combustion Symposium, The Combustion Institute, Pittsburgh, Pa., to be published

FREDERICK KAUFMAN

THRUSH: It seems to me that the outstanding problems are really concerned with the fluorescence of NO_2. I feel that the people who have worked on the air afterglow have really done a rather good job and the ball is very much back with the NO_2 fluorescence people.

F. KAUFMAN: I have unfortunately worked on both the chemiluminescence and the fluorescence and when I toss the ball back I toss it to myself.

THRUSH: One aspect of the air afterflow which I find surprising is that no-one has to my knowledge reported two-body emission from 0 + NO extending beyond the predissociation limit. It should be detectable, as such transitions would have reasonable Franck-Condon factors and the two-body emission constitutes an appreciable fraction of the total emission.

CARRINGTON: There has been one report of emission from 0 + NO extending to wavelengths below the energy threshold. Unfortunately I can't remember at the moment who that was, or whether or not I believed it, but it has been reported, and, as you say, you would expect this effect for at least a couple of kT's above the bond energy. I think I am quoting measurements not predictions, but since I don't remember whose measurement it was, I can't make a very strong case for it.

THRUSH: I believe this was the work of Broida, Schiff and Sugden, and none of them are here to deny it. In fact, the dissociation limit wasn't accurately known then, and they assumed that the emission extended kT above the predissociation limit.

I wish also to make a further comment. The idea that excited NO_2 molecules effectively spend most of their time in high levels of the ground state, and only a small fraction in the excited electronic state (i.e. the total eigen function is largely that of the ground electronic state) is an attractive one. It explains why vibrational relaxation is much more important than electronic quenching. It also explains some of the discrepancy between the apparent radiative lifetime in absorption and emission.

F. KAUFMAN: On this point you are still saying the NO_2 spends most of its time in the ground state, but if you take these numbers seriously and consider the $^2B_1$ state, that at most means three quarters of the time, which isn't enough - we cannot see the number right.

THRUSH: But I think one can get better than a factor of four if one really gets down to it and relies on the convergence of levels towards the dissociation limit, but then you've got the problem of why the radiative lifetime doesn't change as you go to lower energy. Then, of course, nearer the dissociation limit one probably has the $^2B_2$ state coming in as well, which complicates things.

F. KAUFMAN: Yes. The $^2B_1$ and the ground state are part of the same $^2\pi_u$ state in the linear configuration and do converge to the same limit of 0 and NO in their ground states, whereas the B_2 state does not. Nevertheless, there must be a very efficient predissociation coming out of, or going into, the B_2 state.

CARRINGTON: There is one simple thing that one can say about the $^2B_2$ state. We know from Kent Wilson's experiments on the photodissociation, that the predissociation gives the $^2B_2$ state, and the process must be reversible. Therefore, it is possible to get radiative recombination via the $^2B_2$ state. This argument tells you what is possible. On the other hand, it doesn't tell you whether that means 10^{-6} of the total process or 10^{-1} of it.

THRUSH: Another factor lengthening the radiative lifetime would be the need for the $^2B_1 \rightarrow {^2A_1}$ transition moment to go to zero in the linear configuration since, as Dr. Kaufman pointed out, the two states are derived from the same $^2\pi$ state.

OGRYZLO: I feel a bit of a tourist in this 0 + NO stuff at the moment, but I seem to remember that on reading Douglas' paper the impression I gathered was that he did suggest a large change in the transition moment could explain the difference in lifetime. In other words couldn't you say that in absorption the process is Franck-Condon governed, but that in emission this does not occur? Because the species spend a long time in the 'non-emitting' region this could account for a large change in their lifetime.

F. KAUFMAN: Near the continuum limit, the density of states argument must break down. For the B_1 state it certainly must because it will be as near its continuum as the A_1 ground state and the B_2 state, and then one ought to find these rather violent fluctuations in the observed lifetime. I might mention a paper by Zare which is in press in Phys. Rev. Lett. in which amazingly he has found that with excitation near 5900 Å he is not only able to assign some features of the fluorescent spectrum at low pressures clearly to the B_2 state and other features to the B_1, but he also finds that they have different lifetimes. He can separately measure the lifetimes. One is 100 μsec and the other one 30 μsec, which doesn't help us at all. But none of the people who have done careful work with flash excitation have found very non-exponential decays. There may be a slight deviation from the true exponential, but there are no two states of widely differing radiative lifetime. The argument that you've missed the fast state and are looking only at the slow state doesn't seem to stand up: there is no fast state that comes anywhere near the 0.1 μsec or so that we need.

CARRINGTON: At times like this, the best I can do is to remind you that it is more complicated than we think. Perhaps we should consider the density of rotational states manifold, or anharmonic vibrations in the vibrational manifold, which will shift things the way you want.

CHEMILUMINESCENCE REACTIONS INVOLVING METAL VAPORS[*]

C. J. Duthler[†] and H. P. Broida

Department of Physics, University of California

Santa Barbara, California 93106

ABSTRACT

A brief survey of recent experimental work on metal vapors at our laboratories is given. Results of chemiluminescence of group Ia and IIb metal atoms excited by a Lewis-Rayleigh nitrogen afterglow are presented. For group Ia atoms, we have found evidence that the exciting species in the afterglow is $N_2(A)$ rather than $N_2(X)^v$ as previously proposed. A demonstration of spin conservation during collisions of the second kind is seen in the excitation of group IIb atoms and relative quenching of electronically excited nitrogen states is noted.

INTRODUCTION

During the last five years, spectra of metal atoms and diatomic metal oxides have been extensively studied in our laboratories at Santa Barbara. The success of these studies has been a result of the development of systems for producing metal vapors in an inert gas stream for subsequent reactions of metals with active systems to produce light emission from the metals or from metal-oxides or metal-halides. In the first study, the interaction of Ba with long lived excited species in a flowing afterglow was investigated.[1] Since the original experiment, similar systems have been developed for a variety of studies:[2] molecular constants, lifetimes, and quenching cross sections have been measured using laser photoluminescence and more recently microwave-optical double resonance has been used to obtain better molecular constants.[3]

The apparatus used in the present study of the interaction of group Ia and IIb metal vapors is shown in Fig. 1. Several grams of the metal to be evaporated are placed in an alumina crucible which is located inside of an alumina chimney. Heat to vaporize the metal is produced by a tungsten wire heater which is wound around the alumina tube and surrounded by stainless steel heat shields. A stream of nitrogen or other inert gas carries the metal vapor to the reaction chamber which is made from a 10 cm diameter stainless steel cross. In other studies requiring higher temperatures, the tungsten wire heater is in direct contact with the crucible.

Oxidizers or flowing afterglows are introduced into the chamber through an orifice in the tube shown above the heater assembly in Fig. 1. In the case of the Lewis-Rayleigh nitrogen afterglow, excited species are produced by a microwave discharge and flow 30 cm downstream from the discharge and around a right angle through a 14 mm inside diameter pyrex to the reaction chamber. By adjusting the relative flow of the nitrogen (or oxidizer) and the carrier gas through the heater, the chemiluminescence metal atom emission can be centered in the reaction cell. At pressures below 1 torr, light emission is observed to fill the entire cell. At pressures above 10 torr, the flame is confined to an approximately spherical region a few cm in diameter below the inlet orifice. Downstream from the reaction zone, we observe the formation of very small metal oxide particles or in cases where no oxidizer is present we observe very small metallic particles.[4]

Light emission from the cell is viewed normal to the plane of Fig. 1 through a 5 cm diameter quartz window. Relative spectral intensities are measured with a 3/4 meter Fastie-Ebert spectrometer using a sensitive photomultiplier tube either with a grating blazed

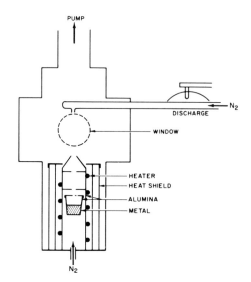

Fig. 1. Apparatus used for producing metal vapors in inert gas stream and for producing chemiluminescence.

at 500 or 200 nm. Preliminary surveys of emission are made using photographic detection with a f/3.5 quartz prism spectrograph for wavelengths less than 500 nm and a f/0.8 grating spectrograph for visible lines. With one hour exposure times, these spectrographs are able to detect lines that are too faint to be measured photoelectrically.

Studies of chemiluminescence emission during the formation of metal oxides and chlorides have not been completed yet, and space does not permit discussion of our present results.[5] The balance of this paper presents results of our studies of the excitation of group Ia and IIb metal atoms by a flowing Lewis-Rayleigh nitrogen afterglow. A more complete account of these experiments is being published elsewhere.[6]

The nitrogen afterglow is a very well known chemiluminescence system and has been studied extensively. Wright and Winkler[7] have reviewed work on active nitrogen prior to 1968 and Bass and Broida[8] have published color pictures of chemiluminescence reactions involving active nitrogen.

In nitrogen afterglows, a flowing stream of N_2 molecules is dissociated by a dc or microwave discharge. Downstream from the discharge dissociated N atoms recombine into electronically excited molecular states or vibrationally excited ground molecular state. Details of the recombination are not yet completely understood, but some of the atoms recombine to form short lived $N_2(B)$ state molecules. Characteristic straw yellow light emission of the Lewis-Rayleigh afterglow then follows:

$$N + N + M \rightarrow N_2(B\ ^3\Pi_g) + M \tag{1}$$

$$N_2(B\ ^3\Pi_g) \rightarrow N_2(A\ ^3\Sigma_u^+) + h\nu\ . \tag{2}$$

Emission from $N_2(A)$ is spin forbidden; we are not able to spectroscopically monitor populations of either $N_2(A)$ or vibrationally excited ground state molecules, $N_2(X)^V$. These three active species [$N_2(A)$, $N_2(B)$ and $N_2(X)^V$] are continually formed downstream from the discharge in the afterglow and are able to transfer energy during collisions to ground state metal atoms.

GROUP Ia METALS

Effective quenching of excited alkali (group Ia) atoms by molecular gases has been known for a long time.[7,9] In quenching of an electronically excited atom by N_2, electronic energy of the atom presumably is transferred to molecular internal vibrational energy. Early evidence for the reverse process was obtained from shock waves in N_2, where it was observed that Na excitation temperatures corresponded to vibrational temperatures of the relaxing N_2 molecules.[10,11] Subsequently, Na and K atomic emissions were observed when alkali atoms were added to nitrogen afterglows and again it was presumed that the electronic excitation energy of the alkali was obtained from vibrational energy of ground state nitrogen molecules.[12-14]

In the present experiment, we have found evidence that the electronically excited alkali atoms obtain their energy from metastable electronically excited $N_2(A)$ molecules rather than from $N_2(X)^v$. This agrees with recent experiments of Krause et al. who observed that vibrational energy can be transferred but requires a larger amount of kinetic energy before the collision than is available in the afterglow.[15] We have recently learned that Gann et al., in a series of experiments concurrent with the present study, have independently concluded that the source of excitation energy of alkali atoms in the afterglow is $N_2(A)$ rather than $N_2(X)^v$.[16]

Atomic emission excited by the Lewis-Rayleigh nitrogen afterglow has been studied for the alkali metal atoms Li, Na, and K. Data for K are typical: 36 spectral lines from 300 to 800 nm are sufficiently intense to be measured photoelectrically with energies to within 800 cm^{-1} of the ionization limit. The emission intensity is pressure dependent with maximum intensity occurring near 0.7 torr and decreasing by two orders of magnitude when the pressure is increased from 0.7 to 10 torr. Except for first positive N_2 band emission $[N_2(B) \rightarrow N_2(A)]$ from the afterglow, no molecular band emission was observed.

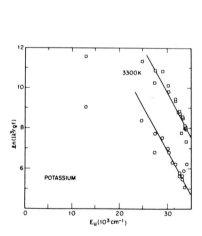

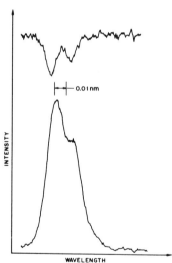

Fig. 2. Excited state K populations as function of excited state energy.

Fig. 3. Li absorption (upper) and emission (lower) line profiles.

Figure 2 presents experimental K emission data at background N_2 pressures of 0.7 and 5.7 torr. Relative excited K state populations are plotted as the ordinate versus excited state energy using the relation

$$N_u \propto I_{ul} \lambda_{ul}^3 / g_l f_{lu} . \qquad (3)$$

If the excited state population has a Maxwell-Boltzmann distribution, a straight line of slope $-1/kT$ is obtained on such a plot.

Excited state K populations with energies greater than 20,000 cm^{-1}, shown in Fig. 2, can be described by a Maxwell-Boltzmann distribution with a temperature of $3300 \pm 500°K$ at both pressures. Similarly, a temperature of $4700 \pm 500°K$ was obtained for upper excited states of Na, which is in reasonable agreement with the result of $4100°K$ previously obtained by Milne.[12]

Departures of the data toward lower intensities from the straight line at high and low excited state energies was observed for both Na and K. Relative intensities were not measured for Li. The two orders of magnitude lower than expected intensity for the lowest energy resonance transition was not a result of quenching of the excited alkali or self-absorption of resonance radiation, both of which are small in our experiment and corrected for in Fig. 2. This is supported by similar results being obtained at all nitrogen pressures and alkali densities used in the experiments.

Instead, we feel that the assumption of efficient but slightly non-resonant transfer from a thermalized distribution of vibrationally excited ground state nitrogen molecules is in error because the distribution of excited state energies and temperature describing the distribution is not degraded with increasing pressure. For excitation of the alkali by $N_2(A)$, the electronic energy transfer would be very non-resonant with the excess energy appearing as relative translational energy and vibrational energy of ground state nitrogen. This can be tested experimentally by measuring the Doppler width of the emission from the excited alkali.

Excess energy released as kinetic energy is partitioned with equal momentum between the alkali atom and the nitrogen molecule so that Li would have a larger velocity and Doppler width than the heavier alkalis such as Na and K.

Emission line widths of the Li doublet at 670.7 nm were measured using a two meter Czerny-Turner spectrometer with photoelectric detection that had a resolution of 0.01 nm. Measured Li emission and absorption line profiles are shown in Fig. 3. Absorption is shown in the upper trace and has Li doublet line widths equal to the instrument width. Linewidths in emission are broadened so that the doublet separated by 0.015 nm is no longer resolved. Radiation entrapment and broadening is not responsible for the emission width as was tested by varying the Li vapor density by a factor of 50 with no observable change in emission width.

The widths of both components of the doublet are 0.02 nm which indicates that 8800 cm^{-1} energy is released as kinetic energy during the transfer. Electronic excitation of the Li requires 14,900 cm^{-1} energy which together with the 8800 cm^{-1} kinetic energy leaves 26,300 cm^{-1} from the original $N_2(A)$ energy for vibrational energy of the ground state nitrogen. This vibrational energy corresponds to the twelfth vibrational level of $N_2(X)$ which we notice has the same internuclear distance at the turning point as N_2 (A, v' = 1) so that the internuclear distance of the nitrogen may be preserved during the energy transfer.

Gann et al. observe 2000 cm^{-1} energy released as kinetic energy in the transfer to Na which is in reasonable agreement with our result of 8800 cm^{-1} with Li. Part of the difference of released kinetic energy could be a result of the preservation of the internuclear distance of the nitrogen molecule during the energy transfer. Excitation of the lowest ^2P level of Na requires 2000 cm^{-1} more energy than for Li so that a correspondingly smaller amount of kinetic energy would be released with Na.

GROUP IIb METALS

Atomic emission was observed when group IIb metal (Hg, Zn, and Cd) vapors were added to the nitrogen afterglow. The most intense emission lines observed for all the metals were the spin forbidden transition from the lowest $^3P_1$ state to the ground $^1S_0$ state with weaker emission being observed from singlet levels and upper triplet levels. Excitation of the group IIb atomic levels is most likely a result of collisions of the second kind with electronically excited $N_2(A\ ^3\Sigma_u^+)$ or $N_2(B\ ^3\Pi_g)$.[17,18] The more intense emission from the triplet levels than from the singlet levels gives evidence of spin conservation during the energy transfer.

Atomic line intensities for the group IIb metals were pressure dependent with maximum intensity occurring near background N_2 pressures of 10 torr which was the highest pressure attainable without quenching the discharge that produced the afterglow. With Hg, the intensity of the 253.6 nm line from the $^3P_1$ level to the ground state increased by a factor of 15 when the pressure was increased from 1 to 10 torr. To the accuracy of our measurements, Hg line intensities from the higher energy ^3S and ^3D levels were pressure independent and were nearly three orders of magnitude less intense than the 253.6 nm line.

In Hg the lowest triplet level is able to be excited by N_2 (A, v' = 0). Excitation of the higher energy ^3S and ^3D levels requires more energy and these levels are probably excited by either high vibrational levels of $N_2(A)$ or $N_2(B)$. The changing relative intensities of Hg lines indicate that excited species in the nitrogen which excite the various Hg levels are degraded with increasing pressure with quenching of high vibrational levels of $N_2(A)$ or quenching of $N_2(B)$ relative to N_2 (A, v' = 0). If we fit a Maxwell-Boltzmann distribution to excited state Hg populations, the temperature at 1 torr is 2800 ± 500°K and decreases by 500°K when the pressure is increased to 10 torr.

Similar results were obtained for Zn atoms excited by the nitrogen afterglow as were obtained for Hg. Emission intensities again were pressure dependent with the intensity of all the Zn lines increasing by a factor of 15 when the pressure was increased from 1 to 10 torr. However, Zn intensity differences were not as large as were observed for Hg with transitions from 5^3S being nearly as intense as the transition from 4^3P_1. In addition, relative Zn intensities were pressure independent to the accuracy of our measurements.

These Zn observations could be a result of preferential population of the 5^3S level which is nearly resonant with N_2 (A, v' = 0). Radiative cascading from 5^3S to 4^3P could effectively couple the populations of these levels and force the intensities to be equal and relative intensities to be pressure independent.

Spectra similar to Hg and Zn were observed with Cd. Maximum emission intensity again was observed near N_2 pressures of 10 torr, but all intensities were weaker so that reliable relative intensities could not be measured.

ACKNOWLEDGMENTS

We are grateful to Drs. Gann, Kaufman, and Biondi for sending us their unpublished results. The assistance of Mr. Steven Zweig during the data acquisition is gratefully acknowledged.

REFERENCES

* Work supported in part by AFOSR Grant No. AFOSR-70-1851.
† Present address: Xonics Inc., 6837 Hayvenhurst Avenue, Van Nuys, California 91406.
1. R. J. Oldman and H. P. Broida, J. Chem. Phys. <u>51</u>, 2764 (1969).

2. For a review of this work see S. E. Johnson, Ph.D. Thesis, University of California, Santa Barbara, 1971.
3. R. W. Field, R. S. Bradford, D. O. Harris, and H. P. Broida, J. Chem. Phys. $\underline{56}$, 4712 (1972); ibid. $\underline{57}$, 2209 (1972).
4. C. J. Duthler, S. E. Johnson, and H. P. Broida, Phys. Rev. Letters $\underline{26}$, 1236 (1971).
5. G. Capelle, R. S. Bradford, and H. P. Broida, J. Chem. Phys. (to be submitted).
6. C. J. Duthler and H. P. Broida, J. Chem. Phys. (submitted).
7. A. N. Wright and C. A. Winkler, <u>Active Nitrogen</u> (Academic, New York, 1968).
8. A. M. Bass and H. P. Broida, J. Res. Natl. Bur. Std. $\underline{67A}$, 379 (1963).
9. A. G. Mitchell and M. W. Zemansky, <u>Resonance Radiation and Excited Atoms</u> (Cambridge University Press, London, 1961).
10. A. G. Gaydon and I. R. Hurle, Symp. Combust. 8th, Pasadena, Calif., 1960 (1962), p. 309.
11. I. R. Hurle, J. Chem. Phys. $\underline{41}$, 3911 (1964).
12. E. L. Milne, J. Chem. Phys. $\underline{52}$, 5360 (1970).
13. W. L. Starr, J. Chem. Phys. $\underline{43}$, 73 (1965).
14. W. L. Starr and T. M. Shaw, J. Chem. Phys. $\underline{44}$, 4181 (1966).
15. H. F. Krause, J. Fricke, and W. L. Fite, J. Chem. Phys. $\underline{56}$, 4593 (1972).
16. R. G. Gann, F. Kaufman, and M. A. Biondi (to be published).
17. W. R. Brennen and G. B. Kistiakowsky, J. Chem. Phys. $\underline{44}$, 2695 (1966).
18. R. A. Young and G. A. St. John, J. Chem. Phys. $\underline{48}$, 2572 (1968).

WAYNE: I am going to start off by choosing a line that goes along with Brian Thrush's. I think that we really shouldn't get too upset by the idea that spin conservation is very important. But I would like to suggest that for systems like mercury it really is not proper to put too much weight on spin conservation. After all spin is not a very good description for mercury. The triplet to singlet transition (2537 Å emission) is after all only perhaps a factor of 100 times less intense than the singlet-singlet (1859 Å) line. So in a collision encounter why should spin be conserved! I find it very hard to go along with that. Organic photochemists talk about sensitization by benzophenone. Why should it be that the benzophenone triplet is populated by a radiationless transition in an isolated molecule since when it makes a collision it is supposed to conserve spin? I find it very difficult to accept that. The other point that I want to make is that in the most recent paper that I have read on the energy transfer from vibrationally excited nitrogen and alkali metals (Sadowski and Schiff) the A state of nitrogen is deliberately excluded and the authors are fairly certain they have only got vibrationally excited nitrogen. They see very efficient energy transfer of the resonant states. They've done it for a series of other levels for sodium other than the first p state and they show that the efficiency of energy transfer depends closely on whether one has a resonant match or not. Now, I wonder whether you would care to comment on the experiments of the York group and on the crossed beam experiments of Krause and Fite?

DUTHLER: I didn't have time to discuss the Krause and Fite experiment. As I understand the Krause and Fite experiment, relative kinetic energy before the collision is necessary to cause energy transfer from the vibrational levels. Such energy is available in Krause and Fite's experiment and they measured the energy dependence. It's also available in the shock tube where energy transfer was also seen from vibrational levels. But it's not available at least in my particular nitrogen afterglow. I am not familiar with the (Schiff) experiment, and I'll just pass that by. All I can say about that, is that I see no evidence for preferential populations and vibrational transfer in my experiment. Perhaps Professor Kaufman would like to comment on this.

You also brought up the point of the spin conservation. Again I can't argue with the theory, all I can do as an experimentalist is fall back on the data. I did see preferential population of the triplet levels and these quantum numbers perhaps are better in the smaller Group IIb metals such as zinc rather than mercury.

I studied the entire set of Group IIb metals. I see evidence for spin conservation in all of them. It's just interesting that spin conservation also holds for mercury, but perhaps not as well.

KAUFMAN: Well, let me first discuss very briefly our experiments on Na excitation, i.e., work by Drs. Gann, Biondi, and myself, which is now in press in Chem. Phys. Letters. Our study shows a broadening of the emitted sodium D-lines under thermal conditions (T 400°K) when active nitrogen and sodium are mixed and we also performed quenching experiments with added NO to titrate out the nitrogen atoms and with added CO, CO_2 and HH_3 to try to unravel the question of the excitation precursor using a simple Stern-Volmer analysis. We were, of course, shocked to find no correlation whatsoever with vibrationally excited nitrogen; that is, introducing a glass wool plug did not reduce the intensity at all but doing anything that took out nitrogen atoms did reduce it, although the atoms themselves were not directly involved in the excitation process, which we could show by adding a little bit of ammonia which strongly quenched the sodium emission without doing anything to the nitrogen afterglow intensity, i.e. the nitrogen atoms were clearly still there, so that another species had to be the precursor. Now that is in direct conflict with the Sadowski experiments. We have been in contact with them and we don't understand at all how the two groups can have experimental results which are so contradictory. I will say that their energy resonance argument would have to be discounted, because (a) they were not aware of the later crossed beam results of Krause et al. which showed that translational energy was required for the Na excitation, and (b) Dugan, also at York University, showed by interferometry that the Na lines were abnormally broad. His results are in excellent agreement with ours and they show that a lot of translational energy does come out in the excitation process. The observed 1500° K Doppler width amounts to about 6 Kcal of translational energy release, which is very much greater than the energy mismatch with N_2 vibrational levels which Sadowski considered significant in Na resonance excitation.

THRUSH: I'd like to stress the importance of making absolute measurements of the intensity and kinetics of these atomic emissions. From such a study of the Hg 2537 Å line Dr. Wild and I have shown that approximately 60% of the recombinations of ground state nitrogen atoms populate the A state of N_2. Hg emission from higher levels of the A state cannot greatly exceed the populations of the corresponding levels of the $B^3\pi_g$ state.

DUTHLER: I did measure the population differences with the zinc and the differences in population were at least a factor of 1000 between single c and triple c.

E.K.C. LEE: I would like to add a comment to what Dr. Wayne said. We have studied the systems in which triplet mercury (6^3P_1) was quenched by simple ketones, and also we have studied quenching of triple benzene ($^3B_{1u}$) by these same ketones. We find that in the case of triplet benzene quenching, about 100% of the product appears as triplet ketones whereas, when we do it with the triplet mercury approximately 15% to 20% appears as a singlet ketone.

Therefore, it is true that when you have heavy atoms there is a substantial amount of mixing between the two states (singlet and triplet) and this is one of the nice examples where, in the intermolecular energy transfer at least 15 to 20% triplet character shows up.

STEDMAN: We have observed the emission of sodium resonance lines on adding metallic sodium to the pure nitrogen ($A^3\Sigma_u^+$). Collisions with excited argon, which are strictly described by jj coupling, give many instances of initial triplet formation.

Can you observe the Hg ($P_{2,1,0}$) sublevel populations?

DUTHLER: I could not observe these populations among the ^3P sublevels.

THE QUENCHING OF SINGLET MOLECULAR OXYGEN

James A. Davidson and E.A. Ogryzlo

Department of Chemistry, University of British Columbia, Vancouver, Canada

The relaxation of electronically excited molecules can occur by the emission of radiation and by several non-radiative pathways. Knowledge of both types of events is essential to the complete understanding of chemiluminescent and photochemical systems since together they determine "quantum efficiencies". Radiative processes are reasonably well understood for most molecules. In the case of oxygen, some novel energy-pooling radiative processes have been observed (1), but these are now reasonably well understood. The major focus of this paper is the non-radiative processes, about which considerably less is known.

Non-radiative processes can be divided into (a) chemical reactions (b) energy-transfer, in which a second species becomes electronically excited, and (c) physical quenching, in which the electronic excitation is degraded into nuclear motion in the system. Chemical reactions of the excited states of oxygen have received a great deal of attention in the last eight years and some of these have been shown to lead to chemiluminescent products (2). Energy transfer is a frequently encountered mode of electronic relaxation. However, in the case of oxygen, where the excitation energy is only 22.5 kcal for the $^1\Delta_g$ state and 37.5 kcal for the $^1\Sigma_g^+$ state, it is seldom observed because few molecules have electronically excited states below this energy. We will only note here that by either consecutive energy transfer (3,4) or some equivalent cooperative pooling process (5) molecules have been produced in states with two or three times excitation energy of the $^1\Delta_g$ state.

The present paper is concerned exclusively with the physical quenching of oxygen in the $^1\Delta_g$ and $^1\Sigma_g^+$ states. In the last few years several laboratories have reported rate constants for the quenching of these species in the gas phase. However, none of these laboratories have presented a theory capable of accounting for the large variations in quenching efficiencies observed. Though a detailed quantitative theory of the quenching process will have to await more accurate and extensive data, we believe that some understanding of the quenching reaction is possible at the moment with a relatively simple model of the process.

EXPERIMENTAL METHODS

The experimental methods used to study the gas phase quenching of $O_2(^1\Sigma_g^+)$ and $O_2(^1\Delta_g)$ can be classified broadly as either photolytic or discharge-flow techniques. The photolytic techniques have used either the vacuum U.V. photolysis of O_2 (9,11,12,13,14,15) or energy transfer from excited sensitizers (7,18,21) to produce singlet oxygen. The most obvious sources of error in these measurements are probably the reactions of oxygen atoms, excited sensitizers or sensitizer fragments. Standard discharge-flow system measurements (6,10,16,17,19,20) are limited to pressures between 0.3 and 10 torr and have been known to yield erroneous results when oxygen atoms are not completely removed (20). A low pressure, very large volume "stopped-flow" variant of this technique (8) has been used to obtain quenching constants for a wide range of both $^1\Sigma_g^+$ and $^1\Delta_g$ quenchers. The effect of atoms and surfaces is uncertain in this last technique.

A few estimates and measurements of these quenching rate constants can be found in the literature earlier than 1968. However, they often differ from the current values by several orders of magnitude, or are obviously erroneous and have not been included in the present collection of data.

RESULTS

Rate constants for the quenching of both $O_2(^1\Sigma_g^+)$ and $O_2(^1\Delta_g)$ in the gas phase have been collected in Table I. An attempt has been made to assess the reliability and accuracy of the various determinations. Under the heading "best value" will be found a somewhat personal estimate of the most probable value of the quenching constant for each species. Rate constants for the removal of $O_2(^1\Delta_g)$ by olefins and dienes have been published and when physical quenching can be separated from chemical reaction, the data may be very interesting. At the moment the two have not been separated, the data is difficult to interpret and hence it is not included in Table I.

TABLE I

Rate Constants for the Quenching of $O_2(^1\Sigma_g^+)$ and $O_2(^1\Delta_g)$

Quencher and ν of highest fundamental	$k_q(^1\Sigma)$ l mole^{-1} sec^{-1} "best" value	$k_q(^1\Sigma)$ l mole^{-1} sec^{-1} reported value	$k_q(^1\Delta)$ l mole^{-1} sec^{-1} "best" value	$k_q(^1\Delta)$ l mole^{-1} sec^{-1} reported value
He	6×10^3	$\sim 6 \times 10^3$ (8) $<6 \times 10^4$ (10) $\sim 6 \times 10^4$ (11) $<7 \times 10^5$ (16)	4.8	≤ 6 (8) <5 (21) 4.8 (22)
Ar	9×10^3	9×10^3 (8) $<6 \times 10^4$ (10) 3.5×10^3 (11) 1.5×10^6 (16)	5.3	≤ 6 (8) <120 (17) <5 (21) 5.4 (22)
Kr			5	5 (22)
Xe			20	20 (22)
H_2 (4405 cm^{-1})	4×10^8	6.6×10^8 (7) 2.4×10^8 (8) 2.4×10^8 (10) 6.6×10^8 (11) 6×10^8 (13) 3.9×10^8 (14)	2.7×10^3	2.2×10^3 (8) 2.7×10^3 (21) 3.2×10^3 (22)
HD (3817 cm^{-1})	1.5×10^8	1.1×10^8 (10) 1.9×10^8 (13)		
D_2 (3118 cm^{-1})	1.2×10^7	1.2×10^7 (10) 1.2×10^7 (11) 1.1×10^7 (13)		
N_2 (2359 cm^{-1})	1.3×10^6	1.3×10^6 (8) 1.2×10^6 (9) 1.8×10^6 (10) 1.1×10^6 (11) 1.3×10^6 (14) 1.4×10^6 (15) 1.5×10^6 (16)	50	≤ 6 (8) ≤ 42 (17) <1.3 (21) <84 (19) 83 (22)
O_2 (1580 cm^{-1})	1×10^5	9.0×10^4 (8) 9.0×10^4 (9) $\sim 6.0 \times 10^5$ (10) 2.7×10^5 (11) 6.6×10^4 (14) $<6.0 \times 10^4$ (15)	1.2×10^3	1.0×10^3 (8) 1.4×10^3 (17) 1.3×10^3 (21) 1.2×10^3 (19)
CO (2168 cm^{-1})	2×10^6	1.8×10^6 (9) 2.6×10^6 (11) 2.0×10^6 (14) 1.5×10^6 (16)		$<4 \times 10^4$ (22)
NO (1904 cm^{-1})	2.6×10^7	3.3×10^7 (8) 2.4×10^7 (10) 2.5×10^7 (11)	2.7×10^4	2.7×10^4 (8)

Table I continued

Quencher and ν of highest fundamental	$k_q(^1\Sigma)$ l mole^{-1} sec^{-1}		$k_q(^1\Delta)$ l mole^{-1} sec^{-1}	
	"best" value	reported value	"best" value	reported value
H_2O (3756 cm^{-1})	2.8×10^9	3.0×10^9 (8) 2.4×10^9 (10) 2.0×10^9 (11) 3.3×10^9 (13) $\sim 1.2 \times 10^{10}$ (14) 6×10^8 (16)	3.1×10^3	2.4×10^3 (8) 8.3×10^3 (17) 3.4×10^3 (21)
D_2O (2789 cm^{-1})	2.4×10^8	2.4×10^8 (10) 2.3×10^9 (13) 4.9×10^8 (16)		
N_2O (2224 cm^{-1})	8.0×10^7	1.1×10^8 (8) 4.2×10^7 (11)	<45	≤ 60 (8) <45 (21)
CO_2 (2344 cm^{-1})	2.0×10^8	2.5×10^8 (7) 1.8×10^8 (8) 1.8×10^8 (9) 9.0×10^7 (10) 2.6×10^8 (11) 1.2×10^8 (14) 2.8×10^7 (16)	10^2	<50 (8) 2.5×10^3 (17) <9 (21)
CS_2 (1532 cm^{-1})	1.7×10^6	1.7×10^6 (7)		
H_2S (2627 cm^{-1})	2×10^8	3.8×10^8 (8) 1.5×10^7 (16)		
NO_2 (1618 cm^{-1})	1.7×10^7	1.5×10^7 (10) 1.9×10^7 (11)	3×10^3	3×10^3 (8)
SO_2 (1362 cm^{-1})	1×10^6	4.0×10^5 (7) 1.8×10^6 (10)		
CH_4 (3020 cm^{-1})	4.5×10^7	4.4×10^7 (6) 4.5×10^7 (8) 6.6×10^7 (11) 8×10^6 (16)	8.4×10^3	8.4×10^3 (8)
C_2H_6 (2995 cm^{-1})	2.0×10^8	1.9×10^8 (6) 2.8×10^8 (7) 2.2×10^8 (11)		
C_3H_8 (2980 cm^{-1})	2.6×10^8	2.7×10^8 (6) 2.2×10^7 (8)	1.45×10^3	1.45×10^3 (8)
n-C_4H_{10}	3.8×10^8	3.8×10^8 (6)		
iso-C_4H_{10}	2.9×10^8	2.9×10^8 (6)		
n-C_5H_{12}	4.5×10^8	4.5×10^8 (6)		
n-C_6H_{14}	5.5×10^8	5.5×10^8 (6)		
n-C_7H_{16}	6.0×10^8	6.0×10^8 (6)		
cyclo-C_3H_6	3.7×10^8	3.7×10^8 (6)		
cyclo-C_6H_{12}	5.8×10^8	5.8×10^8 (6)		
CH_3-cyclo-C_5H_9	5.2×10^8	5.2×10^8 (6)		
CH_3-cyclo-C_6H_{11}	5.8×10^8	5.8×10^8 (6)		

Table I continued

Quencher and ν of highest fundamental	$k_q(^1\Sigma)$ l mole^{-1} sec^{-1}		$k_q(^1\Delta)$ l mole^{-1} sec^{-1}	
	"best" value	reported value	"best" value	reported value
CH_3I	1.8×10^8	1.8×10^8 (10)		
$CHCl_3$	8.8×10^7	8.8×10^7 (16)		
CF_4 (1265 cm^{-1})	1.6×10^6	1.6×10^6 (7)		
CF_2Cl_2	1.8×10^6	1.8×10^6 (8)	2.4×10^2	2.4×10^2 (8)
CCl_4 (791 cm^{-1})	7.5×10^5	2.7×10^5 (7) 1.8×10^6 (10)		
C_2H_2 (3374 cm^{-1})	2.7×10^8	2.7×10^8 (12)		
C_2D_2 (2700 cm^{-1})	4.9×10^7	4.9×10^7 (13)		
C_2H_4 (3272 cm^{-1})	2×10^8	1.2×10^8 (8) 1.9×10^8 (11) 2.7×10^8 (12) 2.6×10^8 (13)	1.2×10^3	1.2×10^3 (8)
C_2D_4 (2304 cm^{-1})	8.4×10^7	8.4×10^7 (13)		
C_2D_6 (2225 cm^{-1})	1×10^8	1×10^8 (7)		
C_2F_6	1.9×10^6	1.9×10^6 (7)		
SF_6 (965 cm^{-1})	3.4×10^5	3.4×10^5 (11)	≤ 6	≤ 6 (8) <7 (21)
NH_3 (3414 cm^{-1})	1.2×10^9	1.3×10^9 (6) 1.1×10^9 (8) 1.8×10^9 (10) 5.2×10^7 (11) 1.6×10^8 (16)	4.2×10^3	4.2×10^3 (8)
CH_3NH_2	1.1×10^9	1.1×10^9 (6)		8.0×10^3 (20)
$C_2H_5NH_2$	8.9×10^8	8.9×10^8 (6)		$? \times 10^3$ (20)
$(CH_3)_2NH$	5.6×10^8	5.6×10^8 (6)		5.6×10^4 (20)
$(C_2H_5)_2NH$				7.3×10^4 (20)
$(CH_3)_3N$	3.6×10^8	3.6×10^8 (6)	1.9×10^6	1.9×10^6 (20)
$(C_2H_5)_3N$	5.8×10^8	5.8×10^8 (6)	2.0×10^6	2.0×10^6 (20)
CH_3SH			2.3×10^3	2.3×10^3 (29)
$(CH_3)_2S$			1.6×10^5	1.6×10^5 (29)
$(C_2H_5)_2S$			4.1×10^5	4.1×10^5 (29)
$(CH_3)_2S_2$			1.0×10^4	1.0×10^4 (29)
CH_3OH (3682 cm^{-1})	2.4×10^9	2.4×10^9 (10) 2.6×10^8 (16)		
CH_3CH_2OH	1.9×10^9	1.9×10^9 (10)		

DISCUSSION

The Quenching of $O_2(^1\Sigma_g^+)$

Several trends have been noted by earlier workers. Hydrogen containing compounds are most effective quenchers (10). This has been interpreted in terms of an increased intermolecular interaction with oxygen for these molecules (2). It has also been pointed out that, since the quenching efficiency tends to monotonically increase with the magnitude of the quencher vibrational frequency, this trend might rather involve a coupling between the $O_2(^1\Sigma_g^+) \rightarrow O_2(^1\Delta_g)$ electronic transition and a vibrational transition in the quenching species (23). Earlier workers (17) proposed a similar mechanism in noting that N_2 was a much more effective quencher than O_2. The extent of such a correlation is illustrated in Figure 1 where $\log k_q(^1\Sigma_g^+)$ is plotted against the highest fundamental vibrational frequency of the quencher.

The general trends noted above are quite evident. The good quenchers are those containing hydrogen, i.e. those with high vibrational frequencies. However, some other correlations are now apparent. The points for the homonuclear diatomic molecules (H_2, D_2, N_2, O_2) fall on a remarkably good straight line. In Figure 1, the line has been curved below 1500 cm^{-1} to pass through the points for He and Ar. We have done this because we feel that the quenching rates for He and Ar represent the basic rate of a process in which the 5240 cm^{-1} excitation energy from the process $O_2(^1\Sigma_g^+ \rightarrow {}^1\Delta_g)$ is transformed into translational motion of the quencher and O_2. The higher quenching rates of the other molecules is then associated with the excitation of vibrational motion in the quencher. This possibility is best considered with the help of Figure 2, where the spacing between the vibrational energy levels of the diatomic quenchers is presented together with the positions of the (0,0), (0,1) and (0,2) bands of the $^1\Sigma_g^+ \rightarrow {}^1\Delta_g$ transition in O_2.

In the case of quenching by H_2, only one process is energetically feasible and this must be the dominant quenching mechanism:

$$H_2(v=0) + O_2(^1\Sigma_g^+\ v=0) \longrightarrow H_2(v=1) + O_2(^1\Delta_g\ v=0) \quad (1)$$

At the other extreme, quenching by O_2 can lead to the excitation of O_2 into the 3rd, 2nd or 1st vibrational level if $O_2(^1\Sigma_g^+)$ relaxes to the 0th, 1st, or 2nd vibrational level of $O_2(^1\Delta_g)$ respectively. In view of the small quenching constant for O_2, it is clear that all these processes are inefficient. The important question now is whether the efficiencies of the various processes are governed by Franck-Condon factors which may be estimated from the overlap of the emission spectrum of the donor, $O_2(^1\Sigma_g^+)$, with the absorption spectra of the acceptors (quenchers). This possibility has been proposed for the quenching of $O_2(^1\Delta_g)$ in solution (24). In the case of homonuclear diatomic molecules where transition between vibrational

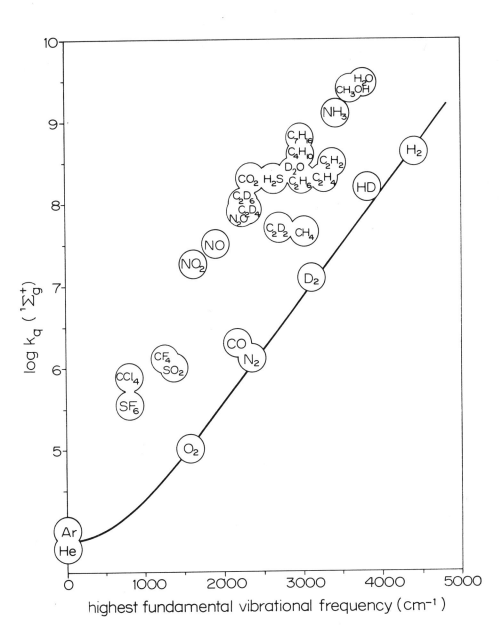

Figure 1 Variation of log $k_q(^1\Sigma_g^+)$ with the Highest Fundamental Vibrational Frequency of the Quencher

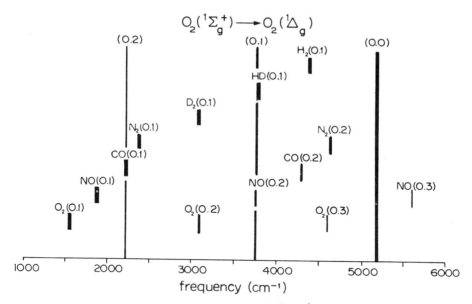

Figure 2 Relative Position of the $O_2(^1\Sigma_g^+ \rightarrow {}^1\Delta_g)$ Transitions and Vibrational Bands of Diatomic Molecules

levels are not observed for the free molecule, the relevant transitions in the quencher would have to be collision-induced (25). It is the positions of these bands which we have shown in Figure 2. To properly assess the data in figure 1, the complete absorption and emission spectra of these molecules in the $1000 cm^{-1}$-$6000 cm^{-1}$ region would have to be determined. Unfortunately most have not been reported and even the $O_2(^1\Sigma_g^+ \rightarrow {}^1\Delta_g)$ transition has not been well characterized (26).

In the absence of such detailed information it is difficult to know whether the overtone bands of the quencher of the 0,1 and 0,2 transitions in $O_2(^1\Sigma_g^+ \rightarrow {}^1\Delta_g)$ contribute significantly to the quenching process. For example, in the case of HD the quenching constant lies slightly above the line in Figure 1. This may be due to the near coincidence of the HD fundamental with the $3750 cm^{-1}$ (0,1) transition in O_2. Similarly, the coincidence of the CO fundamental with the $2300 cm^{-1}$ (0,2) band in O_2 may be why the quenching constant for CO lies above the line as well. However in contrast to the homonuclear diatomics, CO displays a strong infrared absorption and this may be the more important factor. In the case of NO, the rate constant is about two orders of magnitude larger than predicted by the line drawn. Though there is some overlap with the O_2 emission bands and NO has allowed transitions in the infrared, it is difficult to rationalize the great difference between NO and CO and therefore, it is tempting to involve the unpaired electron in NO in a more specific interaction with $O_2(^1\Sigma_g^+)$.

Let us now consider the polyatomic quenchers, most of which have quenching constants one or two orders of magnitude greater than the homonuclear diatomic quenchers with similar fundamental vibrational frequencies. However, for these polyatomic molecules the correlation between the largest fundamental frequency and log $k_q(^1\Sigma_g^+)$ is somewhat more erratic. Both of these effects are no doubt attributable to the presence of additional modes of vibration and the consequent absorptions due to combination bands.

If the quenching of $O_2(^1\Sigma_g^+)$ involves the conversion of the electronic excitation energy of $O_2(^1\Sigma_g^+)$ into vibrational excitation in the quencher, one would expect the quenching rate constant to obey a relationship of the form:

$$k_q(^1\Sigma_g^+) = c\int_0^\infty \epsilon_q(\nu) f_{O_2}(\nu) d\nu \qquad (2)$$

where the integration is carried out over the limits of the $O_2(^1\Sigma_g^+ \to {}^1\Delta_g)$ progression, $fO_2(\nu)$ is the relative emission intensity in that transition, $\epsilon_q(\nu)$ is the quencher extinction coefficient and c is a proportionality constant. Expression (2) may be further broken down into a sum of terms as:

$$k_q(^1\Sigma_g^+) = c\left[FC_1\int \epsilon_q(\nu) f_{O_2}(\nu) d\nu + FC_2\int \epsilon_q(\nu) f_{O_2}(\nu) d\nu + \cdots\right] \qquad (3)$$

where each integration is carried out over the limits of a particular band in the O_2 progression and $fO_2(\nu)$ is the relative O_2 emission intensity within the band. This necessitates the multiplication of each term by the Franck-Condon factor (FC) for the transition involved. In the free O_2 molecule the Franck-Condon factor for the 0,0 transition has been calculated to be about 42 times that for the 0,1 transition and 4200 times that for the 0,2 transition (27). Thus unless the Franck-Condon factors are much more favorable for the induced transition, the second and higher integrals must be quite large or the first integral near zero in order for the higher terms to make a significant contribution to the total rate constant.

In this preliminary investigation we have studied only the overlap in the 5240 cm^{-1} region. As an example, the computer drawn overlap of the NH$_3$ absorption spectrum and the calculated $O_2(^1\Sigma_g^+$ v=0 $\to {}^1\Delta_g$ v=0) emission spectrum is shown in Figure 3. In Table II we have listed the $O_2(^1\Sigma_g^+)$ quenching constants, the relative overlap integrals and the values of c obtained if one truncates (3) after the first term. Note that those compounds for which c deviates markedly from 1 X 10^6 are those which would be expected to show a strong infrared absorption in the region of the $O_2(0,1)$ transition (3750 cm^{-1}). This is illustrated in Figure 4 where the line is drawn to represent the contribution of the overlap in the 5240 cm^{-1} region to the quenching constant.

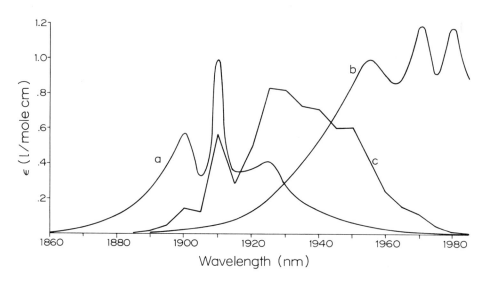

Figure 3 a) Calculated Relative $O_2(^1\Sigma_g^+ \; v=0 \rightarrow {}^1\Delta_g \; v=0)$ Emission Spectrum
b) NH_3 Absorption Spectrum
c) Computer Drawn Relative Overlap Integral

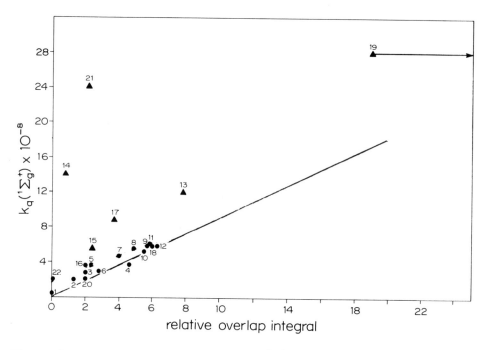

Figure 4 Overlap Integral Versus $O_2(^1\Sigma_g^+)$ Quenching Rate Constants
▲ Designates those compounds containing NH or OH

TABLE II

Comparison of Rate Constants for Quenching of $O_2(^1\Sigma_g^+)$ with Overlap Integration

Number	Compound	k_q "Best" Value	Overlap Integral	c
1	CH_4	4.5×10^7		
2	C_2H_6	2.0×10^8	1.3×10^2	1.6×10^6
3	C_3H_8	2.6×10^8	2.0×10^2	1.4×10^6
4	cyclo-C_3H_6	3.7×10^8	4.6×10^2	8.0×10^5
5	n-C_4H_{10}	3.8×10^8	2.3×10^2	1.7×10^6
6	iso-C_4H_{10}	2.9×10^8	2.8×10^2	1.0×10^6
7	n-C_5H_{12}	4.5×10^8	4.0×10^2	1.1×10^6
8	n-C_6H_{14}	5.5×10^8	4.9×10^2	1.1×10^6
9	cyclo-C_6H_{12}	5.8×10^8	5.7×10^2	1.0×10^6
10	CH_3-cyclo-C_5H_9	5.2×10^8	5.0×10^2	1.0×10^6
11	n-C_7H_{16}	6.0×10^8	5.8×10^2	1.0×10^6
12	CH_3-cyclo-C_6H_{11}	5.8×10^8	6.3×10^2	9.2×10^5
13	NH_3	1.2×10^9	7.8×10^2	1.5×10^6
14	CH_3NH_2	1.1×10^9	1.8×10^2	6.1×10^6
15	$(CH_3)_2NH$	5.6×10^8	2.4×10^2	2.3×10^6
16	$(CH_3)_3N$	3.6×10^8	2.0×10^2	1.8×10^6
17	$CH_3CH_2NH_2$	8.9×10^8	3.7×10^2	2.3×10^6
18	$(CH_3CH_2)_3N$	5.8×10^8	6.0×10^2	9.6×10^5
19	H_2O	2.8×10^9	$>1.9 \times 10^3$	$<1.5 \times 10^6$
20	H_2S	2.0×10^8	2.0×10^2	1.0×10^6
21	CH_3OH	2.4×10^9	2.1×10^2	1.1×10^7
22	CO_2	2.0×10^8	9.5×10^0	2.1×10^7

The Quenching of $O_2(^1\Delta_g)$

Data for the quenching of $O_2(^1\Delta_g)$ from Table I is plotted in Figure 5 as a function of the highest fundamental vibrational frequency of the quencher. Fewer small quenchers have been studied in the case of $O_2(^1\Delta_g)$ and data is therefore much less satisfactory for assessing a mechanism. To determine whether factors similar to those governing the $O_2(^1\Sigma_g^+)$ quenching are important in this case, a line has been drawn through the point for H_2 with a slope identical to that found for the $O_2(^1\Sigma_g^+)$ system. More diatomic molecules need to be studied before any decision is possible. However, the present data is at least not inconsistent with a similar excitation of vibrational modes in the quencher during the relaxation process.

Most quenching constants for $O_2(^1\Delta_g)$ are about 10^5 times smaller than the corresponding constant for $O_2(^1\Sigma_g^+)$ quenching. The origin

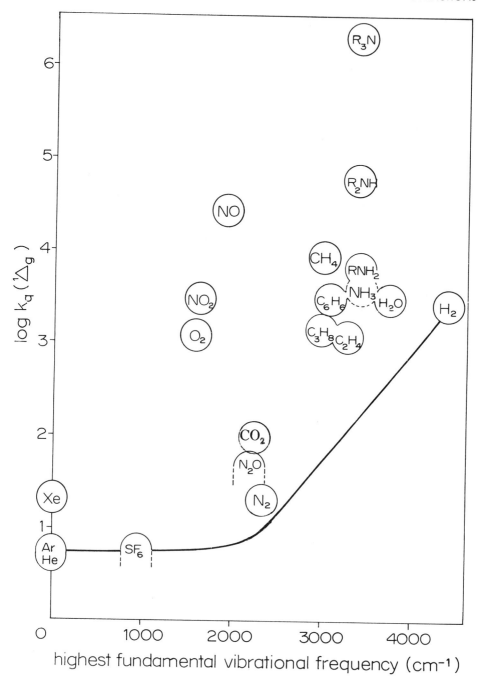

Figure 5 Variation of log $k_q(^1\Delta_g)$ with the Highest Fundamental Vibrational Frequency of the Quencher

of the difference is probably two fold. The energy gap for the $O_2(^1\Delta_g \rightarrow ^3\Sigma_g^-)$ transition is 2640 cm^{-1} greater than for the $O_2(^1\Sigma_g^+ \rightarrow ^1\Delta_g)$ transition and the $O_2(^1\Delta_g \rightarrow ^3\Sigma_g^-)$ transition is "spin forbidden". The larger energy gap places the most important members of the O_2 $(^1\Delta_g \rightarrow ^3\Sigma_g^-)$ progression in a region where few molecules have significant absorptions thus decreasing the value of the overlap integrals and the spin forbidden character of the transition decreases the relaxation probability by an additional several orders of magnitude.

The "anomalously" high quenching constants obtained for molecules with non-zero spin (NO_2, O_2 and NO) are consistent with the spin forbidden nature of the $O_2(^1\Delta_g \rightarrow ^3\Sigma_g^-)$ transition since the transition is made formally allowed in the presence of these species. The higher rate constant for Xe may also be the usual "heavy atom" effect commonly observed in such intersystem crossings.

The varying abilities of amines to quench $O_2(^1\Sigma_g^+)$ and $O_2(^1\Delta_g)$ provide an interesting contrast. In the case of $O_2(^1\Sigma_g^+)$ the rate constants decrease slightly with greater substitution in a manner consistent with their absorption in the 5240 cm^{-1} region, indicating that their quenching efficiencies are related to the presence of vibrational modes which can take up a significant portion of the excitation energy. In the case of $O_2(^1\Delta_g)$ the more substituted amines show a very much greater quenching ability. This is not reflected in any increased absorption in the 7880 cm^{-1} region or in the region of higher members of the $O_2(^1\Delta_g \rightarrow ^3\Sigma_g^-)$ progression. It has been suggested (28) that a "charge-transfer" interaction can account for these results, and a correlation has been noted between quenching efficiencies and ionization energies of the quencher (20). A similar trend has been noted for some sulfides (29). There is little doubt that a different mechanism is required to account for the quenching of $O_2(^1\Delta_g)$ by molecules with very low ionization energies. At the moment the charge-transfer theory provides the only satisfactory interpretation of the data.

Relaxation and Quenching in Solution

We will not attempt a detailed analysis of the quenching data available from studies in the liquid phase. However it is interesting to briefly survey the data noting similarities and differences.

In the condensed phase only $O_2(^1\Delta_g)$ quenching has been studied and the data is not very extensive. When the quenching by a solvent is determined, the characteristic decay rate is reported as a lifetime (τ) or as a first order decay constant ($k_1 = 1/\tau$). Such data is difficult to compare with second order quenching constants since the concentration of the quencher is not a very meaningful quantity and the quenching process may be affected by a "collective" interaction of the solvent. Table III contains data obtained by four different methods (30,24,31,32). Merkel et al.(24) indicated that they have been able to relate the lifetime of $O_2(^1\Delta_g)$, in solution, to the optical densities of the solvents at 1270 and 1590 nm. This would indicate that the quenching of $O_2(^1\Delta_g)$ in solution is governed by factors similar to those found for the quenching of $O_2(^1\Sigma_g^+)$ and $O_2(^1\Delta_g)$ in the gas phase. Some differences in the relative quenching abilities are evident however. Though benzene and water have almost identical $O_2(^1\Delta_g)$ quenching constants in the gas phase, the decay of $O_2(^1\Delta_g)$ is more than ten times as rapid in liquid water as it is in liquid benzene. Such a difference is not unreasonable if both the gas and liquid phase quenching are governed by Franck-Condon factors related to the solvent-absorption at the two frequencies noted. Not only are most bands broadened by condensation, they are frequently shifted by intermolecular forces, especially when these are strong as in hydrogen-bonded solvents.

TABLE III

Rate Constants for the Quenching of $O_2(^1\Delta_g)$ by Solvents

Solvent	τ (μsec)	k (sec^{-1})
Methyl alcohol	5.5	1.8×10^5 (30)
	7	1.4×10^5 (24)
	5	2×10^5 (31)
	11	9.0×10^4 (32)
Ethyl alcohol (95%)	5.5	1.8×10^5 (31)
n-Butyl alcohol	9.1	1.1×10^5 (30)
	19	5.2×10^4 (32)
t-Butyl alcohol	13.5	7.4×10^4 (30)
Benzene	24	4.2×10^4 (24)
	12.5	8×10^4 (31)
Dimethylformamide	7	1.4×10^5 (31)
Acetone	26	3.8×10^4 (24)
CS_2	200	5×10^3 (24)
H_2O	2	5×10^5 (24)
D_2O	20	5×10^4 (24)

TABLE IV

Physical Quenching of $O_2(^1\Delta_g)$ in Solution

Quencher	Solvent	k_q (l moles^{-1} sec^{-1})
Ethylamine	Freon 113	3.1 X 10^4 (33)
Diethylamine	Freon 113	5.7 X 10^5 (33)
Triethylamine	Freon 113	2.1 X 10^6 (33)
	CH$_3$OH	1.2 X 10^7 (32)
DABCO	CH$_3$OH	7.3 X 10^6 (32)
	1-Butanol	4.5 X 10^6 (32)
N-Methylaniline	Freon 113	3.8 X 10^4 (33)
N,N-Dimethylaniline	Freon 113	2.0 X 10^5 (33)

Table IV contains some quenching constants obtained in solution. As in the gas phase, the molecules from these studies show a good correlation of quenching rate with the ionization energy of the quencher (33). The rate constants for quenching by aliphatic amines in freon are in reasonably good agreement with those found in the gas phase. Those obtained in methanol are about an order of magnitude higher. It is possible that with a more polar solvent such as methanol, charge-transfer interactions are enhanced, leading to more effective quenching.

CONCLUSIONS

The physical quenching of $O_2(^1\Sigma_g^+)$ is dominated by the ability of the ($^1\Sigma_g^+ \to ^1\Delta_g$) transition in oxygen to excite vibrational motion in the quencher. More data is required to quantitatively relate the rate constant to the transition probability in the quencher. Though the physical quenching of $O_2(^1\Delta_g)$ in both the gas and liquid phases also appears to be dominated by a similar ability of the quenchers to become vibrationally excited, there are three factors which make the quenching of this species somewhat different. The first factor is the energy gap which for this process is 7880 cm^{-1}. The second is the spin change. These two factors combine to reduce the transition probability by a factor of about 10^5 below that for $O_2(^1\Sigma_g^+)$ quenching. The third factor is the susceptibility of $O_2(^1\Delta_g)$ to charge-transfer interactions with molecules of low ionization energies. This leads to a markedly enhanced rate of quenching for aliphatic amines and sulphides. This is not observed with $O_2(^1\Sigma_g^+)$.

REFERENCES

1. E.W. Gray and E.A. Ogryzlo, Chem. Phys. Lett. $\underline{3}$, 658 (1969)
2. D.R. Kearns, Chem. Rev. 71, 395 (1971)
3. E.A. Ogryzlo and A.E. Pearson, J. Phys. Chem. $\underline{72}$, 2013 (1968)
4. R.G. Derwent and B.A. Thrush, J. Chem. Soc. Far. Trans. II $\underline{68}$, 720 (1972)
5. A.U. Khan and M. Kasha, J. Amer. Chem. Soc. 92, 3293 (1970)
6. J.A. Davidson and E.A. Ogryzlo, Submitted for Publication
7. J.A. Davidson, K.E. Kear and E.W. Abrahamson, Submitted for Publication
8. K.H. Becker, W. Groth and U. Schwath, Chem. Phys. Lett. $\underline{8}$, 259 (1971)
9. J.F. Noxon, J. Chem. Phys. $\underline{52}$, 1852 (1970)
10. R.J. O'Brien Jr. and G.H. Myers, J. Chem. Phys. $\underline{5}$, 3832 (1972)
11. S.V. Filseth, A. Zia and K.H. Welge, J. Chem. Phys. $\underline{52}$, 5502 (1970)
12. F. Stuhl and H. Niki, Chem. Phys. Lett. $\underline{5}$, 573 (1970)
13. F. Stuhl and H. Niki, Chem. Phys. Lett. $\underline{7}$, 473 (1971)
14. F. Stuhl and K.H. Welge, Can. J. Chem. $\underline{47}$, 1870 (1969)
15. T.P.J. Izod and R.P. Wayne, Proc. Roy. Soc. (London) $\underline{A308}$, 81 (1968)
16. S.J. Arnold, M. Kubo and E.A. Ogryzlo, Adv. Chem. Ser. $\underline{77}$, 133 (1968)
17. T.D. Clark and R.P. Wayne, Proc. Roy. Soc.(London) $\underline{A314}$, 111 (1969)
18. F.D. Findlay, C.J. Fortin and D.R. Snelling, Chem. Phys. Lett. $\underline{3}$, 204 (1969)
19. P.D. Steer, R.A. Ackerman and J.N. Pitts, J. Chem. Phys. $\underline{51}$, 843 (1969)
20. K. Furukawa and E.A. Ogryzlo, J. Photochem. In Press
21. F.D. Findlay and D.R. Snelling, J. Chem. Phys. $\underline{55}$, 545 (1971)
22. R.J. Collins, D. Husain and R.J. Donovan, J. Chem. Soc., Far. Trans. II (1972) In Press
23. J.A. Davidson, PhD. Thesis October, 1971
24. P.B. Merkel, R. Nilsson and D.R. Kearns, J. Am. Chem. Soc. $\underline{3}$, 1030 (1972)
25. See for example E.J. Allin, A.D. May, B.P. Stoicheff, J.C. Stryland and H.L. Welsh, Appl. Optics $\underline{6}$, 1597 (1967) and references therein
26. J.F. Noxon, Can. J. Phys. $\underline{39}$, 1110 (1961)
27. R.W. Nicholls, J. Res. NBS $\underline{69A}$, 369 (1965)
28. C. Ouannes and T. Wilson, J. Am. Chem. Soc. 90, 6527 (1968)
29. R.A. Ackerman, I. Rosenthal and J.N. Pitts Jr., J. Chem. Phys. $\underline{54}$, 4960 (1971)
30. R.H. Young, K. Wehrly and R.L. Martin, J. Am. Chem. Soc. $\underline{93}$, 5774 (1971)
31. D.R. Adams and F. Wilkinson, J. Chem. Soc., Far. Trans. II $\underline{4}$, 586 (1972)
32. R.H. Young and R.L. Martin, J. Am. Chem. Soc. 94, 5183 (1972)
33. I.B.C. Matheson and J. Lee, J. Am. Chem. Soc. $\underline{94}$, 3310 (1972)

WAYNE: There are several things that I would like to ask you. You said rather fleetingly that all of the evidence points toward the quenching of O_2 singlet sigma going down to the singlet delta state and then you left it at that and proceeded on the assumption that was the case and then demonstrated some beautiful correlations on that assumption, and I would like to ask you: what reason do you really have for supposing that's what happens, what direct evidence is there rather than indirect evidence? And also I think that you said singlet sigma singlet delta transition had not been observed in emission, which is not the case because John Noxon saw it back in 1961.

OGRYZLO: That is the one we looked at in fact. It is a rather poor spectrum. There is no structure to it, a very noisy thing, the line width, etc. was impossible to determine.

WAYNE: But I believe that you have to recognize that that this rather difficult experiment has in fact been done, rather than to say that it hasn't been done.

OGRYZLO: That's right and I should have said that it was Noxon's work that we used initially but it was Nichol's calculations that we used to get the synthetic spectrum, because it wasn't very clear. Noxon only observed the 0,0 band and the spectrum we didn't have and we would have liked, was a 0,1 line.

WAYNE: But to go back to the first question which is the more substantial one.

OGRYZLO: The spin change effect of O_2 and NO is the indirect effect that we mentioned. And I have always considered that a very good argument. In other words O_2 is totally ineffective or at least lies on that line for singlet sigma. And for singlet delta it lies many orders of magnitude above. And so it is clear therefore that for forbidden transitions such as the quenching of singlet delta, the paramagnetic species are very important so it can't be a transition down to ground state, or it does not look as if it is.

WAYNE: The reason I asked this question is because that we have all been thinking about this possibility and in fact I remember our Chairman asking this question many years ago, whether singlet sigma goes down to singlet delta or not and we have repeatedly tried to demonstrate this explicitly. Every experiment that we have done to try to demonstrate that in the gas phase, has not been positive experiment. It neither says that it does nor it doesn't produce singlet delta.

KEARNS: One experiment which I think very strongly supports the quenching of sigma to delta is Evans' experiment where he irradiated directly in the sigma, the triplet sigma to singlet sigma absorption and was able to get photochemical reactions characteristic of delta. I think that it would be very hard to expect that sigma would have given those reactions. We know that the lifetime is too short, therefore it must have been delta. Sigma goes down to delta.

WAYNE: This is closer but it is still slightly indirect. But still, I think that there is a definite need to demonstrate explicitly the formation of delta in that system. There is just one niggly point if I may just hold this thing one moment longer. I saw on your graph for the delta quenching. Before I go on here, just let me say that there is something about the quenching by O_2 which is interesting. It's some work from Jones and Bayes from UCLA about the energy transfer from O_2 singlet delta to O_2. And it appears that in almost every gas kinetic collision between O_2 singlet delta and O_2 the energy is handed on. By doing an isotopic experiment you can demonstrate that the energy is just passed on from one O_2 molecule to the next and it is transferred on virtually every collision. One wonders whether perhaps the inefficiency of quenching O_2 singlet sigma might not be due to some similar effect where the energy is just held up. That is to say collisions lead to energy transfer so to the change in the ownership of the excitation and not to a loss of the energy.

OGRYZLO: How another process could actually decrease the probability of quenching is a little hard to see. But as you saw on our correlations we don't need any such explanation.

WAYNE: I recollected at Sheffield we had the beginnings of the same discussion about whether energy could migrate to the walls more rapidly by this kind of "passing on" collision of electronic energy or not. But the last thing that I was going to ask about the graph showing the singlet delta quenching rate I observed with some dismay, first, there were two points of N_2O, I think that Dr. Thrush pointed that out, and secondly, with greater dismay that CO_2 has an exceedingly small value, which you also said was a limit and there are those of us who might not go along entirely with the rate constants you have chosen to put in for the CO_2 quenching of singlet delta. Do you want the number here and now? It is at least two orders of magnitude bigger than that.

OGRYZLO: There are three determinations that indicate that it is less than, one has it less than 50 the other one less than 9 and it is true that one is about two orders of magnitude greater but because of the other two values, personally I decided that the upper limit determined by the other two was probably more accurate.

WAYNE: Well, that is a very difficult one because the absolute determinations, where quenching is being looked for, do give a number that is quite higher. Now the experiments are the ones where there is some doubt anyway whether one would expect to see an effect at all because there may not be enough CO_2 in the system. What are those two N_2O points doing on the graph? There is one immediately above the N_2 and there is one to the immediate left of the H_2.

OGRYZLO: Yes, I have another slide that was made in Cambridge which in fact has the correct order but it is not here.

J. LEE: Concerning the solvent effect you pointed out between freon and methanol, it was of interest to us to determine the quenching rate by beta carotene which is assumed generally to be diffusion controlled. We have looked at this using our laser technique for generating singlet oxygen and we found that it was about an order of magnitude less than diffusion controlled. We were concerned about this but in fact if there is a solvent effect as you pointed out in the charge transfer process this would explain it.

OGRYZLO: Yes, I have noticed these results and have thought perhaps that there was an explanation. You seem very worried about the discrepancies and looked for experimental errors and I thought to myself that it might be a very real effect.

HOYTINK: I think that the quenching of singlet sigma oxygen by ground state oxygen is a very interesting case and I really am surprised that Dr. Wayne still argues about whether the singlet sigma goes to the delta or to the ground state. Because the tremendous difference in Franck-Condon factors it must go to the singlet delta state. The other rate would be much much lower because of the fairly unfavorable Franck-Condon Factors.

KEARNS: Furthermore it is a spin allowed process. Singlet sigma to singlet delta is still allowed but singlet sigma to the ground state would be forbidden.

WAYNE: Yes, but it must be recognized that the rate constant is pretty damn small. I am not arguing that it doesn't go to singlet delta, don't misunderstand that, but what I am arguing is that there is no direct evidence that it does either. And to argue that this is a spin allowed process and it should go rapidly or that the Franck-Condon effect allows it to be. This is an exceedingly slow quenching rate the O_2 singlet sigma quenched by O_2. The strength of the argument there seems to be that it is something like 7 orders of magnitude below the collision frequency. So a factor of 10^7 isn't to be regarded as evidence for a spin forbidden process, if I wanted to make that argument. It could be either way, I don't think we can argue on the basis of Franck-Condon with a factor of 10^7.

HOYTINK: I don't want to act as a die-hard but why is it spin forbidden?

KEARNS: I am sorry for oxygen quenching. No.

M. KAUFMAN: Can you get better agreement between the polyatomics and diatomics if you sum the contributions from different vibrations rather than just consider the highest frequency vibration?

OGRYZLO: Yes, we have taken everything into account. We have just taken the absorption spectrum of the polyatomics in that region; these are overtone regions and combination band regions for polyatomics. We don't know what they are but we have just taken the absorption spectrum.

M. KAUFMAN: On your graphs you plotted just the highest frequency vibrations.

OGRYZLO: Yes, that's right. That was the initial graph. We then went on to see why the anomaly occurred and took the whole absorption spectrum.

Chemiluminescence of Perhydroxyl- and Carbonate-Radicals

J. Stauff, U. Sander, and W. Jaeschke

Institut für Physikalische Biochemie

der Universität Frankfurt am Main, Germany

I.

Many oxidations and reductions of hydrogenperoxide in aqueous solution are accompanied by chemiluminescence of mostly very low intensity. Its origin might be sought in reactions of the hydroxyl- and perhydroxyl-radicals, which appear as intermediates. In the absence of excitable molecules of inorganic or organic character the following radical recombinations are to be discussed:

$$\dot{O}H + \dot{O}H \rightarrow H_2O_2 \qquad (1)$$

$$\dot{O}_2H + \dot{O}_2H \rightarrow H_2O_2 + O_2 \qquad (2)$$

$$\dot{O}_2H + \dot{O}_2^- \rightarrow HO_2^- + O_2 \qquad (3)$$

$$\dot{O}_2H + \dot{O}H \rightarrow H_2O + O_2 \qquad (4)$$

Reaction (1) does not lead to an excited state because the excitation energy of H_2O_2 is much higher than the energy of reaction. The reactions (2), (3) and (4) produce O_2 altogether, and each of them more energy than necessary for the excitation of the $^1\Delta_g$-state (22,4 kcal/mole) or the $^1\Sigma_g^+$-state (37,5 kcal/mole) of oxygen. Because the transitions of the states of O_2 emit light at wavelength of 760 or 1260 nm, they are usually not registered by a normal photomultiplier as light detector. Therefore the observable chemiluminescence of such reactions were believed to be closely associated with the known emissions of excited O_2-dimers. Possible transitions are

$$^1\Delta_g\ ^1\Delta_g \rightarrow 2\ ^3\Sigma_g^-$$
$$^1\Delta_g\ ^1\Sigma_g^+ \rightarrow 2\ ^3\Sigma_g^-$$
$$^1\Sigma_g^+\ ^1\Sigma_g^+ \rightarrow 2\ ^3\Sigma_g^-$$

Because of the weakness of the chemiluminescence of the radical redox reactions of H_2O_2 it has not been possible to obtain completely accurate spectra of the emitted light[1]. Otherwise, the only reaction of H_2O_2 (with ClO^- or BrO^-) with a quantum yield, sufficient for spectroscopic investigation is surely nonradical[2]. The difficulties with radicals do not only arise from their very low stationary concentrations, which combined with a low quantum efficiency shifts the observability of the emitted light to its lower limit they also arise from special conditions of their kinetics.
For the study of the chemiluminescence of the reactions (2) and (3) a system should be used which has not the disadvantages of the aqueous hydroperoxide redox system. Several authors reported on the relative high stability of solutions of the superoxide anion O_2^- in aprotic solvents[3,4] generated by electrolytic reduction of O_2. These observations could be confirmed by experiments[2] at which the superoxide anions were produced by different methods. They were: Reduction of dissolved molecular O_2 at a Hg-cathode in dimethylformamide, acrylonitrile and hexamethylphosphoramide (=HMPA), treatment of solutions of solvated electrons in HMPA with dry (!) molecular oxygen and dissolution of KO_2 in HMPA.

The chemiluminescence from electrolysis of O_2 in aprotic solvents has been reported already by several authors[5,6]. But they used tetraalkylammoniumperchlorates as carrier electrolyte which leads to erroneous results, because the organic cation is attacked and forms peroxides which luminesce for themselves. Only when $NaClO_4$ is used as electrolyte one obtains a solution of O_2^- (NaO_2) the e.s.r.-spectrum of which is identical with that already known[7]. A current-voltage curve on a Hg-cathode with a 0.1 m solution of $NaClO_4$ in acetonitrile saturated with streaming O_2 has a half-wave-potential of about -0.85 V. When a solution electrolysed at -1 V is examined by a photomultiplier a very weak luminescence can be detected. Addition of traces

of water to this solution gives an outburst of light
which decays in some seconds. The same emission is observed
when acids or alcohols are added. This observation
reveals that protons are necessary for the production
of an excited species out of O_2^--anions. Fig.1
shows the dependence of luminescence intensity from
time of O_2^--solutions with different amounts of $HClO_4$
in acetonitrile. The decay follows 2nd order kinetics.

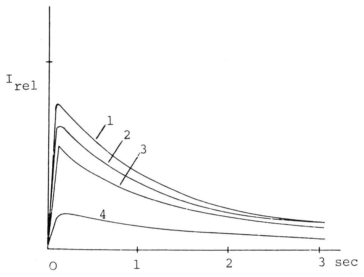

Fig.1. Chemiluminescence of NaO_2 solutions in acetonitrile
with different amounts of $HClO_4$. NaO_2-concentration
0.01 m, $HClO_4$-concentration 1: 0.01 m, 2: 0.004 m,
3: 0.0025 m, 4: 0.001 m.

Hexamethylphosphoramide (($Me_3N)_3PO$: HMPA) dissolves alkali
metals to alkali cations and solvated electrons[8].
The treatment of such solutions with extremely dry oxygen
leads to colourless solutions of O_2^--ions detectable
by their e.s.r.-spectra. Chemiluminescence is observable
only when proton donors are added e.g. when humid O_2
is bubbled through a solution of solvated electrons.
Very high intensities of the luminescence are obtained
when the solvent contains solid pieces of alkali metal
because solvated electrons are continuously formed at
the metal surface and can react at once with O_2 and

H_2O [(9)] (The light is detectable by the dark adapted eye). Fig.2 shows the spectra of the light emitted from systems containing Na, K or Li as sources for solvated electrons. In all cases no emissions from eventually excited states of the alkali metals were observed but only a main maximum at 620 - 640 nm and two secondary maxima at about 590 and 680 nm. (Na emits at 589 nm, Li at 670 nm and K at 770 nm.)

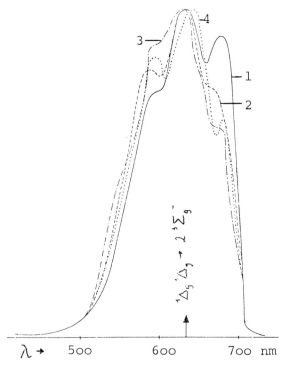

Fig.2. Chemiluminescence spectra of solutions of alkali metals in HMPA with water saturated O_2, the solid alkali metals being present at the bottom of the vessel. 1: Na, 2: K, 3: Li. Spectrum Nr. 4 is due to the emission of a Na-mirror on Raschig-rings treated with humid O_2.

To exclude the possibility of an excitation of the solvent metallic Na was deposited as mirror on Raschig-rings in vacuo in a special device. This system also generates luminescence with O_2 only when the gas contains H_2O; its spectrum is drawn as curve 4 in fig.2. This experiment proves that the solvent is not involved in the chemiluminescence. These results reveal that the formation of O_2^- or MeO_2 alone is not sufficient to form excited oxygen. The reason is to be sought in the unreactivity of O_2^- in the absence of protondonors. Though being radicals, they do not recombine according to

$$2\,\dot{O}_2^- \rightarrow O_2^{2-} + O_2$$

to detectable amounts because this reaction is highly endergonic(7). To start the production of singlet oxygen perhydroxy-radicals have to be formed first. In the presence of protons or proton donors the equilibria

$$\dot{O}_2^- + H^+ \rightleftarrows \dot{O}_2H \qquad (8)$$

$$\dot{O}_2^- + H_2O \rightleftarrows \dot{O}_2H + OH^- \qquad (9)$$

$$\dot{O}_2^- + H_2O_2 \rightleftarrows \dot{O}_2H + O_2H^- \qquad (10)$$

will adjust themselves very rapidly. The O_2H radicals can react at once according to reaction (2) and/or (3) which both produce O_2 at least partially in an excited state. As it is revealed by the spectra the excited state is an $^1\Delta_g$-state because only transitions from the $^1\Delta_g\,^1\Delta_g$-dimer could be observed. The absence of the transitions $^1\Delta_g\,^1\Sigma_g^+ \rightarrow 2\,^3\Sigma_g^-$ at 478 nm and $^1\Sigma_g^+ \rightarrow\,^3\Sigma_g^-$ at 760 nm prove that no O_2 in the $^1\Sigma_g^+$-state is formed or that the amounts are too small to be detected (See e.g. Khan and Kasha[2]).
Because both reactions (2) and (3) develop more energy than necessary ($\Delta H > 40$ kcal/mol in both cases) and the formation of $^1\Sigma_g^+ - O_2$ needs only 37.5 kcal/mol the reason for the absence of this species cannot be the lack of energy of formation. More probable is the conception that the energy of activation for a reaction path leading to $^1\Delta_g - O_2$ is lower than for one leading to $^1\Sigma_g^+ - O_2$.

II.

In many reactions at which perhydroxyl radicals are known as intermediates a weak blue-green chemiluminescence has been observed[1]. In default of other possibly exitable species this luminescence was ascribed to combined transitions of O_2 molecules in the $^1\Delta_g$- and $^1\Sigma_g^+$-states. This assumption is found not to hold up in view of the results of the preceding chapter. Therefore a search for a contaminant was started which ended with the result that atmospheric CO_2 was responsible for that effect. It could easily be confirmed by careful exclusion of CO_2 from the used solutions and apparatuses which diminished the blue green luminescence to very small values. Otherwise, the addition of carbonate or bicarbonate to the solutions under investigation produced increasing luminescence intensities with increasing carbonate concentration, (Fig. 3) especially in the spectral region of 450 - 550 nm[18].

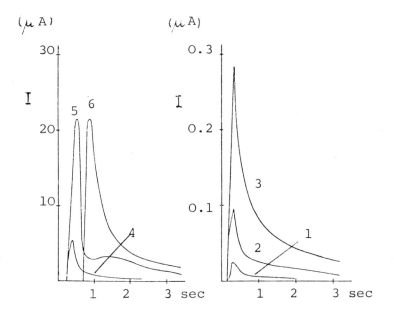

Fig.3. Intensity-time curves of the reaction KIO_4/H_2O_2. 1: without additive, 2: with 0.1 m K_2CO_3, 3: 0.1 m $KHCO_3$, 4: +0.00001 acridineorange, 5: 0.1 m K_2CO_3 + acridineorange, 6: 0.1 m $KHCO_3$ + acridineorange.

In the reaction Fe^{2+} + H_2O_2 addition of 0.2 m $KHCO_3$ to the H_2O_2 before mixing and using the light filter BG 25 (Schott) the intensity increases to 43 times the value without $KHCO_3$, in the reaction KIO_4 + H_2O_2 the increase with 0.0475 $KHCO_3$ is 69. The luminescence is much amplified by fluorescers like eosine Y, acridine orange, and fluoresceine even by substances like quininesulfate and quinoline which are not excited by the system

without carbonate. (Compare the intensity scale in fig. 3a and b.)
Also to be considered is the observation that the light producing intermediate is still present in small amounts even after 50 sec from the mixing time of the components. When some fluorescer is added after this time a new outburst of luminescence is to be seen.

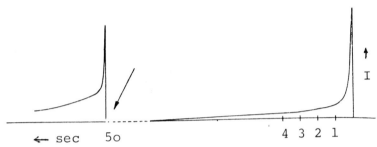

Fig.4. Intensity-time course of the chemiluminescence KIO_4 + H_2O_2 containing 1 m $KHCO_3$, pH = 10.8. Addition of 1 ml 0.001 m quininesulfate 50 sec after the starting point (arrow).

The generation of the blue-green luminescence can be explained by the formation of carbonate or bicarbonate radicals according to the reactions

$$HCO_3^- + \dot{O}H \rightarrow H\dot{C}O_3 + OH^- \quad (11)$$

$$HCO_3^- + \dot{O}_2H \rightarrow H\dot{C}O_3 + O_2H^- \quad (12)$$

$$H\dot{C}O_3 \rightleftarrows \dot{C}O_3^- + H^+ \quad (13)$$

$$2\ H\dot{C}O_3 \rightarrow X \quad (14)$$

The radicals recombine to a product X which acts as energy rich precursor of an excited molecule. Reactions (11) and (12) are known from radiolytic and photochemical processes (11, 12).
At first sight the recombination of the carbonate radicals should lead to mono- or diperoxalic acid which could undergo similar reactions like those found by Rauhut et al.(13). Out of this reasons some experiments were made with the best known reaction of oxalic acid

and $KMnO_4$. This reaction shows normally no chemiluminescence but develops increasing amounts of light if Mn^{2+}-ions are present from the beginning of the reaction. Its maximal intensity is proportional to the Mn^{2+}-concentration in the region up to the equality of Mn^{2+} and oxalic acid. The luminescence will be amplified by the same fluorescers like that originating from the carbonate radical reactions, the same is valid for the lifetime of the precursor[10].

These experiments lead directly to the speculation on the formation of dioxetanedione, discussed by Rauhut et al. and others[14]. It could be formed after recombination of CO_3^--radicals by internal conversion and splitting off of H_2O_2 or by two-electron oxidation of oxalic acid within the Mn(IV)-complex. However, the following experiments could demonstrate that the substance X is not identical with the dioxetanedione (C_2O_4) of Rauhut's system.

Out of the liquid reaction mixture, dinitrophenyloxalic-diester + H_2O_2 in dimethylphthalate, a gaseous product can be extracted by bubbling argon or nitrogen through the mixture or by evacuation. The gas excites diphenylanthracene to a bright fluorescence. But the gas alone also luminesces at wavelengths of approximately 540 nm. Its luminescence decays with 1st order, the decay constant being 0.166 sec^{-1} at 20°C. At -60°C a product can be condensed which after being evaporated again shows the same luminescence characteristics as the original gas[16]. If at all, this product should contain C_2O_4 as far as one could say so at the momentary state of investigation. When this gas is passed into water one observes a weakening of the luminescence and a faster decay as in the gaseous state. In solutions of acridine orange, fluoresceine or quinine sulfate no excitation of the fluorescers takes place in contrast to the reactions discussed above. (Chemical analysis of the aqueous solution showed only CO_2 and H_2O_2 but no oxalic or formic acid.) Therefore, if the gas of the Rauhut reaction contains C_2O_4, this is not identical with the energy rich and long living species from CO_3^--radical recombination.

In cause of these experimental results and in consideration of the most probable structure of the CO_3^-- or HCO_3^--radicals with the lone electron on one of the O-atoms the radical recombination should lead to dicarboxylperoxide

$$2 \; HO-\overset{\overset{\displaystyle \cdot}{\|}}{\underset{O}{C}}-O \longrightarrow O = \underset{OH}{\overset{|}{C}}-OO-\underset{OH}{\overset{|}{C}} = O$$

which will split into 2 CO_2 and H_2O_2 in the end. Otherwise, with CO_2 as the only endproduct of very different reactions altogether being producers of chemiluminescence one has to assume or at least consider an excited state which may be formed in different ways. The excited state cannot be one of CO_2 or CO because none of this molecules has an excited electronic state in the environment of 3 eV (blue-green spectral region)[15]. Tentatively, the following scheme could serve as basis for discussion:

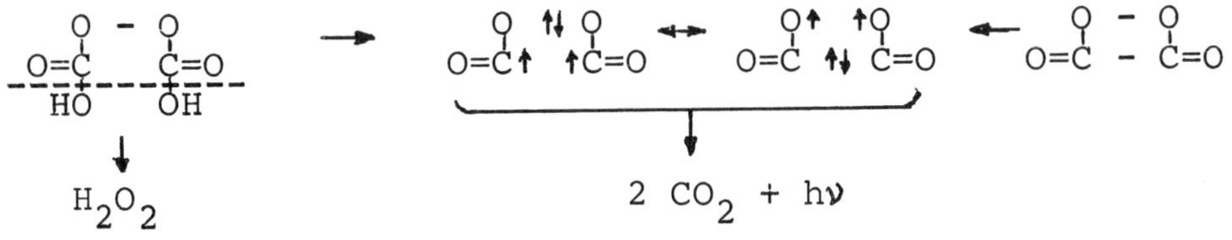

$$\underset{(I)}{\underset{\downarrow}{\underset{H_2O_2}{\overset{O=C\overset{O-O}{\underset{HO\ \ OH}{\rule{0pt}{0pt}}}C=O}{}}}} \longrightarrow \underset{(IIa)}{O=C\!\uparrow\ \overset{O\ \uparrow\downarrow\ O}{\uparrow C=O}} \longleftrightarrow \underset{(IIb)}{O=C\ \overset{O\uparrow\ \uparrow O}{\uparrow\downarrow}\ C=O} \longleftarrow \underset{(III)}{O=C\overset{O-O}{-}C=O}$$

$$\underbrace{\rule{4cm}{0pt}}_{\downarrow}$$
$$2\ CO_2 + h\nu$$

Dicarboxylperoxide (I) transforms under abstraction of H_2O_2 into the biradical (IIa) which could be regarded as an excited triplet dimer of two CO_2-molecules. It is possibly in resonance with substance (IIb) which should be formed out of (III) more rapidly than (IIa) out of (I). Substance (IIa) or (IIb) or both of them can go over to the ground states of two CO_2 molecules only by changing the spin of one single electron of the system. This has to be performed under emission of radiation and is to be regarded as triplet-singlet transition the probability of which could be much increased by the neighbourhood of the radical electrons.

Synopsis

Solutions of O_2^--radical anions in aprotic solvent react to perhydroxyl radicals (O_2H) which recombine with O_2^- or themselves to give O_2 in the excited $^1\Delta_g$-state. The chemiluminescence spectrum shows only transitions from the $^1\Delta_g\ ^1\Delta_g$-dimer of O_2. No transitions out of the $^1\Sigma_g^+$-state could be observed.
The blue-green chemiluminescence of many redox reactions of H_2O_2 in aqueous solutions is due to CO_2 as contaminant. From experiments with bicarbonate ions in hydroxyl or perhydroxyl radicals producing reactions a new luminescing species could be recorded which originates from the recombination of intermediary formed carbonate-radicals. Because similar results were obtained

with the oxidation of oxalic acid with permanganate in the presence of Mn (II)-sulfate the formation of an CO_2-dimer as excited species is discussed.

References

1) J.Stauff and F.Lohmann, Z.Physik.Chem.(Frankfurt) 40, 123 (1964)
2) A.U.Khan and M.Kasha, J.Amer.Chem.Soc. 92, 3293 (1970)
3) M.E.Peover and B.S.White, Electrochim.Acta 11, 1061 (1966)
4) D.L.Maricle and W.C.Hodgson, Anal.Chim. 37, 1562 (1965)
5) J.M.Bader and I.KUWANA, J.Electroanal.Chem. 10, 104 (1965)
6) K.D.Legg and D.M.Hercules, J.Amer.Chem.Soc. 91, 1902 (1969)
7) U.Sander, Dissertation Frankfurt a.M., 1971
8) T.Cuvigny, J.Normant and H.Normant, C.R.hebd.Sceances acad.Sci. 258, 3503 (1964)
9) U.Sander and J.Stauff, Anales Asoc.Quim.Argentina 59, 149 (1971)
10) J.Stauff and U.Bergmann, Z.Physik.Chem.(Frankfurt) 78, 263 (1972)
11) J.P.Keene, Y.Raef and A.T.Swallow in "Pulse Radiolysis", London, 1965, p.99 ff.
12) A.K.Chibisov, V.A.Kuz'min and A.P.Vinogradov, Dokl. Akadem.Nauk SSSR, 187, 142 (1969)
13) M.M.Rauhut, B.G.Roberts and A.M.Semsel, J.Amer. Chem.Soc. 88, 3604 (1966)
14) F.McCapra, Pure Appl.Chem. 1970, 611
15) G.Herzberg, Molecular Spectra, I.Diatomic Molecules, van Nostrand New York, 1950
16) J.Stauff, W.Jaeschke and G.Schlögl, Z.f.Naturforsch. Teil b, in press

J. LEE: In your measurement of the spectrum of singlet oxygen emission in the reaction of the metal ions was there any bubbling around the metal itself? Could the emission come from gas bubbles or gas bubbles formed by some reaction with the metal? To my knowledge no singlet oxygen emission from solution--in true solution has ever been observed. The Khan-Kasha emission obviously comes from the gas bubbles. I think they keep emphasizing that point.

STAUFF: Solutions of solvated electrons with sodium metal present show luminescence after having been shaken with air. The light can be seen by eye. It is not developing from the metal surface alone but also from cloud-like regions where solvated electrons react with the dissolved oxygen and water from the air. Weak chemiluminescence is observed also when solutions of 10^{-4}m KO_2 in hexamethylphophoramide are mixed with 10^{-4}m solutions of water in the same solvent. The singlet oxygen generated this way should have a solubility of less than 10^{-4} m - that is less than 1/10 of the solubility of O_2 in the ground state - if it is produced in the gaseous state. If not, the luminescence should be due to a dissolved species.

KEARNS: It is possible to see what appears to be rotational structure on oxygen dissolved in solution.

J. LEE: Well, you don't see much of a rotation struction in absorption.

KEARNS: You do see rotational structure. It is blurred out.

J. LEE: It is blurred out. But the Khan-Kasha emission has a very distinct rotational structure, from which they assumed it was coming from gas bubbles.

KEARNS: Right, agreed.

THRUSH: You mentioned in your paper that the afterglow is from carbon dioxide, which is the same as that produced in a microwave discharge. Now this afterglow is known to be electronic emission by quite a higher excited state of CO_2. It is the singlet C_2 state which has an energy of 120 Kcal and the reason you see emission from that in the blue is because of the Franck-Condon factors involved, if you are going from a bent molecule to a high vibrational level of the ground state of CO_2. And this emission we know comes from the combination of ground state oxygen atoms and CO. My question is, do you think that your emission spectrum in solution is any way related to the afterglow of CO_2, in the microwave discharge?

STAUFF: There is no correlation.

CHEMILUMINESCENT REACTIONS OF RADICAL-IONS

Edwin A. Chandross

Bell Laboratories

Murray Hill, New Jersey 07974

Chemiluminescent reactions of radical-ions, or chemiluminescent electron transfer reactions as I prefer to call them, represent the newest and most general area of chemiluminescence. In spite of the fact that the field is less than ten years old the reaction mechanisms are quite well understood in the major area of electrogenerated chemiluminescence (ECL) and reasonable hypotheses can be advanced for other processes.

The four papers which follow this introduction review various areas. It seems most appropriate to devote this section to a brief introduction to the field. Detailed reviews have been published by Hercules[1,2] and Zweig[3] and my last paper[4] contains an overview of the work done at Bell Laboratories.

The adiabatic annihilation of a hole by an electron to produce an excited state is the fundamental step in electron transfer chemiluminescence. This process is well known in the solid state, e.g., light-emitting diodes (GaAs, GaP) and the elegant demonstration of electron-hole recombination in crystalline anthracene by Helfrich and Schneider.[5] The most direct analogy in an organic system is the recombination luminescence originally observed in ultraviolet irradiated rigid glassy solutions, of species such as the potassium salt of carbazole, thirty years ago by Lewis and Bigeleisen.[6,7] This process, which requires two photons, leads to photoionization of the solute and produces radical-cations and trapped electrons. Luminescence is produced by the recombination of these species and it characteristically lasts much longer than the normal excited state lifetimes, sometimes for several hours. Further, allowing the matrix to soften produces a burst of light.

The best understood solution analog of recombination luminescence is the reaction between aromatic hydrocarbon radical-cations and radical-anions. The history of this discovery has been described. As so often

$$Ar_1^+ + Ar_2^- \longrightarrow {}^1Ar^* \text{ and/or } {}^3Ar^*$$

happens, several people thought of approximately the same idea at about the same time. It seems clear that G. J. Hoytink was the first to consider the experiment and theory in detail but experimental problems prevented him from realizing success.

Radical-anions also yield chemiluminescence upon treatment with a variety of oxidants and these reactions have been explained in terms of

$$Ar^- + R \longrightarrow {}^1Ar^* \text{ and/or } {}^3Ar^* + R^-$$

which is similar to the cation-anion annihilation.

Finally there is the solution analog of recombination luminescence-- the reaction of solvated electrons with organic species, which is suggested to involve two steps--the generation of a radical and

$$Rx + e^- \longrightarrow R + x^-$$

$$R + e^- \longrightarrow (R^-)^*$$

its reduction to an electronically excited anion by a second electron.[4]

All of these processes are fundamentally the same. The theoretical treatments of Hoytink and Marcus[8] deal with the creation of excited states in some detail. Some years ago I suggested that one might qualitatively think of this as an analog of a Franck-Condon transition.[9] The energy liberated by the reaction appears rapidly (compared to vibrational frequencies) as the electron is transferred and this favors an adiabatic transition to a state of similar geometry, which is an electronically excited state, rather than a highly excited vibrational state of the lowest electronic state. In fact the work of Faulkner and Freed,[10] using a triplet counting technique, suggests that the yield of exicted states can approach one hundred percent.

Electrochemiluminescence has some promise for application in display devices. Many people have discussed this as well as the possibility of an ECL laser but nothing practical has yet appeared. It will be interesting to see what happens in the future.

1. D. M. Hercules, Accts. Chem. Res., **2**, 301 (1969).

2. D. M. Hercules, "Techniques of Chemistry," Vol. I, Pt. IIB, Ed. by A. Weisseberger, Wiley-Interscience, New York, 1971.

3. A. Zweig, "Advances in Photochemistry," Vol. 6, Interscience, New York, 1968.

4. E. A. Chandross, Trans. N.Y. Acad. Sci., II, **31**, 571 (1969).

5. W. Helfrich and W. G. Schneider, J. Chem. Phys., **44**, 2902 (1966).

6. G. N. Lewis and J. Bigeleisen, J. Am. Chem. Soc., **65**, 2424 (1943).

7. H. Linschitz, M. G. Berry and D. Schweitzer, ibid., **76**, 5833 (1954).

8. R. Marcus, J. Chem. Phys., **43**, 2654 (1965).

9. E. A. Chandross and F. I. Sonntag, J. Am. Chem. Soc., **86**, 3176 (1964).

10. D. J. Freed and L. R. Faulkner, ibid., **94**, 4790 (1972).

CATION-ANION ANNIHILATION OF NAPHTHALENE, ANTHRACENE, AND TETRACENE

G. J. Hoytink

Department of Chemistry, The University

Sheffield, S3 7HF, England

Abstract

A qualitative study has been made of the cation-anion annihilation in polar solvents for the hydrocarbons naphthalene, anthracene and tetracene. The reaction leads to a final state in which one of the two molecules occurs in the triplet state or the lowest singlet excited state. The crossing from the initial to the final state may take place by a resonance or a non-resonance electron transfer depending on the location of the excited state of the molecule. For naphthalene one finds two very efficient reactions both governed by resonance transfer one of which leads to the triplet and the other to the lowest excited singlet. For anthracene and tetracene the most efficient reaction involves a non-resonance transfer to a final state in which one of the two molecules occurs in the triplet state.

The quenching of triplet molecules by the cations and anions can have a very unfavourable effect on the quantum yield of the annihilation process.

During the past decade several investigators have observed the emission of light arising from the annihilation of positive and negative ions of aromatic molecules[1]. A few years ago the author made a qualitative study of this process taking the molecules benzene, naphthalene, anthracene and tetracene as examples[2]. This study was largely focussed on the thermodynamics of the annihilation and the role the polar solvent plays in the kinetics of the primary reaction. The aim of the present paper is to consider the kinetic aspects in some more detail including the quenching of the excited molecules by the hydrocarbon ions which under certain conditions may have an unfavourable effect on the quantum yield of the annihilation process. In order to keep the discussion as simple as possible the benzene molecule will not be considered in this paper. The molecules naphthalene, anthracene and tetracene show a sufficient variety of possibilities to give an impression of the process in general.

The thermodynamic data for the annihilation process as far as relevant for the present discussion are listed in Table 1 together with the energies of the lower excited singlet and triplet states. The reaction appears to be sufficiently exothermic to produce one of the two molecules in the triplet state and in one case even in

Table 1

Free energies and heats of reaction of the cation-anion annihilation and energies of the lower excited singlet and triplet states*

Molecule	$-\Delta G^{o}$ [a]	$-\Delta H^{o}$	E_S	E_T
Naphthalene	35	34	31.8	21.3 [c] 30.6 [d]
Anthracene	27	25	26.4	14.9 [c] 26.0 [e]
Tetracene	21	20	21.2	10.3 [c]

* in $cm^{-1} \times 10^{-3}$

a) from Table 1, ref 2; values converted from eV's into cm^{-1}

b) from Table 1, ref 3; ⎫
c) from Table 1, ref 4; ⎬ where further references are given

d) ref 5

e) ref 6

the lowest excited singlet state. These data, of course, do not give any information about the efficiency of the annihilation. For that purpose we have to make a more detailed study of the kinetics of the reactions which play an essential role in the overall process.

The kinetics of the cation-anion annihilation

Let us consider the reaction between a positive and negative ion of the same parent hydrocarbon molecule in a polar solvent like acetonitrile or dimethylformamide which are most commonly used in the experimental investigations.

$$^2M_o^+ + {}^2M_o^- \rightarrow {}^{1,3}M_1 + {}^1M_o \qquad (1)$$

In the initial state the presence of two charged species will give rise to solvent-dipole orientation, whereas in the final state the solvent dipoles will be randomly orientated. For fast reactions the distance of closest approach will be determined by the two ions separated by two solvent layers ($R \simeq 10\text{Å}$). Any closer approach will require a substantial increase in free energy because of a drastic change in solvation. In order to proceed from the initial to the final state a point must be reached at which the nuclear configurations, including that of the surrounding solvent, is the same for both states. As far as the solvent is concerned this means that both states must have the same solvent-dipole orientation. Using Marcus' dielectric theory[7] the free energy required for reorientation of the solvent dipoles follows from

$$G_{pol} = N\, m^2 e^2 \left(\frac{1}{r_+} + \frac{1}{r_-} - \frac{1}{R}\right)\left(\frac{1}{D_o} - \frac{1}{D_s}\right) \qquad (2)$$

where r_+ and r_- are the radii of the positive and negative ions, R is the inter-ionic distance, D_o and D_s denote the optical and static dielectric constants and e represents the electronic charge. The variable m indicates the fraction of charge transferred from the negative to the positive ion. For any value of m formula (2) gives the free energy for the orientation of the solvent dipoles in the corresponding equilibrium situation relative to the free energy for $m = o$. Actually this means that the electron spends a fraction of time m on the positive ion and $1-m$ on the negative ion. The electronic polarisation of the surrounding medium follows the electron motion instantaneously and will therefore be the same for any value of m. It is obvious that m represents the reaction co-ordinate for solvent-dipole reorientation. Instead of (2) we will use the empirical formula, first suggested by Marcus[8]:

$$G_{pol} = m^2 \lambda \qquad (3)$$

in which λ is a constant depending on the nature of the solvent and the ions considered. For large aromatic systems λ will be approximately constant and throughout this paper we will use the rounded-off value

$$\lambda = 5000 \text{ cm}^{-1} \qquad (4)$$

which has been derived from the rate of electron transfer between tetramethylphenylene diamine and its positive ion[8].

Because of the relatively large dielectric constant the coulombic interaction between the ions at the distance of closest approach will be of the order of kT at room temperature and may therefore be neglected.

Using (3) and (4) and the data listed in Table 1 we have drawn the free energy curves for the initial states and all relevant final states for the annihilation reactions of naphthalene, anthracene and tetracene. In the diagrams shown in Figure 1 the curves near the crossing points have not been rounded off (as was done in our previous paper[2]), because solvent-dipole reorientation as such can never lead to a mixing of the initial and final states. In order to achieve that, we have to move along the reaction co-ordinates determining the separation of the two reacting species. The value of m at the crossing point follows from

$$(1 - 2m^*) \lambda = \Delta G^o \qquad (5)$$

By substituting m^* into (3) we then find ΔG^*_{pol}.

In principle one would have to follow the same procedure for the change in nuclear configuration of the two reacting species. However, for large rigid aromatic hydrocarbons the change in nuclear configuration on the addition or removal of an electron is very small, so that this effect can safely be left out of consideration.

The rate constant for the cation-anion annihilation can now be expressed by

$$k = g \cdot \rho \cdot k_d \exp(-\Delta G^*_{pol}/RT) \qquad (6)$$

where k_d is the diffusion encounter rate constant, g is the statistical factor for the electron spin and ρ the probability for crossing from the initial to the final state in the "transition complex". If during an encounter of the two ions the solvent orientation satisfies the condition for crossing (the probability for such an encounter is given by the exponential in (6)) the initial and final states will interact and the extent of interaction "during the encounter" will determine the magnitude of ρ. The statement "during the encounter" is somewhat vague and needs some further explanation. In principle the interaction can take place at any separation between the two ions. However, since it falls off very rapidly with increasing distance, far the largest contribution comes from interactions within a range of a few Å from the distance of closest approach. The overall range during which a significant interaction takes place will therefore be of the order of 10 Å. In solvents like acetonitrile and dimethyl formamide at room temperature this corresponds to a "duration of the encounter" of about $10^{-10} sec^2$.

The interaction between the initial and final states

In order to obtain some idea about the kind of interaction between the initial and final states we will make use of the conventional configurational description based on the two sets of molecular orbitals of the hydrocarbon molecules in their ground states:

$$\ldots m\ n\ n'\ m' \ldots$$
$$\ldots \mu\ \nu\ \nu'\ \mu' \ldots \qquad (7)$$

The latin and greek characters refer to the molecular species on the left and on the right. The orbitals n and ν are the highest bonding orbitals and the primes indicate the anti-bonding orbitals. The notation already indicates that we are dealing with alternant hydrocarbons for which the well-known pairing exists between bonding and anti-bonding orbitals.

The encounter of the two ions which both have doublet ground states can lead to either a singlet or a triplet transition complex both of which have the electron configuration:

$$^{3,1}\phi_i = m^2\ n\ \mu^2\ \nu^2\ \nu' \qquad (8)$$

The configurations for the final states can be defined most easily if we start from the situation with the two molecules in their closed shell ground states:

$$m^2 n^2 \mu^2 \nu^2 \qquad (9)$$

By promoting an electron from the highest bonding n (or ν) to the lowest anti-bonding orbital n' (or ν') the corresponding singlet configuration approximately describes the final state in which one of the two molecules occurs in the ^1L_a state, which is the lowest singlet excited state for anthracene and tetracene and the second excited singlet state for naphthalene. The corresponding triplet configuration approximately describes one of the two molecules in the lowest triplet state for all three hydrocarbons considered.

$$^{3,1}\phi_f = m^2 n\, n' \mu^2 \nu^2 \qquad (10)$$

$$^{3,1}\phi_f = m^2 n^2 \mu^2 \nu\nu' \qquad (11)$$

If H is the Hamiltonian operator for the transition complex and we omit all terms which are second order in intermolecular overlap the matrix elements for the interaction between the initial and final states become:

$$(8) + (10): H_{if} = \langle n'|H|\nu'\rangle \qquad (12)$$

$$(8) + (11): H_{if} = \langle n|H|\nu\rangle \qquad (13)$$

The matrix elements (12) and (13) are the same for both the initial and final states being singlets or triplets.

Starting again from (9) we now promote an electron from m to n'. Because of the pairing properties of the m.o.'s the configuration obtained this way has the same energy as the one arising from the promotion of an electron from n to m'. Configuration interaction leads to an in phase and an out of phase combination of which the latter has the lower energy for the singlet and approximately describes the final state in which one of the two molecules occurs in its ^1L_b state. This is the lowest singlet excited state for the naphthalene molecule. For the triplet state the in phase combination has the lower energy and for all three hydrocarbons considered approximately describes the final state in which one of the two molecules occurs in its second triplet state. By repeating the same procedure for the other molecule ($\mu \to \nu'$: $\nu \to \mu'$) we arrive at the following description for the final states:

$$^{3,2}\phi_f = m\, n'\, n^2\, \mu^2\, \nu^2 \pm m^2\, n\, m'\, \mu^2\, \nu^2 \tag{14}$$

$$^{3,1}\phi_f = m^2\, n^2\, \mu\, \nu'\, \nu^2 \pm m^2\, n^2\, \mu^2\, \nu\, \mu' \tag{15}$$

Interaction with the initial state here leads to

$$(8) + (14) : H_{if} = \frac{1}{\sqrt{2}} <m'|H|\nu'> \tag{16}$$

$$(8) + (15) : H_{if} = \frac{1}{\sqrt{2}} <n|H|\mu> \tag{17}$$

These matrix elements in which terms of second order in intermolecular overlap have again been omitted are the same for both combinations and for interactions between singlets and triplets. These results are typical for electron transfer reactions for which one expects the interaction to be first order in intermolecular overlap. It should be emphasized, however, that this does not necessarily happen for every electron transfer leading to an excited species. As we will see later on in this paper there certainly are configurations for which the interaction is second order in intermolecular overlap.

The matrix elements (12), (13), (16) and (17) show an interesting result. In (12) and (16) the two orbitals are anti-bonding, whereas in (13) and (17) they are bonding orbitals. Since anti-bonding orbitals will be more extended into space than bonding orbitals we may expect (12) > (13) and (16) > (17). Since (12) and (13) hold for the same crossing and consequently both contribute to the same factor ρ a comparison between (10) and (11) indicates that the chance for the positive ion to end up in the excited state is higher than for the negative ion. From (16) and (17) we arrive at exactly the same conclusion for the reaction leading to the final states described by (14) and (15).

Although the above conclusions are based on a simple configurational description, a more extended configuration interaction would not alter them qualitatively because the configurations (8), (10), (11), (14) and (15) still would give the major contributions to the wave functions describing the initial and final states.

For two species separated by two solvent layers it is reasonable to assume that the energy of interaction H_{if} is of the order of 1 cm^{-1}. According to the uncertainty principle this requires a time of interaction of $\frac{1}{3} \times 10^{-10}$ sec. for the crossing to occur with unit probability. This is even less than the estimated "duration of the encounter" and hence we can safely put $\rho = 1$ for each of the annihilation reactions considered before.

Crossing to the lowest triplet state

From Figure 1 we see that all crossings from the initial state to the final state in which one of the molecules occurs in its lowest triplet state are at negative values of the reaction coordinate m. Since the corresponding values for ΔG^*_{pol} are fairly large one wonders if not a more efficient crossing might be possible from the initial state at m = o to the vibrationally excited final state. Within the Born-Oppenheimer approximation the initial state can be described by

$$\Phi_i = \psi_i \, \chi_o(^2M_o^+) \cdot \chi_o(^2M_o^-) \qquad (18)$$

where ψ_i is the electronic part of the wave function for the initial state and $\chi_o(^2M_o^+)$ and $\chi_o(^2M_o^-)$ are the vibrational functions for the zero vibrational states of the two ions. Similarly we can write for the final state

$$\Phi_f^{j,k} = \psi_f \, \chi_j(^3M_1) \, \chi_k(^1M_o) \qquad (19)$$

where j and k are sets of vibrational quantum numbers for the various modes of the triplet and ground state molecule with the restriction that the total vibrational energy must equal

$$\varepsilon_j + \varepsilon_k = \Delta E \pm \tfrac{1}{2} \alpha \qquad (20)$$

Here ΔE is the energy gap between the vibrationless initial and final state for which we can write

$$\Delta E = \Delta G^o - \lambda \qquad (21)$$

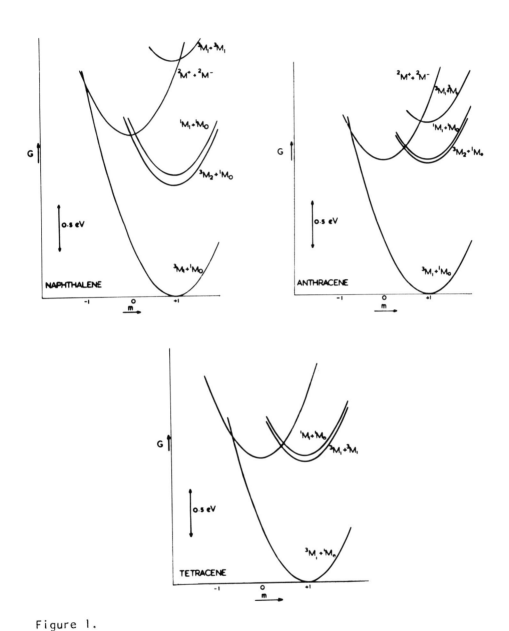

Figure 1.

Plot of the free energy for the initial and final states of the cation-anion annihilation against the reaction co-ordinate for solvent-dipole orientation m.

Because of the interaction with the environment the vibrationally excited states will be mutually coupled and the initial state will "see" a manifold of N vibrationally excited final states spread over an energy interval α. The value of α follows from the rate of vibrational relaxation and approximately equals 10 cm^{-1}. The matrix element between the initial state (10) and one the final states (19) becomes:

$$\beta = H_{if} <\chi_o(^2M_o^+)|\chi_j(^3M_1)><\chi_o(^2M_o^-)|\chi_k(^1M_o)> \qquad (22)$$

Since the energy gap ΔE is a few thousands of wave numbers the two factors for the vibrational overlap will be much smaller than unity. With H_{if} = 1 cm^{-1}, this means $\beta<<\alpha<<\Delta E$. Under this condition we can use the formula derived by Robinson and Frosch[9] for the probability per unit time, which for our case reads:

$$P = \frac{2H_{if}^2}{\alpha\hbar} \sum_{j,k}^{N} <\chi_o(^2M_o^+)|\chi_j(^3M_1)>^2 <\chi_o(^2M_o^-)|\chi_k(^1M_o)>^2 \qquad (23)$$

(For a qualitative explanation of this formula we refer to[10]). The probability for a crossing from the initial to the final state directly follows from (23), since

$$\rho = P \times \Delta t \qquad (24)$$

where Δt is the time during which the interaction takes place, i.e. the "duration of the encounter".

Unfortunately no data are available to calculate the summation in (23), which may be called the Franck-Condon factor for the conversion considered here. In order to obtain some idea about the magnitude of (23) we will assume that all the excess vibrational energy will be on the ground state molecule. In that case we obtain:

$$P' = \frac{2H_{if}^2}{\alpha\hbar} <\chi_o(^2M_o^+)|\chi_o(^3M_1)>^2 \sum_{\ell}^{N'} <\chi_o(^2M_o^-)|\chi_\ell(^1M_o)>^2 \qquad (25)$$

Since the summation in (25) accounts for only part of the summation in (23) we know that P' < P.

If one looks at the electronic absorption spectra of the aromatic hydrocarbon molecules, their positive and negative ions and their triplet-triplet spectra one finds in general that the strongly allowed transitions give rise to an absorption band with a Franck-Condon progression, arising from the C-C stretching modes, which covers only a relatively small frequency range. The intensity falls off very rapidly with increasing vibrational quantum number and usually the o-o transition accounts for at least 50% of the total intensity of the electronic transition. (For weakly allowed transitions the Franck-Condon pattern is more complicated because of the presence of antisymmetric progressions which arise from borrowing of intensity from strongly allowed transitions by vibronic interactions.) One also finds a very strong similarity between the electron spectra of the hydrocarbon dipositive and dinegative ions and the triplet-triplet spectra of the corresponding molecule[11]. All these facts indicate that for large aromatic hydrocarbons a change in the π-electronic structure causes very little change in the nuclear configuration and gives only a slight alteration in the force constants of the C-C stretching modes. It seems therefore reasonable to assume that the vibrational overlap $<\chi_o(^2M^+)|\chi_o(^3M_1)>$ will be very close to unity and in the further discussion we will use a value of 0.5 for its square. An estimate of the summation in (25) may be obtained in the following way:

If we replace $\chi_o(^2M^-)$ by $\chi_o(^3M_1)$ the summation corresponds to the Franck-Condon factor F for the conversion from the lowest triplet state to the ground state for the hypothetical case that the energy gap between the two states equals ΔE given by (21). Siebrand and Williams[4,12] have made a detailed study of these Franck-Condon factors and their results are very useful for our present discussion. Although, as we have mentioned already, a change in the π-electronic structure involves only small changes both in the nuclear configuration and in the force constants of the vibrational modes we may expect these changes to be somewhat larger if we add an electron to the ground state molecule than if we excite it to its lowest triplet state. According to the theory[4] this means that for the same energy separation the Franck-Condon factor given in (25) will be somewhat larger than for the triplet-singlet conversion. If we therefore make use of Siebrand's results[12] we will at least arrive at a lower limit of (25) and hence of (23). For all three molecules considered the energy gap following from (21) exceeds 4000 cm^{-1} which implies that the main contribution to the Franck-Condon factors comes from the C-H stretching modes[4]. Using Siebrand's plot of F^{-1} vs $(\Delta E-4000)/\eta^{12}$, where $\eta = N_H/(N_H + N_C)$ and N_H and N_C are the number of hydrogen and carbon atoms in the molecule we arrive at the approximate values for F listed in Table 2.

Inserting these and the values mentioned before into (25) and using (24) and (6) we obtain the lower limits for the relative rate constants listed in the first column of Table 3. The data listed in the second column are derived from (5) and (3) using the ΔG^o values given in Table 1.

Table 2

Energies of separation and lower limits of the Franck-Condon factors for the cation-anion annihilation to the lowest triplet state.

Molecule	ΔE (cm^{-1})	F
Naphthalene	8,700	10^{-2}
Anthracene	7,100	3×10^{-2}
Tetracene	5,700	5×10^{-2}

Table 3

Estimated relative rates of formation of the lowest triplet state by cation-anion annihilation k/gk_d (g = 3/4).

Molecule	non-resonance $\Delta G^*_{pol} = 0$	resonance $\rho = 1$
Naphthalene	$> 2 \times 10^{-2}$	2×10^{-8}
Anthracene	$> 6 \times 10^{-2}$	10^{-6}
Tetracene	$> 10^{-1}$	10^{-3}

The non-resonance transfer appears to be much faster than the resonance transfer. Because of the very fast vibrational relaxation the excess vibrational energy will be dissipated to the environment during the encounter so that the total reaction can be described as:

$$^2M_o^+ + {}^2M_o^- \to {}^3({}^2M_o^+, {}^2M_o^-) \to {}^3({}^3M_1^*, {}^1M_o^*) \to {}^3({}^3M_1, {}^1M_o) \begin{smallmatrix} \nearrow {}^3(M_2)_1 \\ \searrow {}^3M_1 + {}^1M_o \end{smallmatrix}$$

(26)

where the asterisks indicate the excess vibrational energy. Part of the encounters may lead to the formation of triplet excimers, which may dissociate again into the triplet and ground state molecule or into two ground state molecules. The role of triplet excimers in the quenching of triplet molecules by collisions with ground state molecules has been discussed in a previous paper[13].

A comparison with other annihilation reactions

From Figure 1 it appears that the crossings to the other possible final states all occur through resonance interaction. For these reactions the relative rates can be derived in the same way as before, taking $\rho = 1$ and using the data in Table 1 to calculate ΔG_{pol}^*. A comparison of the relative rates of cation-anion annihilation for the molecules naphthalene, anthracene and tetracene is given in the Tables 4, 5 and 6.

Table 4

Estimated relative rate constants for the cation-anion annihilation of naphthalene

final state	g	k/gk_d
${}^3M_1 + {}^1M_o$	3/4	$> 2 \times 10^{-2}$
${}^3M_2 + {}^1M_o$	3/4	≈ 1
${}^1M_1 + {}^1M_o$	1/4	≈ 1

Table 5

Estimated relative rate constants for the cation-anion annihilation of anthracene

final state	g	k/gk_d
$^3M_1 + {}^1M_0$	3/4	$> 6 \times 10^{-2}$
$^3M_2 + {}^1M_0$	3/4	$\approx 10^{-3}$
$^1M_1 + {}^1M_0$	1/4	3×10^{-3}

Table 6

Estimated relative rate constants for the cation-anion annihilation of tetracene

final state	g	k/gk_d
$^3M_1 + {}^1M_0$	3/4	10^{-1}
$^3M_2 + {}^1M_0$	3/4	unknown
$^1M_1 + {}^1M_0$	1/4	3×10^{-3}

For naphthalene there are two very efficient reactions leading to the lowest excited singlet and to the second excited triplet. In the latter case the final product, of course, will be the lowest triplet because internal conversion and vibrational relaxation proceed so fast that during the encounter the $^3M_2$ will already decay to the $^3M_1$. The overall process will be

$$2M_o^+ + {}^2M_o^- \to {}^3({}^2M_o^+, {}^2M_o^-) \to {}^3({}^3M_2, {}^1M_o) \to {}^3({}^3M_1, {}^1M_o) \nearrow {}^3(M_2)_2 \atop \searrow {}^3M_1 + {}^1M_o \qquad (27)$$

How much the contribution of the first reaction will be is difficult to say because we only know the lower limit of the rate constant.

The overall reaction for the production of the lowest excited singlet will be:

$$^2M_o^+ + {}^2M_o^- \to {}^1({}^2M_o^+, {}^2M_o^-) \to {}^1({}^1M_1, {}^1M_o) \nearrow {}^1(M_2)_1 \atop \searrow {}^1M_1 + {}^1M_o \qquad (28)$$

Here part of the encounters may give rise to the formation of excimers.

As we see from Tables 5 and 6 for anthracene and tetracene the cation-anion annihilation predominantly leads to the formation of the lowest triplet states and the singlet excited monomer and the excimer will be produced only indirectly by triplet-triplet annihilation. The latter reaction is diffusion controlled which because of the spin statistical factor implies that one of every nine collisions leads to a singlet excited molecule or excimer. Every one out of three collisions leads to quenching of one of the two triplets. If this were the only efficient quenching reaction the quantum yield of light emission from cation-anion annihilation would still be fairly high. However, as we will see in the following section the positive and negative hydrocarbon ions also quench the triplet molecule and consequently lower the quantum yield of light emission.

Quenching of triplet molecules by hydrocarbon ions

Since the positive and negative ions behave practically in the same way and more experimental data are available for the negative ions, in the following discussion we will consider the quenching of triplets by negative ions. The following two reactions are possible

$$^3M_1 + {}^2M_o^- \rightarrow {}^2M_i^- + {}^1M_o \tag{29}$$

and

$$^3M_1 + {}^2M_o^- \rightarrow {}^1M_o + {}^2M_i^- \tag{30}$$

where the index i indicates that the ion may end up in its ground state or in one of its lower excited states. Reaction (29) involves electron transfer so that solvent-dipole reorientation has again to be taken into consideration. Its description is therefore very similar to that of the cation-anion annihilation. In Figure 2 we have drawn the free energy curves for the initial and relevant final states. For one of these final states the ion occurs in its first (anthracene and tetracene) or its second excited state (naphthalene). Naphthalene has a lower excited state the transition to which from the ground state is forbidden[14]. The three excited states considered are all of the same kind and appear in the absorption spectrum as a weak short axis polarized absorption band located at 9,500 cm^{-1} for naphthalene and at 11,500 cm^{-1} for both anthracene and tetracene[15].

For a spin allowed reaction the encounter complex must be a doublet and the configuration for the initial state can be described by:

$$^2\phi_i = \{n\,n'\,\nu^2\,\nu'\}\,\frac{1}{\sqrt{6}}\{2\alpha\alpha\beta - \alpha\beta\alpha - \beta\alpha\alpha\} \tag{31}$$

where the spin part gives the proper combination for the three electrons moving in half filled orbitals (for $S_z = +\frac{1}{2}$).

In the final state in which the ion is in its ground state the electron configuration becomes:

$$^2\phi_f = n^2\,n'\,\nu^2 \tag{32}$$

Omitting again terms of second order in intermolecular overlap we obtain for the interaction between (31) and (32)

$$H_{if} = \sqrt{\frac{3}{2}}\,<n|H|\nu'> \tag{33}$$

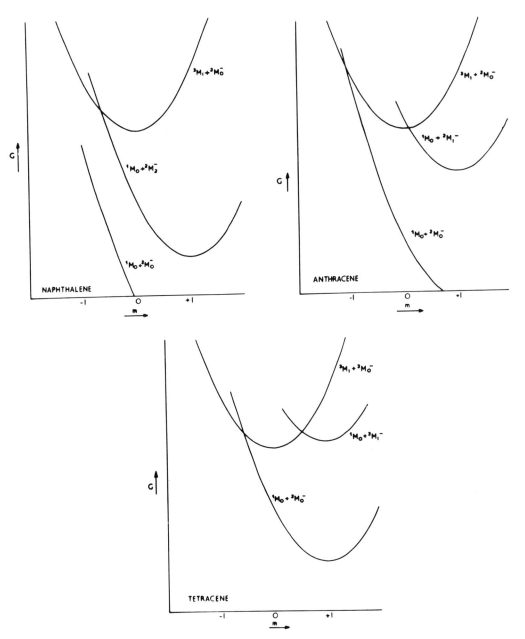

Figure 2.

Plot of the free energy for the initial and final states of the quenching of triplets by negative ions against the reaction co-ordinate for solvent-dipole orientation m.

As has been discussed in previous papers[14, 15] the excited states of the ions considered cannot be described by one single configuration. There are two electron configurations which both contribute strongly to the wave function for the excited state. These are

$$\phi_{f_1} = n^2 \, m' \, \nu^2 \qquad (34)$$

and

$$\phi_{f_2} = n(n')^2 \, \nu^2 \qquad (35)$$

A comparison between (31) and (34) shows that the two configurations differ in two molecular orbitals and consequently the interaction will be second order in intermolecular overlap. The configurations (31) and (35) differ only in one orbital and omitting again higher order terms we find

$$H_{if} = \sqrt{\tfrac{3}{2}} \, <n'|H|\nu'> \qquad (36)$$

A large fraction of (36) will contribute to the overall interaction and as far as the order of magnitude is concerned we therefore assume both for (33) and (36) the value $H_{if} = 1$ cm^{-1} which has been used before. Following the same procedure as in previous sections we find two kinds of reactions: a resonance transfer where $\rho = 1$ and the rate constant is determined by ΔG^*_{pol} and a non-resonance transfer for which $\Delta G^*_{pol} = 0$ and the probability for transfer depends on the Franck-Condon factor for the energy gap between the initial and final state at m = o. The most efficient quenching reactions are listed in Table 7. For naphthalene the state $^2M_2$ will exist only for a very short time. Because of a fast internal conversion and vibrational relaxation it will decay to the lowest excited state during the encounter.

Reaction (30) in which the negative charge stays on the same species does not require solvent reorientation so that the closest distance of approach will be smaller than in reaction (29). The interaction, however, is determined by electron exchange, i.e. second order in intermolecular overlap. The overall effect is probably about the same as in reaction (29). Since solvent reorientation plays no role the energy gaps between the initial and final states will be 5000 cm^{-1} larger and consequently the Franck-Condon factors will be smaller. On the other hand transition can take place to one of the higher excited doublet states of the ion.

Table 7

Estimated relative rate constants for the quenching of triplet molecules by negative ions ($g = 1/3$)

Molecule	Final state	k/gk_d
Naphthalene	$^1M_o + {}^2M_2^-$	> 0.1
Anthracene	$^1M_o + {}^2M_1^-$	1
Tetracene	$^1M_o + {}^2M_o^-$	> 0.2

Only one of these states is known experimentally. The absorption spectra of the ions of naphthalene, anthracene and tetracene show a second absorption band separated from the first one by about 1000 cm^{-1} which arises from a long-axis polarized transition[14, 15]. From the theory one expects a few more low-lying excited doublet states. The transitions from the ground state to these states are symmetry forbidden which makes it very difficult if not impossible to observe them spectroscopically. Because of the lack of experimental data we cannot make any estimate about the rate constants of these reactions, although they probably will be as efficient as reactions (29).

The estimates given in Table 7 show that the quenching of the triplet by the ion is very fast and almost or entirely diffusion controlled.

The effect of quenching of excited molecules on the quantum yield of the annihilation process.

For the most favourable case that the cation-anion annihilation is diffusion controlled $3/4$ of the encounter will lead to a triplet molecule and $1/4$ leaves the ions unchanged. Since cations and anions have the same effect the ratio triplet: ion in the "reaction zone"* will be 3:2. If we take into account the spin statistical factors and consider both the triplet-triplet and triplet-ion reactions diffusion controlled we obtain a yield of 20% of excited singlet molecules. Actually this yield will be appreciably lower because we did not account for the fact that a very large fraction of the triplet will diffuse away from the reaction zone and subsequently

become quenched by encounters with positive or negative ions. The
singlet excited molecules are much less vulnerable because of their
much shorter lifetime. A similar discussion as has been given before
for the triplet molecule will show that the singlet excited molecule
can be rapidly converted into the corresponding triplet molecule during
an encounter with a positive or negative ion. In those cases where
the cation-anion annihilation is not very efficient the concentration
of ions in the reaction zone may become very large and quenching of
singlet excited molecules may be expected as well. This can be
prevented by keeping the bulk concentration of ions low enough so
that the concentration of ions in the reaction zone stays below the
critical limit for quenching of the singlet molecules.

As has been mentioned in the previous paper[2] the diagrams as
shown in Figure 1 reveal that the encounter of two triplet molecules
can also give rise to the formation of the cation and anion. In
some cases the energy barrier is so low that this reaction may
compete very favourably with the familiar triplet-triplet annihilation
reactions. Jarnagin et al.[16] have found experimental evidence for
this reaction.

Although the qualitative study presented in this paper has been
restricted to the molecules naphthalene, anthracene and tetracene
it can be extended to other aromatic molecules as well as to reactions
in which the cation and anion originate from different molecular species.
The latter reactions have been investigated in detail by Weller and
Zachariasse[17] using hydrocarbon negative ions as donors and tetramethyl
phenylene diamine or tri-paratolyl amine cations as acceptors. As
solvents they used tetrahydrofuran, 2-methyltetrahydrofuran and
1,2-dimethyoxy ethane. The dielectric constants of these solvents
(THF, $D_s \simeq 7.6$)[18] are appreciably lower than those of acetonitrile
and dimethylformamide ($D_s \simeq 37.6$)[18]. From equation (2) (with $D_o \simeq 2$)
it follows that for these solvents one would have to use a value of
$\lambda = 3,900$ cm^{-1}. When applied to the reactions considered in this
paper this lower value of λ would imply a smaller Franck-Condon
factor for a non-resonance transfer. On the other hand because of
the relatively low dielectric constant the coulomb interaction
between the two ions will increase the duration of the encounter
and this in turn will contribute favourably to the probability of
the transition.

As Brocklehurst and others[19] have shown luminescence can also
be observed from the annihilation of cations and anions produced
by γ irradiation of aromatic hydrocarbons in rigid solvents. These
experiments were carried out in non-polar solvents and at low
viscosities so that the main conclusions arrived at in this paper
do not hold for this kind of annihilation reaction.

References

1. T. C. Werner, J. Chang and D. M. Hercules, J. Amer. Chem. Soc., 92, 763 (1970).
   D. M. Hercules, "Physical Methods of Chemistry", Vol. 1, Part IIB, A. Weissberger and B. Rossiter, eds. (Wiley, New York, 1971), page 257.
   E. A. Chandross and F. I. Sonntag, J. Amer. Chem. Soc., 88, 1089 (1966).
   E. A. Chandross, Trans. N. Y. Acad. Sci., series II, 31, 571 (1969)
   A. Zweig, Advances in Photochemistry, 6, 425 (1968).
   A. Zweig, D. L. Maricle, J. S. Brinen and A. H. Maurer, J. Amer. Chem. Soc., 89, 473 (1967).
   A. J. Bard, K. S. C. Santhanam, S. A. Cruser and L. R. Faulkner, "Fluorescence: Theory, Instrumentation and Practice", G. G. Guilbault, ed., (Marcel Dekker, New York, 1967) Chapter 14.
   L. R. Faulkner and A. J. Bard, J. Amer. Chem. Soc., 90, 6284 (1968).
   C. A. Parker and G. D. Short, Trans. Faraday Soc., 36, 2618 (1967).
   M. Sano and F. Egusa, Bull. Chem. Soc. Japan, 41, 1490 (1968).
   K. Mori, N. Yamamoto and H. Tsubomura, Bull. Chem. Soc. Japan, 44, 2661 (1971).
   and related papers cited in the above references

2. G. J. Hoytink, Discuss. Faraday Soc., 45, 14 (1968).

3. W. Siebrand and D. F. Williams, J. Chem. Phys., 49, 1860 (1968).

4. W. Siebrand, J. Chem. Phys., 47, 2411 (1967).

5. R. L. de Groot and G. J. Hoytink, J. Chem. Phys., 46, 4524 (1967).

6. P. E. Kellogg, J. Chem. Phys., 44, 411 (1966).

7. R. A. Marcus, J. Chem. Phys., 43, 679 (1965).

8. R. A. Marcus, J. Chem. Phys., 43, 2654 and references cited therein.

9. G. W. Robinson and R. P. Frosch, J. Chem. Phys., 37, 1962 (1962); ibid. 38, 1187 (1963).

10. G. J. Hoytink, Accounts Chem. Res., 2, 114 (1969).

11. G. J. Hoytink, J. Pure Appl. Chem., 11, 393 (1965).

12. W. Siebrand, J. Chem. Phys., 44, 4055 (1966).

13. J. Langelaar, G. Jansen, R. P. H. Rettschnick and G. J. Hoytink, Chem. Phys. Letters, 12, 86 (1971).

14. P. Balk, S. de Bruijn and G. J. Hoytink, Rec. Trav. Chim. Pays Bas., 76, 907 (1957).

15. K. H. J. Buschow and G. J. Hoytink, J. Chem. Phys., 40, 2501 (1964), and papers cited therein.

16. K. de Groot, L. P. Cary and R. C. Jarnagin, J. Chem. Phys., 48, 5280 (1968).

17. A. Weller and K. Zachariasse, Chem. Phys. Letters, 10, 161 (1971). K. Zachariasse, Thesis, Free University, Amsterdam (1972), and papers cited therein.

18. Landolt Börnstein II. Band, 6. Teil I (1959).

19. B. Brocklehurst and R. D. Russell, Trans. Faraday Soc., 65, 2159 (1969), with references to papers by other authors.

*A detailed kinetic analysis of the entire process of electrochemiluminescence has recently been made by R. Bezman and L. R. Faulkner, J. Amer. Chem. Soc., 94, 3699 (1972). This is a revised version of an earlier study by S. W. Feldberg, J. Phys. Chem., 70, 3928 (1966).

CHEMILUMINESCENCE FROM RADICAL ION RECOMBINATION

VI. REACTIONS, YIELDS, AND ENERGIES

A. Weller and K. Zachariasse

Max-Planck-Institut für biophysikalische Chemie

Abt. Spektroskopie, Göttingen, Germany

INTRODUCTION

Our studies [1 - 8] of chemiluminescent radical ion recombination reactions of the general type

$$^2A^- + {}^2D^+ \longrightarrow A + D + h\nu \qquad (1)$$

were initiated by the question whether singlet hetero-excimers, $^1(A^-D^+)$, can be formed from the corresponding radical ions in their doublet ground states. The results obtained with more than 170 different A,D-systems [6] revealed however that electron transfer processes leading to molecular excited states of A and/or D are effectively competing with direct hetero-excimer formation and that subsequent reactions such as triplet-triplet annihilation [1 - 3]

$$^3A^* + {}^3A^* \longrightarrow {}^1A^* + A \qquad (2)$$

$$^3D^* + {}^3D^* \longrightarrow {}^1D^* + D \qquad (3)$$

$$^3A^* + {}^3D^* \longrightarrow {}^1(A^-D^+) \qquad (4)$$

and hetero-excimer dissociation [5 - 8]

$$^1(A^-D^+) \longrightarrow {}^1A^* + D \qquad (5)$$

$$^1(A^-D^+) \longrightarrow A + {}^1D^* \qquad (6)$$

leading, with the yield γ_d, to the lowest molecular

excited singlet state of the A,D-system have to be taken into account.

In order to elucidate the kinetic possibilities involved and to obtain more information about the mechanistic aspects an attempt is made here to evaluate the general kinetic scheme in terms of reaction yields and to calculate the free energy differences that are associated with the competitive and sequential processes taking place in these systems. It is on the basis of the generalized results of this attempt that the chemiluminescence spectra of various A,D-systems, their temperature dependence and quantum yields can be discussed and interpreted.

REACTIONS AND YIELDS

It is clear that in order to produce chemiluminescence in radical ion recombination reactions (1) the energy stored chemically in the radical ions must be transformed into excitation energy of the A,D-system. Evidently, this transformation cannot occur until the solvated radical ions have formed an encounter complex with a center-to-center distance, a, between the ionic charges of, say, 6 or 7 Å. In view of its comparitively slow diffusional separation ($\sim 10^6 \text{sec}^{-1}$ in the low dielectric constant solvents ($\varepsilon \approx 7$) used in this research) this encounter complex, $A^-..D^+$, between the solvated radical ions represents a convenient origin of the competitive processes that eventually lead to the chemiluminescence observed.

Apart from the diffusional separation of the radical ion encounter complex there are all together six competing reactions which have to be considered in these systems:

$$A^-...D^+ \xrightarrow{\gamma_s^A} {}^1A^* + D \qquad (7)$$

$$A^-...D^+ \xrightarrow{\gamma_s^D} A + {}^1D^* \qquad (8)$$

$$A^-...D^+ \xrightarrow{\gamma_c} {}^1(A^-D^+) \qquad (9)$$

$$A^-...D^+ \xrightarrow{\gamma_t^A} {}^3A^* + D \qquad (10)$$

$$A^{-}\ldots D^{+} \xrightarrow{\gamma_t^D} A + {}^3D \quad (11)$$

$$A^{-}\ldots D^{+} \xrightarrow{\gamma_g} A + D \quad (12)$$

Their yields γ_x are related to the corresponding free energy differences ΔG_{ipx} as will be discussed below.

Reactions (10) and (11) followed by triplet-triplet annihilation give rise to luminescence from the molecular excited singlet states of A and D, respectively, with the chemiluminescence yields

$$\chi^A = \gamma_t^A p^A \phi^A \quad \text{and} \quad \chi^D = \gamma_t^D p^D \phi^D \quad (13)$$

where ϕ^A and ϕ^D are the fluorescence quantum yields and p^A and p^D the respective annihilation probabilities which are proportional to the square of the actual triplet lifetimes (triplet quenching by radical ions included). It should be mentioned that even under the most favorable experimental conditions $\gamma_t p$ never exceeds 10^{-3} [6]. The yield of the hetero-excimer chemiluminescence generated by the mixed triplet-triplet annihilation process (4) is given by

$$\chi' = \gamma_t^A \gamma_t^D p' \phi' \quad (14)$$

where ϕ' is the hetero-excimer fluorescence quantum yield and p' the mixed triplet-triplet annihilation probability which is proportional to the product of the actual triplet lifetimes of A and D in the chemiluminescent system.

One might suppose that molecular triplet state formation could also occur by intermediate triplet hetero-excimer formation followed by rapid internal conversion within the hetero-excimer triplet manifold This process which would necessarily and exclusively lead to the lowest triplet state of the A,D-system can, however, be ruled out on the basis of the results obtained with systems where both molecular triplet states, $^3A^*$ and $^3D^*$, are energetically accessible from the radical ion pair. The chemiluminescence of these systems brought about by triplet-triplet annihilation, invariably, shows both molecular fluorescence components [2]. Triplet energy transfer experiments corroborate these results [6,8].

Reaction (9) may be followed by thermal heteroexcimer dissociation (5) or (6) with the yield γ_d depending on ΔG_d, the amount of free energy required in this process (see below), and may thus lead to chemiluminescence from the lowest molecular excited singlet state of the A,D-system (cf. [5,7]). Taking into account contributions due to direct production of the molecular excited singlet states (reactions (7) and (8)) and due to triplet-triplet annihilation (equations (13) and (14)) one obtains now for the total chemiluminescence yield of molecular A or D fluorescence

$$\chi = (\gamma_s + \gamma_c \gamma_d + \gamma_t p) \phi \tag{15}$$

and for that of hetero-excimer emission

$$\chi' = (\gamma_c(1 - \gamma_d) + \gamma_t^A \gamma_t^D p') \phi' \tag{16}$$

These expressions, which in any particular case may be simplified considerably by energy considerations, form the basis of interpretation of the chemiluminescence quantum yields that have been determined experimentally [4,6]. The highest chemiluminescence yield observed so far (0.075 Einstein/mole(anion)) was with the system: A = 9,10-dimethylanthracene, D = tri-p-tolylamine in methyltetrahydrofuran at room temperature and could be shown to be due to $\chi^A + \chi' = \gamma_c \gamma_d^A \phi^A + \gamma_c(1-\gamma_d)\phi'$ with $\gamma_c = 0.10 \pm 0.01$, $\gamma_d \approx 0.8$, $\phi^A = 0.90$, $\phi' = 0.15 \pm 0.10$ [4,5].

An upper limit of the reaction yields is, of course, given by

$$\gamma_s^A + \gamma_s^D + \gamma_c + \gamma_t^A + \gamma_t^D + \gamma_g = 1 \tag{17}$$

For spin-statistical reasons equation (17) may be split up into

$$\gamma_s^A + \gamma_s^D + \gamma_c + \gamma_g = 0.25 \tag{18}$$

and

$$\gamma_t^A + \gamma_t^D = 0.75 \tag{19}$$

It should be pointed out, however, that equations (18) and (19) can only be valid as long as the corresponding reaction rates fulfil the conditions

$$k_s^A + k_s^D + k_c + k_g > \frac{1}{\tau_{spin}} \qquad (20a)$$

$$k_t^A + k_t^D > \frac{1}{\tau_{spin}} \qquad (20b)$$

where τ_{spin} is the spin relaxation time of the radical ion pair in the encounter complex which may be of the order of or even smaller than 1μsec.

ENERGY CONSIDERATIONS

Fig. 1 schematically shows the free energy differences which have to be taken into account when dealing with chemiluminescent radical ion recombination reactions in ether solvents with dielectric constant $\varepsilon \approx 7$.

$\Delta G(\overline{A}...\overset{+}{D})$, the amount of free energy stored chemically in the radical ion pair and available in subsequent reactions is associated with the equilibrium $A + D \rightleftarrows \overline{A}...\overset{+}{D}$ and can be calculated according to equation (21) derived in [2]

$$\Delta G(\overline{A}...\overset{+}{D}) = E_{\frac{1}{2}}(D/D^+) - E_{\frac{1}{2}}(A^-/A) + 0.20 \text{ eV} \qquad (21)$$

from the oxidation and reduction potentials of D and A, respectively, measured in acetonitrile or dimethylformamide. For the A,D-systems investigated $\Delta G(\overline{A}...\overset{+}{D})$ varies between 3.7 and 1.7 eV and since

$$\Delta G(\overline{A}...\overset{+}{D}) = -\Delta G_{ipg} \qquad (22)$$

it is this amount of energy which under the prevailing isothermal conditions has to be transformed into heat in reaction (12).

It has been tacitly assumed that reaction (12) because of the very unfavorable Franck-Condon factors involved should be much slower than those leading to excited states. However, the results presented in table I as well as other evidence [9] seem to indicate

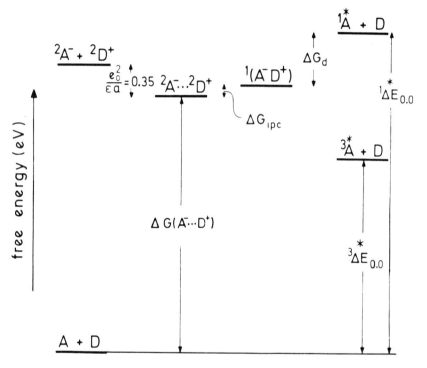

Fig. 1. Generalized energy diagram in ether solution ($\varepsilon \approx 7$).

that γ_g is by no means negligible and that k_g increases as $\Delta G(A^-...D^+)$ becomes smaller.

Information as to which molecular singlet or triplet excited states of the A,D-system are accessible in radical ion recombinations is provided by the free energies ΔG_{ips} and ΔG_{ipt} (associated with reactions (7), (8) and (10), (11), respectively) that can be calculated according to

$$\Delta G_{ips} = {}^1\!\Delta E^* - \Delta G(A^-...D^+) \qquad (23a)$$

$$\Delta G_{ipt} = {}^3\!\Delta E^* - \Delta G(A^-...D^+) \qquad (23b)$$

from the zero-zero transition energies, ${}^1\!\Delta E^*$ and ${}^3\!\Delta E^*$, between the ground and the excited singlet and triplet states, respectively. It is clear and has been confirmed with all systems investigated that only those molecular excited singlet and triplet states are likely to be populated to any appreciable extent for which ΔG_{ips} and ΔG_{ipt}, respectively, are < 0.1 eV.

The above procedure implies that only negligible entropy changes are involved in electronic excitation. Therefore, equations (23a) and (23b) are applicable to aromatic hydrocarbons and probably also to their derivatives but not to hetero-excimers.

In order to estimate (rather than calculate) the free enthalpy change, ΔG_{ipc}, associated with reaction (9) use was made of the thermodynamic quantities which are associated with the hetero-excimer dissociation process (5) or (6) and have been determined for about forty A,D-systems in n-hexane [10].

In contrast to the dissociation entropy, for which within the limits of error of these measurements a constant value $\Delta S_d(hex) = 18.5 \pm 1.5$ e.u. so that at room temperature

$$T\Delta S_d(hex) = 0.23 \pm 0.02 \text{ eV} \qquad (24)$$

was obtained, the dissociation enthalpy was found to vary with the donor and acceptor properties of the molecules involved. The latter result was interpreted on the basis of the relation (cf.[11,12])

$$\Delta H_d(\text{hex}) = {}^1\Delta E^* - E(A^-D^+) - (U_{dest} - U_{stab}) \quad (25)$$

where $E(A^-D^+)$ is the energy of the hetero-excimer charge-transfer state (with respect to the separated ground state molecules) which can be expressed in terms of the polarographic oxidation and reduction potentials (measured in acetonitrile). $(U_{dest} - U_{stab})$ accounts for specific destabilizing and stabilizing interactions (affecting the hetero-excimer energy) of the hetero-excimer charge-transfer state with the no-bond ground state and with locally excited states, respectively. With

$$U_{dest} - U_{stab} = \pm 0.10 \text{ eV} \quad (26)$$

and in excellent agreement with the experimental results equation (25) then becomes

$$\Delta H_d(\text{hex}) = {}^1\Delta E^* - (E_{\frac{1}{2}}(D/D^+) - E_{\frac{1}{2}}(A^-/A) + 0.13) \pm 0.10 \text{ eV} \quad (27)$$

Combining equations (24), (26) and (27) one obtains for the dissociation free enthalpy in hexane

$$\Delta G_d(\text{hex}) = {}^1\Delta E^* - (E_{\frac{1}{2}}(D/D^+) - E_{\frac{1}{2}}(A^-/A))$$

$$- (U_{dest} - U_{stab}) - 0.36 \pm 0.02 \text{ eV} \quad (28)$$

In order to find the corresponding thermodynamic data in the ether solvents one has to calculate the free enthalpy and entropy changes, ΔG_{hex}^{eth} and ΔS_{hex}^{eth}, associated with the equilibrium

$${}^1A_h^* + D_h + {}^1(A^-D^+)_e \rightleftharpoons {}^1A_e^* + D_e + {}^1(A^-D^+)_h \quad (29)$$

where the subscripts h and e refer to the solvents hexane and ether, respectively. Assuming that solvation of the hetero-excimer (with dipole moment μ) is by far the predominating effect in this equilibrium one can write

$$\Delta G_{hex}^{eth} = \frac{\mu^2}{\rho^3}\left[\left(\frac{\varepsilon-1}{2\varepsilon+1}\right)_{eth} - \left(\frac{\varepsilon-1}{2\varepsilon+1}\right)_{hex}\right] \quad (30)$$

and

$$T\Delta S_{hex}^{eth} = \frac{\mu^2}{\rho^3}\left[\left(\frac{d \ln \varepsilon}{d \ln T} \frac{3\varepsilon}{(2\varepsilon+1)^2}\right)_{hex} - \left(\frac{d \ln \varepsilon}{d \ln T} \frac{3\varepsilon}{(2\varepsilon+1)^2}\right)_{eth}\right] \quad (31)$$

with ρ being the equivalent sphere radius in the Kirkwood-Onsager continuum model which is applied here.

Values of μ^2/ρ^3 have been determined [11,13] for hetero-excimers to be in the range 0.75 ± 0.25 eV. The dielectric constant (ε) and its temperature dependence (d ln ε / d ln T) are, respectively, 7 and -1.2 for the ethers and 1.9 and -0.23 for hexane, so that one obtains for the ether solvents at room temperature with equations (24) and (31)

$$T\Delta S_d = T\Delta S_d(hex) + T\Delta S_{hex}^{eth}$$
$$= 0.27 \pm 0.03 \text{ eV} \quad (32)$$

and with equations (28) and (30)

$$\Delta G_d = \Delta G_d(hex) + \Delta G_{hex}^{eth}$$
$$= {}^1\Delta E^* - (\underline{E_1(D/D^+)} - \underline{E_1(A^-/A)})$$
$$ - (U_{dest} - U_{stab}) - 0.20 \pm 0.07 \text{ eV} \quad (33)$$

Inspection of fig. 1. shows that

$$\Delta G_{ipc} = {}^1\Delta E^* - \Delta G_d - \Delta G(A^-\!\ldots D^+) \quad (34)$$

which with equations (21), (26) and (33) leads to

$$\Delta G_{ipc} = (U_{dest} - U_{stab}) \pm 0.07 = \pm 0.17 \text{ eV} \quad (35)$$

This result shows that direct hetero-excimer formation from radical ions (reaction (9)), in principle, may be possible (if $\Delta G_{ipc} \leq 0$), but because of the high uncertainties involved it does not allow to make predictions as to the yield of that process for any specific A,D-system. It is, however, on the other hand possible by comparing experimentally determined heteroexcimer chemiluminescence quantum yields χ' to arrive at conclusions with respect to the relative magnitude of ΔG_{ipc} in different systems (cf. [7]).

SINGLET POPULATION FROM RADICAL ION ENCOUNTERS

Results obtained with tri-p-tolylaminium perchlorate (TPTA$^+$ ClO$_4^-$) and the radical anions (A$^-$) of coronene, rubrene and pentacene in tetrahydrofuran at room temperature are collected in table I.

Table I.

Yields and energies of some radical ion recombination reactions with D=tri-p-tolylamine

A	ϕ^A	$\Delta G(A^{\cdot\cdot}...D^+)$ (eV)	$\Delta E(^1A^*)$ (eV)	χ^A	$\gamma_s^A + \gamma_c$	γ_g
Coronene	0.23	2.98	2.95	0.02	0.10	0.15
Rubrene	1.0	2.41	2.39	0.05	0.05	0.20
Pentacene	0.08	2.30	2.12	0.0015	0.02	0.23

The chemiluminescence observed consists only of the hydrocarbon fluorescence ($^1A^*$) which according to the energy data of table I can be generated by reaction (7) and/or by reaction (5) following reaction (9), so that one has according to equation (15) for the chemiluminescence quantum yield since contributions due to triplet-triplet annihilation can be neglected

$$\chi^A = (\gamma_s^A + \gamma_c \gamma_d) \phi^A \qquad (36)$$

With $\gamma_d = 1$ (because no hetero-excimer chemiluminescence could be observed even at low temperatures and ΔG_d calculated according to equation (33) turns out to be negative for these systems) and since $\gamma_s^D = 0$ for energetic reasons ($\Delta E(^1D^*) = 3.51$ eV) one obtains with equation (18)

$$\gamma_g = 0.25 - \frac{\chi^A}{\phi^A} \qquad (37)$$

The values of γ_g (cf. last column of table I) are surprisingly high and increase as expected with decreasing energy $\Delta G(A^-\ldots D^+)$.

CONCLUDING REMARKS

Analogous considerations as given here with respect to the reactions, yields and energies involved in heteromolecular radical ion recombinations, when applied to the homomolecular case, will lead to the same equations as (15) and (16) for the molecular and excimer chemiluminescence yields, respectively and, also, to similar expressions for the free energy differences. Unlike hetero-excimers, however, excimers have $\mu = 0$ and $U_{dest} - U_{stab}$ always negative [11,12].

REFERENCES

[1] A. WELLER and K. ZACHARIASSE, J.Chem.Phys. 46(1967) 4984 (part I of the series on chemiluminescence from radical ion recombination).

[2] A. WELLER and K. ZACHARIASSE, in: Molecular Luminescence (E.C. Lim ed.) W.A. Benjamin Inc., New York (1969) p. 895 (Part II).

[3] A. WELLER and K. ZACHARIASSE, Chem.Phys.Letters, 10 (1971) 197 (part III).

[4] A. WELLER and K. ZACHARIASSE, Chem.Phys.Letters, 10 (1971) 424 (part IV).

[5] A. WELLER and K. ZACHARIASSE, Chem.Phys.Letters, 10 (1971) 590 (part V).

[6] K. ZACHARIASSE, thesis, Free University, Amsterdam (1972)

[7] A. WELLER and K. ZACHARIASSE, subsequent paper (part VII).

[8] A. WELLER and K. ZACHARIASSE, J.Lumin. in print (part VIII).

[9] H. SCHOMBURG, H. STAERK and A. WELLER, unpublished results

[10] H. KNIBBE, D. REHM and A. WELLER, Ber.Bunsenges. physik.Chem. 73 (1969) 839 and unpublished results.

[11] H. BEENS and A. WELLER, Acta Phys. Polon. 34 (1968) 593.

[12] D. REHM and A. WELLER, Z.physik.Chem. NF, 69 (1970) 183.

[13] H. BEENS, H. KNIBBE and A. WELLER, J.Chem.Phys. 47 (1967) 1183 and unpublished results

CHEMILUMINESCENCE FROM RADICAL ION RECOMBINATION

VII. HETERO-EXCIMER CHEMILUMINESCENCE YIELDS

A. Weller and K. Zachariasse

Max-Planck-Institut f. biophysikalische Chemie
Abt. Spektroskopie, Göttingen, Germany

INTRODUCTION

The study of the chemiluminescence [1 - 6] produced in recombination reactions between radical anions ($^2A^-$) and radical cations ($^2D^+$) has shown [5] that hetero-excimers ($^1(A^-D^+)$) can be formed directly from the radical ions according to

$$^2A^- \ldots ^2D^+ \xrightarrow{\gamma_c} {}^1(A^-D^+) \qquad (1)$$

whereby the free energy, $\Delta G(A^-\ldots D^+)$, stored chemically in the radical ion pair (=encounter complex of the solvated radical ions) is adiabatically and isothermally transformed into excitation energy of the A,D-system.

The chemiluminescence quantum yield, χ', of the hetero-excimer emission is given by (cf. [5,6])

$$\chi' = \gamma_c (1-\gamma_d) \phi' \qquad (2)$$

where γ_c is the yield of process (1), ϕ' the hetero-excimer fluorescence quantum yield and γ_d the yield of hetero-excimer dissociation

$$^1(A^-D^+) \xrightarrow{\gamma_d} A + {}^1D^* \qquad (3)$$

leading to $^1D^*$ the lowest molecular excited singlet state of the A,D-systems investigated here.

An estimate of the free energy change, ΔG_{ipc}, associated with reaction (1) which was based on spectroscopic and thermodynamic data of hetero-excimer systems in hexane and ether solvents [6] led to

$$\Delta G_{ipc} = (U_{dest} - U_{stab}) \pm 0.07 \text{ eV} \qquad (4)$$

where U_{dest} and U_{stab} are the energy changes (affecting the hetero-excimer energy) due to the destabilizing and stabilizing interactions of the hetero-excimer charge-transfer state (of zero-order energy E_{ct}) with the no-bond ground state (E_g) and locally excited states (E^*) respectively.

The amount of $(U_{dest} - U_{stab})$, according to perturbation theory, can be correlated with the energy differences between the interacting states and will be changing from negative to positive values as

$$Q = \frac{E^* - E_{ct}}{E_{ct} - E_g} \qquad (5)$$

increases. On account of the corresponding change of ΔG_{ipc} from negative to positive values γ_c is expected to decrease strongly and approach zero in A,D-systems with high values of Q.

Recent chemiluminescence studies [2b] which were carried out with radical anions of anthracene and radical cations of a series of structurally related p-substituted triphenylamines seem to indicate that χ' decreases as $\Delta G(A^{\bar{}}...D^+)$ the free energy available in the radical ion pair becomes smaller. As $\Delta G(A^{\bar{}}...D^+)$ can be calculated [2,6] according to

$$\Delta G(A^{\bar{}}...D^+) = E_{\frac{1}{2}}(D/D^+) - E_{\frac{1}{2}}(A^{\bar{}}/A) + 0.20 \text{ eV} \qquad (6)$$

from the polarographic oxidation and reduction potentials of D and A, respectively, it is closely related to $E_{ct} - E_g$. Correspondingly, $E^* - E_{ct}$ can be taken as the difference $^1\Delta E - \Delta G(A^{\bar{}}...D^+)$ where $^1\Delta E$ is the energy of the lowest molecular excited singlet state of the A,D-system. Thus, instead of equation (5) one can write

$$Q = \frac{^1\Delta E - \Delta G(A^-\ldots D^+)}{\Delta G(A^-\ldots D^+)} \quad (7)$$

Clearly, as $\Delta G(A^-\ldots D^+)$ becomes smaller Q will increase and this, by making $(U_{dest}-U_{stab})$ and thus ΔG_{ipc} more positive, will lead to smaller values of γ_c.

In order to gain more insight into these relationships the chemiluminescence spectra and intensities of a number of A,D-systems are compared in which D is either tetramethyl-p-phenylenediamine (TMPD) or tris(p-dimethylaminophenyl)amine (TPDA) whose oxidation potentials differ by only 0.08 V. However, since their $^1\Delta E$ values are 3.52 and 2.99 eV, respectively, the Q values for systems with the same A differ considerably, while the values of $\Delta G(A^-\ldots D^+)$ are almost equal.

EXPERIMENTAL

The purification of the ether solvents dimethoxyethane (DME), tetrahydrofuran (THF) and methyltetrahydrofuran (MTHF), the preparation of the radical ions, the flow-apparatus in which the chemiluminescence was generated and the methods to measure the spectral distribution and to determine approximate quantum yields are described in [1,3,4].

RESULTS AND DISCUSSION

Fig. 1 and fig. 2 show the chemiluminescence spectra obtained in ether solvents at room temperature with the radical anions of the compounds indicated in the figures and the radical cations TMPD$^+$ and TPDA$^+$. These spectra, whose absolute intensities vary within about a factor of 200, exhibit the hetero-excimer as well as the molecular fluorescence spectra of A and D with different relative intensities.

The value of $\Delta G(A^-\ldots D^+)$ decreases from the top of fig. 1 to the bottom of fig. 2 (cf. tables I and II). The hetero-excimer chemiluminescence quantum yields, χ', given in table I for the TMPD$^+$-systems also decrease from top to bottom, while those of the TPDA$^+$-systems (table II) show a maximum value of 0.005 Einstein/mole (anion) for the system with A = naphthalene.

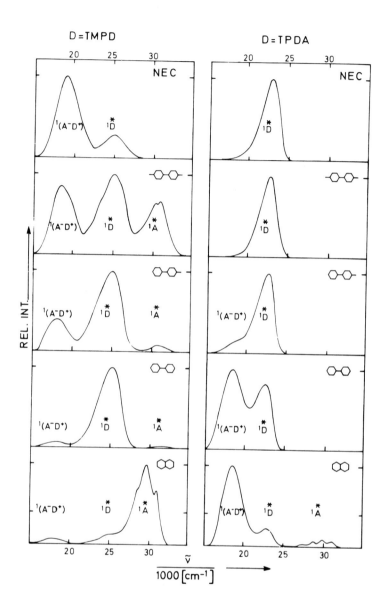

Fig.1. Chemiluminescence spectra at room temperature of
A^-,D^+-systems. The molecules A are indicated in
the spectra;
l.h.s. D=tetramethyl-p-phenylenediamine (TMPD)
r.h.s. D=tris(p-dimethylaminophenyl)amine (TPDA).

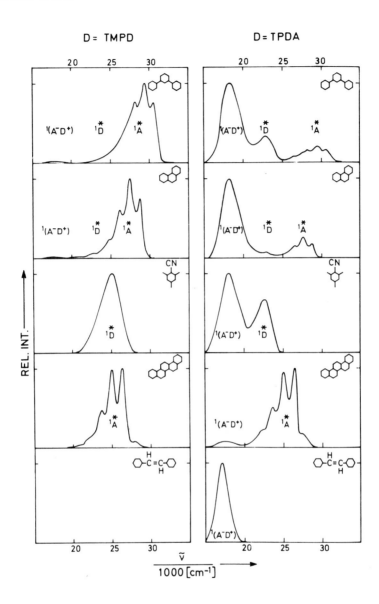

Fig. 2. Chemiluminescence spectra at room temperature of A^-,D^+-systems. The molecules A are indicated in the spectra;
l.h.s. D=tetramethyl-p-phenylenediamine (TMPD)
r.h.s. D=tris(p-dimethylaminophenyl)amine (TPDA).

Table I. Energy data and quantum yields for chemiluminescent radical ion recombination reactions in ether solvents at room temperature with D=tetramethyl-p-phenylenediamine (TMPD)
$E_{\frac{1}{2}}(D/D^+) = 0.16$ V (vs.SCE); $\Delta E(^3D^*) = 2.83$ eV; $\Delta E(^1D^*) = 3.52$ eV

no.	A	$\Delta G(A^{\bar{\cdot}}\ldots D^+)$ (eV)	Q	$\chi' \times 10^5$ (a)
1	N-ethylcarbazole (NEC)	3.16	0.11	20
2	4,4'-dimethyl-biphenyl	3.13	0.12	3
3	4-methylbiphenyl	3.07	0.15	1
4	biphenyl	3.01	0.17	0.2
5	naphthalene	2.94	0.19	0.4
6	m-terphenyl	2.87	0.23	0.1
7	phenanthrene	2.84	0.24	0.1
8	2,4,6-trimethyl benzonitrile	2.83	0.24	0.01
9	picene	2.64	0.25	0 (b)
10	trans-stilbene	2.55	0.38	0
11	anthracene	2.32	0.41	0

(a) hetero-excimer chemiluminescence quantum yield in Einstein/mole(anion). (b) 0 means $\chi' < 10^{-8}$.

Table II. Energy data and quantum yields for chemiluminescent radical ion recombination reactions in ether solvents at room temperature with D=tris(p-dimethylaminophenyl)amine(TPDA)
$E_{\frac{1}{2}}(D/D^+)=0.08$ V (vs.SCE); $\Delta E(^3D^*)=2.63$ eV; $\Delta E(^1D^*)=2.99$ eV

no.	$\Delta G(A^-\ldots D^+)$ (eV)	Q	$\chi' \times 10^5$ (a)	$\Delta E(^3A^*)$ (eV)	$\Delta E(^1A^*)$ (eV)
1	3.08	(-0.03)	(b)	3.04	3.66
2	3.05	(-0.02)	(b)	2.95	4.03
3	2.99	0.00	100	2.98	4.07
4	2.93	0.02	300	3.01	4.13
5	2.86	0.05	500	2.64	3.95
6	2.79	0.07	100	2.79	3.90
7	2.76	0.08	100	2.69	3.58
8	2.75	0.09	20	3.12	4.53
9	2.56	0.17	10	2.49	3.30
10	2.47	0.21	7	2.20	3.79
11	2.24	0.34	0 (c)	1.82	3.28

(a) hetero-excimer chemiluminescence quantum yield in Einstein/mole(anion). (b) see text. (c) 0 means $\chi' < 10^{-8}$.

Comparing the χ' values of corresponding systems in table I and II one finds that in general the hetero-excimer fluorescence quantum yields of the TPDA$^+$-systems are greater by about three orders of magnitude than those of the corresponding TMPD$^+$-systems. This difference can be understood on the basis of the values of Q given in table I and II. χ' clearly decreases as Q increases and approaches zero for Q values greater than about 0.3 so that in the case of anthracene (no. 11; spectra not shown) even with TPDA no hetero-excimer chemiluminescence is observed. Here as well as with other systems of fig.1 and 2 the emission from $^1A^*$ is brought about by triplet-triplet annihilation [1].

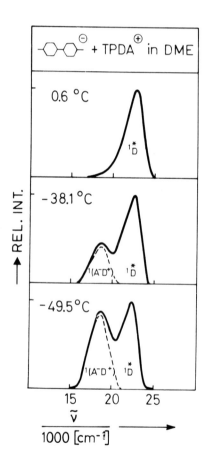

Fig. 3. Chemiluminescence spectra of the system bitolyl$^-$, TPDA$^+$ in dimethoxyethane (DME) at different temperatures.

The decrease in χ' as one goes from no. 5 to no. 1 in table II is, clearly, due to thermal hetero-excimer dissociation (reaction (3)) which for the systems no. 1 and 2, by γ_d approaching unity (cf. equation (2)), makes the hetero-excimer chemiluminescence yield virtually zero at room temperature. On lowering the temperature, however, the hetero-excimer emission returns as shown in fig. 3

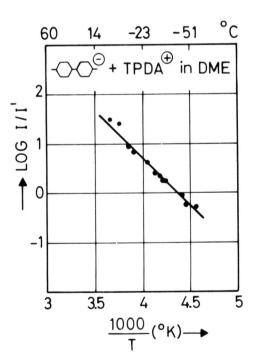

Fig. 4. Temperature dependence of the chemiluminescence intensity ratio $I(^1D^*)/I(A^-D^+)$ in dimethoxyethane of the system A = bitolyl, D = TPDA.

In the systems no. 1 - 4 of table II the chemiluminescence quantum yield of the TPDA fluorescence is given by

$$\chi^D = \gamma_c \gamma_d \phi^D \qquad (8)$$

where ϕ^D is the fluorescence quantum yield of TPDA which between room temperature and -50° was found to be virtually independent of temperature. From equations (2) and (8) one obtains for the quantum yield ratio

$$\frac{\chi^D}{\chi'} = \frac{\gamma_d}{1-\gamma_d} \frac{\phi^D}{\phi'} = \frac{k_d}{k'_f} \phi^D \qquad (9)$$

where k_d is the rate constant of the dissociation reaction (3) and k'_f is the radiative rate constant of the hetero-excimer. As the latter can be assumed to be independent of temperature a semilogarithmic plot of the corresponding intensity ratio versus 1000/T, as shown in fig. 4, gives a slope that corresponds to the activation energy E_d of reaction (3) which according to

$$E_d = \Delta H_d + E_{diff} \qquad (10)$$

is the sum of the dissociation enthalpy and the activation energy of diffusion. From fig. 4 one obtains $E_d = 0.38$ eV so that with $E_{diff} = 0.10$ eV for DME one has $\Delta H_d = 0.28$ eV. With $T\Delta S_d = 0.27 \pm 0.03$ eV (cf. [6]) one finds for the bitolyl, TPDA-system $\Delta G_d = 0.01 \pm 0.03$ eV.

Biphenyl, TMPD is another system for which more information is available. Its free energy of dissociation has been determined in n-hexane at room temperature as $\Delta G_d(hex) = 0.21$ eV. From this, according to a procedure applied in [6] one obtains for the ether solvents $\Delta G_d = 0.37 \pm 0.07$ eV.

These data allow better estimates of ΔG_{ipc} (cf. [6]) than equation (4), which indicate (cf. table III) that for the TMPD systems $U_{dest} - U_{stab}$ is positive (in agreement with large values of Q) and likely to be negative (small values of Q) for the TPDA systems.

Table III. Energy data at room temperature.

A	D	ΔG_d (eV)	ΔG_{ipc} (eV)	$U_{dest}-U_{stab}$
biphenyl	TMPD	0.37 ± 0.07	$+0.14 \pm 0.07$	> 0
bitolyl	TPDA	0.01 ± 0.03	-0.07 ± 0.03	≤ 0

ACKNOWLEDGEMENT

Part of this work was supported by the Netherlands Foundation for Chemical Research (S.O.N.) with financial aid from the Netherlands Organization for the Advancement of Pure Research (Z.W.O.)

REFERENCES

1. A. WELLER and K. ZACHARIASSE, J.Chem.Phys. 46 (1967) 4984 (part I of this series); Chem.Phys.Letters, 10 (1971) 197 (part III).

2. A. WELLER and K. ZACHARIASSE, (a) Molecular Luminescence (E.C. Lim ed.) W.A. Benjamin Inc. New York (1969) 895 (part II); (b) J.Luminescence, in print (part VIII).

3. K. ZACHARIASSE, thesis, Free University, Amsterdam (1972)

4. A. WELLER and K. ZACHARIASSE, Chem.Phys.Letters, 10 (1971) 424 (part IV).

5. A. WELLER and K. ZACHARIASSE, Chem.Phys.Letters, 10 (1971) 590 (part V).

6. A. WELLER and K. ZACHARIASSE, preceding paper (part VI).

7. H. KNIBBE, D. REHM and A. WELLER, Ber.Bunsenges. physik. Chem. 73 (1969) 839.

On The Efficiency of Electrogenerated Chemiluminescence

Allen J. Bard, Csaba P. Keszthelyi, Hiroyasu Tachikawa, and Nurhan E. Tokel

Department of Chemistry, The University of Texas at Austin, Austin, Texas 78712

INTRODUCTION

Electrogenerated chemiluminescence (ECL) occurs when redox reactions between oxidizing and reducing species generated electrochemically result in the production of light. Since the reactant species are frequently radical ions, this luminescence is sometimes called radical ion chemiluminescence, although a number of cases of ECL have now been described where at least one of the reacting partners is not a radical ion. Because a number of reviews (1-6) on ECL have appeared since the initial experiments described in 1964, only a brief outline of the methods and recent results will be presented here. A review of the experimental methods in ECL and an outline of the results of the as yet unreviewed research of 1970-1972 will appear shortly. (7)

ECL can be generated at a single electrode by switching the potential to regions where first one of the reacting species and then the other is generated. For example, consider the ECL obtained with 9,10-diphenylanthracene (DPA) in dimethylformamide (DMF) solutions containing 0.1 $\underline{M}$ tetrabutylammonium perchlorate (TBAP) as a supporting electrolyte. In this medium DPA is reduced to the radical anion (DPA$^{\overline{\cdot}}$) at potentials more negative than -2.0 V (all potentials vs. an aqueous S.C.E.) and is oxidized to the radical cation (DPA$^{\dot{+}}$) at potentials more positive than +1.4 V. If an electrode is first switched to the negative potential region, the electrode reaction

$$DPA + e \rightarrow DPA^{\overline{\cdot}} \qquad [1]$$

occurs, and DPA$^{\overline{\cdot}}$ diffuses from the electrode. On switching the electrode to the positive potential region, $DPA^{\bar{\cdot}}$ and DPA in the vicinity of the electrode are oxidized

$$DPA \rightarrow DPA^{\dagger \cdot} + e \qquad [2]$$

$$DPA^{\bar{\cdot}} \rightarrow DPA^{\dagger \cdot} + 2e \qquad [3]$$

The $DPA^{\dagger \cdot}$ diffusing away from the electrode encounters $DPA^{\bar{\cdot}}$ diffusing back toward the electrode, and a redox reaction between them occurs

$$DPA^{\bar{\cdot}} + DPA^{\dagger \cdot} \rightarrow DPA^* + DPA \qquad [4]$$

$$DPA^* \rightarrow DPA + h\nu \qquad [5]$$

A light pulse, with the light intensity, I, decaying more or less exponentially with time, t, during the oxidation step is observed. Digital simulation methods have been used to give theoretical I vs. t curves in ECL. (8,9)

An important consideration in ECL is the standard enthalpy of the redox process [4], since that governs whether sufficient energy is available to form a singlet excited state in this step. This enthalpy can be estimated from the cyclic voltammetric peak potentials (E_p) for systems unperturbed by decomposition or strong ion pairing of the reactants, and assuming an entropy term, $T\Delta S°$, of about 0.1 eV, using the equation (10-12)

$$-\Delta H° = E°(R'^{\dagger \cdot}/R') - E°(R/R^{\bar{\cdot}}) - T\Delta S° \qquad [6]$$

$$-\Delta H° = E_p(R'^{\dagger \cdot}/R') - E_p(R/R^{\bar{\cdot}}) - 0.16 \text{ eV} \qquad [7]$$

When $-\Delta H°$ is greater than the energy of the emitted photon (E_s), the reaction is called "energy sufficient". Since for DPA, $E_p(R/R^{\bar{\cdot}})$= -1.89 V, $E_p(R/R^{\dagger \cdot})$ = +1.35 V, and E_s = 3.00 eV, DPA ECL would be an example of an energy sufficient system.

ECL can also be obtained where the reactants in the redox reaction [4] arise from different parent molecules (so-called "mixed systems"). For example, if N,N,N',N'-tetramethyl-p-phenylenediamine (TMPD) is added to the DPA solution, it will oxidize well before DPA does ($E_p(TMPD/TMPD^+)$ = +0.24V) so that on cycling between -2.0 and +0.3 V, $DPA^{\bar{\cdot}}$ and $TMPD^{\dagger \cdot}$ are generated. The emitted photon in this case is still that characteristic of DPA fluorescence. This system, for which $-\Delta H° < E_s$, is called an "energy deficient" system. Note also that photons characteristic of DPA^* are emitted when the reacting species are $DPA^{\bar{\cdot}}$ and $TMPD^{2+}$ generated at potentials of about +0.9 V. (13)

ECL can also be produced by generating the reactants at two separate electrodes by using a rotating ring-disk electrode (RRDE) system. (13,14) The RRDE consists of a planar disk electrode and a concentric ring electrode separated from the disk by a very thin insulating gap. When this electrode is rotated, solution flows normal to the electrode surfaces from the bulk solution and radially from disk to ring at the electrode surface. Both electrodes can be maintained at constant potentials and disk-generated reactant will be swept into the zone of ring-generated reactant; reaction there causes the appearance of a ring of light. In this case a steady light intensity is observed, at least for several minutes.

Recent research has been concerned with the details of the mechanism of ECL for different systems. The simplest system involves direct generation of an excited singlet in the redox step [the S-route (9)].

$$R'^{+\cdot} + R^{-\cdot} \rightarrow R' + {}^1R^* \qquad [8]$$

For energy deficient systems the proposed pathway involves generation of an excited triplet state in the redox step, followed by triplet-triplet annihilation (TTA).

$$R'^{+\cdot} + R^{-\cdot} \rightarrow R' + {}^3R^* \qquad [9]$$

$${}^3R^* + {}^3R^* \rightarrow R + {}^1R^* \qquad [10]$$

This path has been named (9) the T-route. The possibility of simultaneous production of excited singlet and triplet state molecules in the redox step (the ST-route) also exists. In considering the mechanism of ECL processes, quenching of the excited states must also be considered,

$${}^1R^* + Q \rightarrow R \qquad [11]$$

$${}^3R^* + Q \rightarrow R$$

where Q represents radical ions, electrolyte ions, adventitious impurities, etc.

The study of magnetic field effects on ECL has been a useful tool in determining the mechanism of the reactions and uncovering the participation of triplets in the ECL process. (12,15) The presence of a magnetic field effect has been taken as evidence of the intermediacy of triplets in the light-producing process, since a magnetic field will affect both TTA (reaction [10]) and quenching of triplets by paramagnetic species (e.g., reaction [12] with Q = $R^{+\cdot}$). Studies of delayed fluorescence in solution in the absence and presence of radical cations (16) have independently demonstrated these effects. Some recent results of this effect are show in Figure 1 for the ECL of a solution containing 0.5 mM tetracene and 0.5 mM TMPD in DMF - 0.1 M TBAP. (17) Since $E_p(R/R^{-\cdot})$ = -1.62 V (tetracene), $E_p(TMPD/TMPD^{+\cdot})$ = +0.22 V, and E_S (tetracene) = 2.53-2.58 eV, the tetracene-TMPD system is clearly energy deficient. The large effect points to a T-route in this ECL process, and additionally, demonstrates that TTA of tetracene triplets is energetic enough to produce excited singlet tetracene. The pronounced effect of radical cation quenching on TTA is illustrated by the delayed fluorescence studies shown in Figure 2. (17) These results, a continuation of those previously reported, (16) were performed by coulometrically generating $TMPD^{+\cdot}$ in a solution containing 1 mM anthracene and (initially 0.6-2 μM) TMPD in CH_2Cl_2-0.1 M TBAP. Since intensity

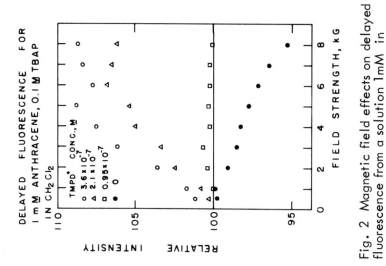

Fig. 2 Magnetic field effects on delayed fluorescence from a solution 1mM in anthracene, 0.1 M in TBAP and different amounts of TMPD$^{+}_{-}$ radical cation.

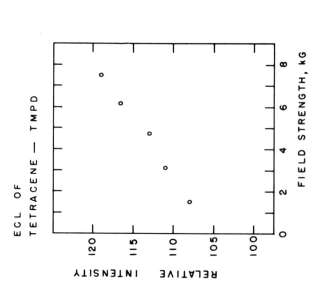

Fig. 1 Magnetic yield effects on ECL from tetracene–TMPD system. The solution contained 0.5 mM tetracene, 0.5 mM TMPD and 0.1 M TBAP in DMF.

increases with increasing magnetic field are always seen in ECL studies of energy deficient systems, triplet quenching by radical ions probably makes an important contribution to the overall reaction scheme of systems following the T-route.

Additional persuasive evidence of the involvement of triplets in energy-deficient ECL processes are the triplet interception and quenching experiments described by Freed and Faulkner (18) and other studies of energy-deficient systems. (11,19) Taken together these investigations show that ECL techniques provide a useful method of producing triplet states without involving photoexcitation; e.g., observation of TTA in DPA and rubrene, not easily observed by spectroscopic methods, is possible in ECL. Moreover, triplet energies can be estimated from peak potential measurements for mixed systems showing ECL proceeding via the T-route.

EFFICIENCY MEASUREMENTS

One of the most important parameters for chemiluminescence reactions is the efficiency of the process, in terms of excited states produced per redox event, photons emitted per coulomb consumed, etc. In measuring the efficiencies of ECL processes, a number of unique problems arise, both in the definition of the type of efficiency measured and in deriving these efficiencies from the experimental electrical and photometric values. Some of these problems have been considered previously, (9,13,20-21) and the notation below follows previous practice as far as possible.

In the measurement of ECL efficiency by transient (double pulse, triple pulse or continuous) methods the light intensity, I (einsteins/sec), and current, i (coulomb/sec), both changing rapidly with time during each pulse, are measured. Depending upon the type of efficiency reported (see below) it may be necessary to convert from total current, i_t, to faradaic current, i_f, by subtracting the current (or coulombs) necessary to charge the electrical double layer (the capacitive current) i_c, and any background current; e.g., due to impurities in solution which do not produce reactants participating in a light-emitting process. The total number of reactive species undergoing the redox reaction, N, must then be extracted from i_f, and is a function of the relative concentrations of reacting species in mixed systems as well as any decomposition of the reactants in side paths (9). In efficiency measurements at the RRDE, since an electrochemical steady state is reached quickly (about 1 second or less), the problem of obtaining N is smaller, since i_c does not contribute and, for most rotation rates and practical RRDE's, the conditions can be established where essentially all of the disk-produced species reacts in the redox process with ring-generated species.

In the measurement of I, aside from the usual problems of absolute luminescence measurements (calibration, geometry corrections etc.) one must correct the total collected intensity, I_t, for photons absorbed in the solution between the electrode and the cell wall and attempt some correction for the number of photons not reflected from the electrode, to yield the intensity at the redox reaction site, I_e.

Several different types of efficiencies can be measured and reported. Since the efficiency itself may be a function of the environment where the redox reaction [8] and emission occurs, and this environment changes with time, t, during a transient experiment (or with rotation rate in an RRDE experiment) in a general sense one can define a differential (or instantaneous) efficiency

$$\phi = IF/i \qquad [13]$$

(where F is the faraday) and an integral (or overall) efficiency

$$\overline{\phi} = F\int_0^t I dt / \int_0^t i dt \qquad [14]$$

where t would represent the pulse length in a transient experiment. Depending upon the extent of correction of I and i for the effects mentioned above, several subscripts can be associated with the ϕ-values (Table I).

Table I. Types of Efficiencies in ECL

Type	i =	I =	ϕ	Comments
Practical	i_t	I_t	ϕ_{prac}	$\dfrac{\text{Total Emitted Light}}{\text{Total Current}}$
Electrical	i_t	I_c	ϕ_{elec}	Correction of I_t for solution losses, reflectivity...
Coulombic	i_f	I_c	ϕ_{coul}	Correction of i for background and capacitive current.
ECL	N	I_c	ϕ_{ecl}	Correction for number of electrogenerated species which are lost.

ϕ_{prac} may be of interest in ECL devices, but would be critically dependent upon the actual geometry, cell path length, etc. ϕ_{ecl} comes closest to being the function of fundamental chemical interest, and can lead to a determination of the yield of excited states (ϕ_s, the excited singlet yield or ϕ_t, the excited triplet yield) if the fluorescence efficiency, ϕ_f, is known, quenching effects can be taken into account, the TTA efficiency is determined, etc. In the simplest case of an S-route reaction and no quenching by radical ions (5,9) $\phi_{ecl} = \phi_s \phi_f$; obtaining ϕ_t for T-route reactions is obviously more difficult. (9)

A number of ECL efficiency measurements have been reported (Table II). The large variation between some of these values testifies to the numerous difficulties in the measurement but also probably reflects differences in experimental procedures and solution conditions in the vicinity of the redox couple (R'^{+} and R^{-}) at the time of the measurement. In addition to the problems mentioned previously, other effects may cause variations. In transient methods the solution conditions at the first pulse may be different than those after a number of repetitive pulses because of build-up of radical ions and decomposition products in the vicinity of the electrode; however, it is possible that small amounts of quencher (e.g., oxygen) initially present are removed by radical ions or excited states during the early pulses. In transient measurements the emitted light must pass through a layer of radical ion and may cause some decrease in I_c. It is also possible that excited states produced very near the electrode surface are quenched by direct interaction with the electrode; such effects have been noted in fluorescence measurements. [28]

RECENT EFFICIENCY MEASUREMENTS

We will outline here some recent investigations we have carried out which have some bearing on ECL efficiency measurements. The experimental methods and details of results of this work will be published elsewhere. These results, however, provide evidence for solvent and supporting electrolyte effects in ECL and give preliminary results of the first direct actinometric efficiency measurements in ECL.

The TPTA-DMA System

The large efficiency, ϕ_{cl}, found by Weller and Zachariasse [27] for the reaction of dimethylanthracene (DMA) radical anion in THF solution with solid tri-p-tolylaminium perchlorate (TPTA$^{+}\cdot$ClO$_4^{-}$) (see Table II) suggested ECL measurements of this system. Results are summarized in Table III and Figure 3. Briefly it was found that ϕ_{ecl} of the process (determined at the RRDE as described in Table IV) for 1.0 mM each DMA and TPTA in THF containing 0.2 $\underline{M}$ TBAP was 0.07×10^{-2}. Moreover, a strong magnetic field effect was observed for the ECL, suggesting reaction via the T-route. However, as shown in Fig. 3, the ECL intensity depends strongly on the TBAP concentration so that the value obtained for 0.1 mM each DMA and TPTA in THF with 0.01 $\underline{M}$ TBAP (the lowest electrolyte concentration for which meaningful electrochemical measurements could be made) was 10 times that for solutions with 1.0 mM reactants containing 0.2 $\underline{M}$ TBAP. Based on these results the extrapolated value ϕ_{ecl} ([$\overline{TBAP}$]$\rightarrow$0) is 7×10^{-2}, in excellent agreement with the chemiluminescence results. The cause of this strong dependence of electrolyte concentration is not clear. In this low dielectric constant medium ion pairing is certainly of importance.

Table II. Reported ECL Efficiencies

Compound (concentration)	Solvent[c]	Method	Efficiency × 10²	Ref.
MADI (1mM)	DMF	Trans.-2 elec. 60 hz. sq. wave IS-PM (s.s.)	ϕ_{prac} = 0.1-0.2	Zweig, et al. (21)
Rubrene (?)	BZN	Trans. (?)	$\overline{\phi}_{coul}$(?) ⩽1	Watne (22) quoted in (5,6)
DPA (2mM)	DMF	RRDE-Direct PM	ϕ_{ecl} = 1.5[a] (0.08)	Maloy & Bard (13)
DPA-TMPD (1mM-1mM)	DMF	"	ϕ_{ecl} = 0.1 (0.006)	"
Rubrene (1mM)	DMF	"	ϕ_{ecl} = 0.1 (0.007)	"
Pyrene-TMPD	DMF	"	ϕ_{ecl} = 0.1 (0.007)	"
Rubrene (2mM)	BZN	Trans.-2 elec. 3-25 hz. sq. wave (s.s.)	ϕ_{prac} = 8.7	Schwartz, et al. (23)
DPA (0.27-2.7mM)	DMF	Trans-3 elec. 1 sec. pulses (1st pulse) IS-PM F-Plot	$\overline{\phi}_{ecl}$ = 0.3-0.8[b] (0.05-0.15) [$\overline{\phi}_{ecl}$(extrap.) =1.4]	Bezman & Faulkner (24)
Rubrene (0.4-1mM)	DMF	"	$\overline{\phi}_{ecl}$ = 0.3 (0.05)	" (25)
Rubrene (0.6-1.2mM)	BZN	"	$\overline{\phi}_{ecl}$ = 0.5 (0.1)	"
FA-10MP (0.2-1.5mM)	DMF	"	$\overline{\phi}_{ecl}$ = 0.03-0.05 (0.006-0.01)	" (26)
DMA⁻-TPTA⁺ (0.3mM) (solid)	THF	Direct Actinometry	$\overline{\phi}_{cl}$ = 7.5	Weller & Zachariasse (27)

Footnotes to Table II.

[a] Re-evaluation of the luminescence calibrating procedure yields values about 13 times those originally reported (original values in parenthesis); consideration of electrode reflectivity adds a factor of 1.3-1.5.

[b] Re-evaluation of the luminescence calibrating procedure yields values 5.4 times those originally reported (original values in parenthesis) (L. R. Faulkner, private communication).

[c] In all cases except the CL measurement in THF the supporting electrolyte was 0.1 M TBAP.

Abbreviations: Compounds: MADI, N-methyl-1,3-p-anisyl-4,7 isoindole; DPA, 9,10-diphenylanthracene; TMPD, N,N,N',N'-tetramethyl-p-phenylene diamine; FA, fluoranthene; DMA-dimethylanthracene; TPTA-tri-p-tolylamine. Solvents: DMF, N,N-dimethylformamide; THF, tetrahydrofuran; BZN, benzonitrile; TBAP, tetra-n-butylammoniam perchlorate. Other: RRDE, rotating ring-disk electrode, Trans., transient method; PM, photomultiplier; PD, silicon photodiode, F-plot, Feldberg plot; s.s., steady state; IS - integrating sphere.

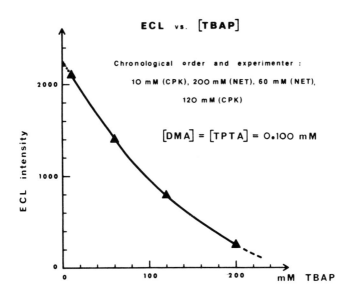

Fig. 3 The effect of supporting electrolyte (TBAP) concentration on ECL intensity in the DMA/TPTA system in THF.

Table III. Electrochemical and Spectroscopic Data for DMA-TPTA System

Compound	Solvent	Electrolyte Conc., M	$E°$(V vs. S.C.E.)[c]
DMA	DMF[a]	0.100	(R/R^-) -1.95
	THF[a]	0.200	-1.95
	THF[b]	0.0100	-2.00
TPTA	DMF[a]	0.100	$(R'/R'^{+\cdot})$ +0.84
	THF[a]	0.200	+0.98
	THF[b]	0.0100	+1.05

E_s (DMA) = 3.09 eV E_T (DMA) = 1.80 eV

$-\Delta H°$ (0.2 M TBAP-THF)[a] = 2.83 eV

$-\Delta H°$ (0.01 M TBAP-THF)[b] = 2.95 eV

$-\Delta H°$ (0.1 M TBAP-DMF) = 2.69 eV

Concentration of DMA and TPTA [a] 1.00 mM, [b] 0.100 mM

[c] Obtained from cyclic voltammetry measurements by averaging E_p (cathodic) and E_p (anodic)

The $E°$ measurements suggest that ion pairing of the reactant TPTA$^{+\cdot}$ and DMA$^{-\cdot}$ with supporting electrolyte ions occurs (although these measurements may be slightly perturbed by uncompensated iR-effects). The value for $-\Delta H°$ for the ion paired radical ions would be less than that for the free ions, and perhaps in this system with marginal energy sufficiency ion-pairing tips the reaction from S-route to T-route. The possibility that ion-paired radical ions have additional paths for energy dissipation during the redox step leading to a lower yield of excited states also exists.

Efficiency Measurements Based on the DPA-TH System

We have recently discovered that the DPA-thianthrene (TH) system, especially in a solvent of a mixture acetonitrile (AN)-benzene (BZ)-toluene (TOL) yields very intense ECL when the reactant species are DPA$^{-\cdot}$ and TH$^{+\cdot}$. The appropriate peak potentials for the electrode reactions are E_p (DPA/DPA$\bar{\cdot}$) = -1.89 V, E_p (DPA/DPA$\overset{+}{\cdot}$) = 1.35 V, and E_p (TH/TH$\overset{+}{\cdot}$) = 1.25 V, so that the electron transfer reaction producing excited states involves reaction of DPA$\bar{\cdot}$ with both TH$\overset{+}{\cdot}$ and DPA$\overset{+}{\cdot}$. The high intensity of this system allowed direct actinometric measurements of ECL at an RRDE to be carried out. The apparatus employed is shown in Fig. 4. Briefly, ECL was carried out for a total electrolysis time of 16 minutes with TH$\overset{+}{\cdot}$ and DPA$\overset{+}{\cdot}$ generated at the disk and DPA$\bar{\cdot}$ generated at the ring for 2-3 minutes, followed by a rest period of 30 seconds. Conventional actinometry was carried out (13) and corrections were made for photons lost through the monitoring window at the bottom of the cell (a 7% correction) as well as emission, primarily in the green spectral region, escaping through the actinometry solution (a 30% correction). Additional corrections which are needed in obtaining $\int I_c dt$ involve the number of photons absorbed in the solution without re-emission and the number not reflected back from the electrode. Correction for these effects introduce the largest uncertainty in the luminescence measurement. The absorbance losses depend upon the ϕ_f-value for the emitter, the presence or build up of absorbers in the solution and the path length between cell wall and electrode. The reflectivity of an electrode in solution has been considered by several authors (see e.g., (29,30)) and depends upon electrode material, wavelength, (changing from 0.2 in UV-region to 0.7 in visible for platinum electrodes in aqueous solutions (30)) and, at least for RRDE's, on losses from the Teflon insulating portions. In the absence of definitive data for the conditions of our experiments, we have corrected the measured integrated number of photons by a factor of 3 (i.e., loss of 1/2 of the emitted photons by solution absorption and loss of 1/2 of the photons emitted in the direction of the electrode). The disk current integrated for the electrolysis time yields directly, to a very good approximation, the number of redox events. The efficiency for the process under these conditions is listed in Table IV.

An alternate approach to efficiency determinations involved direct measurement of luminescence with a calibrated photodiode (PD), correcting for the relative area of the photodiode to the total emission area. The PD employed was a UDT(UV)-500 PD (United Detector Technology, Santa Monica, Cal.) with a 1.00 cm^2 active area and a flat response (in terms of incident power) over a spectral region of 200 nm to 1100 nm. The manufacturer's nominal sensitivity figure at specified operating conditions is 5 μV/pwatt. The detector was calibrated using a He-Ne laser and 0.99% neutral density filter, with the laser in turn calibrated with an Eppley thermopile and a Spectraphysics Model 401C power meter to yield a response of 2.5 μV/pwatt at the specified operating conditions. From this measurement, the detector could be calibrated in terms of response per photon flux in all regions of the spectrum. Direct measurements with the RRDE using this procedure for the DPA-TH systems and others are also shown in Table IV.

Table IV. ECL Efficiency Measurements

Compounds (conc.)	Solvent[a]	Method	Efficiency x10^2
DPA (1.0mM)	DMF	RRDE-Calibrated PD	$\phi_{ecl} = 3$
DPA (2.20mM)	ACN-BZ	"	$\phi_{ecl} = 10$
DPA (7.77 mM)	ACN-BZ-TOL	"	$\phi_{ecl} = 8$
Rubrene (2.00 mM)	BNZ	Trans.-3 elec. 60 hz (s.s.)-calib. PD	$\overline{\phi}_{coul} = 2$
DPA-TH (7.77-11.11mM)	ACN-BZ-TOL	RRDE-Direct actinometry (expl. duration 16 min.)	$\overline{\phi}_{ecl} = 8$
		RRDE-Calib. PD	ϕ_{ecl}(max. value) = 14
		Trans-3 elec. 60 hz (s.s.)-calib. PD	$\overline{\phi}_{coul}$(1-2 min.) = 7
			$\overline{\phi}_{coul}$(2-16 hr.) = 6
			$\overline{\phi}_{coul}$(max. value) = 16-20
		" 300 hz	$\overline{\phi}_{coul}$(2 hr.) = 3
		Trans-3 elec., 60 hz Direct actinometry for 75 min.	$\overline{\phi}_{coul} = 5$
DMA-TPTA (1.00-1.00mM)	THF[b]	RRDE-calib. PD	$\phi_{ecl} = 0.07$
			ϕ_{ecl}(extrap. [TBAP]$\to$0) = 7

Footnotes for Table IV

Abbreviations are the same as in Table II.

[a] Supporting electrolyte 0.1 M TBAP, except as noted ACN-BZ, 1:1 acetonitrile:benzene; ACN-BZ-TOL, acetonitrile: benzene: toluene: 0.5:0.33:0.17.

[b] [TBAP] = 0.2 M

[c] Parentheses indicate duration of experiment; (max. value) denotes largest value obtained during an experiment.

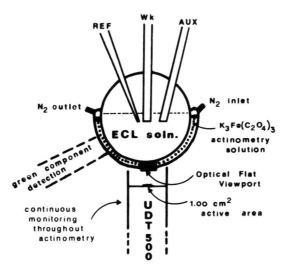

Fig. 4 Apparatus used to determine ECL quantum efficiency by actinometry.

Additional measurements were made of the TH-DPA system in the transient mode. Reported values (Table IV) represent measurements at "steady state", i.e., for pulses after a number of square wave cycles have been passed. In the actinometric experiment the corrected integrated photon flux was obtained in the apparatus in Figure 4; the total coulombs passed were determined by integrating the i-t curve for a single cathodic pulse, correcting for the coulombs needed to charge the double layer and multiplying by the total number of cathodic pulses in the experimental interval. In transient measurements with the calibrated PD, the area under the I-t curves during an anodic and cathodic pulse pair were used to determine the integrated luminescence. This value was divided by the area of a cathodic i-t pulse, corrected for charging, to obtain ϕ_{coul}. Finally, the DPA-TH system, which appears to have the highest ϕ_{ecl} of any reported system, shows no magnetic field effect. The reasonable agreement among efficiencies measured by different techniques provides some confidence in these reported values.

The most recent results of ECL efficiencies show that electron-transfer reactions of the type reported here, especially for systems which proceed by the S-route, can effectively produce excited states and ultimately photons. Indeed, the recent triplet interception experiment of Freed and Faulkner [18] demonstrating near unity efficiency in production of triplet fluoranthene in the $FA^{\bar{\cdot}}$-10-phenylphenothiazine radical cation redox reaction is further evidence of effective excited state production in a T-route system. The effect of solvent, electrolyte, and reactant structure on the efficiency of these processes may provide new insights into the paths by which excited states are produced and react.

ACKNOWLEDGMENT

The support of this research by the U.S. Army Research Office-Durham and an Electrochemical Society Joseph W. Richards Fellowship to one of us (CPK) is gratefully acknowledged.

REFERENCES

1. T. Kuwana in "Electroanalytical Chemistry", A. J. Bard, Ed., Vol. 1, Marcel Dekker, N. Y., 1966, Chap. 3

2. A. J. Bard, K. S. V. Santhanam, S. A. Cruser and L. R. Faulkner in "Fluorescence", G. G. Guilbault, Ed., Marcel Dekker, N. Y., 1967, Chap. 14.

3. A. Zweig, Advances in Photochemistry, 6, 425 (1968).

4. E. A. Chandross, Trans. N. Y. Acad. Sci. Ser. 2, 31, 571 (1969).

5. D. M. Hercules, Accounts Chem. Res., 2, 301 (1969).

6. D. M. Hercules in "Physical Methods of Organic Chemistry", 4th ed., Part II, A. Weissberger and B. Rossiter, Eds., Academic Press, N. Y., 1971.

7. A. J. Bard and L. R. Faulkner in "Creation and Detection of the Excited State", W. R. Ware, Ed., Vol. 3, Marcel Dekker, N. Y., (in preparation).

8. S. W. Feldberg, J. Amer. Chem. Soc., 88, 390 (1966); J. Phys. Chem., 70, 3928 (1966).

9. R. Bezman and L. R. Faulkner, J. Amer. Chem. Soc., 94, 3699 (1972).

10. G. J. Hoytink, Discuss. Faraday Soc., 45, 14 (1968).

11. A. Weller and K. Zachariasse, J. Chem. Phys., 46, 4984 (1967).

12. L. R. Faulkner, H. Tachikawa and A. J. Bard, J. Amer. Chem. Soc., 94, 691 (1972).

13. J. T. Maloy and A. J. Bard, J. Amer. Chem. Soc., 93, 5968 (1971).

14. J. T. Maloy, K. B. Prater and A. J. Bard, J. Amer. Chem. Soc., 93, 5959 (1971).

15. L. R. Faulkner and A. J. Bard, J. Amer. Chem. Soc., 91, 209 (1969).

16. L. R. Faulkner and A. J. Bard, J. Amer. Chem. Soc., 91, 6495, 6497 (1969).

17. H. Tachikawa and A. J. Bard, unpublished results.

18. D. J. Freed and L. R. Faulkner, J. Amer. Chem. Soc., 93, 2097, 3565 (1971); 94, 6324 (1972).

19. J. Chang, D. M. Hercules and D. K. Roe, Electrochim. Acta, 13, 1197 (1968).

20. R. Bezman and L. R. Faulkner, Anal. Chem., 43, 1749 (1971).

21. A. Zweig, A. K. Hoffman, D. L. Maricle and A. H. Maurer, J. Amer. Chem. Soc., 90, 261 (1968).

22. B. M. Watne, Sc. B. Thesis, Massachusetts Institute of Technology, 1967.

23. P. M. Schwartz, R. A. Blakeley and B. B. Robinson, J. Phys. Chem., 76, 1868 (1972).

24. R. Bezman and L. R. Faulkner, J. Amer. Chem. Soc., 94, 6317 (1972).

25. Ibid., p. 6324.

26. Ibid., p. 6331.

27. A. Weller and K. Zachariasse, Chem. Phys. Letters, 10, 424 (1971).

28. H. Kuhn, J. Chem. Phys., 53, 101 (1970).

29. J. Horkans, J. Electrochem. Soc., 118, 38C (1971).

30. J. D. E. McIntyre, Paper No. 94, Electrochemical Society Meeting, Houston, Texas, May, 1972.

ZACHARIASSE: I have a question on the possibility of an S or a T route on a system like TPTA and the amine. It is fairly easy to make a decision as to where we have an S or a T route from a standard. In a T route you will see an excimer emission from a molecule like 9,10-dimethylanthracene and in an S route you will not. So there is an easy experimental means of determining an S or a T route. What have you seen about this?

BARD: Well in the high concentration of electrolyte I don't believe we see any exciplex (a hetero excimer).

ZACHARIASSE: Even in 9,10-dimethylanthracene?

BARD: No we don't. I think the magnetic field effect shows this also.

WELLER: What was the solvent you studied the salt concentration effect in?

BARD: DMF.

WELLER: This points in the same direction as I always suspected namely, that increasing salt concentration apparently increases the effective dielectric constant. That's why all the potentials, polarographically measured, for instance in THF with salt present, are about the same order of magnitude as those measured in acetonitrile or dimethylformamide. The point is you cannot measure it without salt present. If you could do that then you would get different results.

BARD: Well I think there are two effects. There is perhaps a dielectric constant effect and an ion-pairing effect.

WELLER: I do not distinguish between the two.

BARD: Yes, but there are differences. The data I showed, that is, in the measurement of DMF with 0.2 M TBAP and THF with 0.2 M TBAP, the ΔG and ΔH values differ by 0.14 volts. Now if you cut this down to 10mM, and that's not now too far different from your conditions, your conditions are lower yet, then the difference goes up by another 60 mV or so. So I think there is a difference, that's right.

CHEMILUMINESCENCE OF DIAZAQUINONES AND RELATED COMPOUNDS

Karl-Dietrich Gundermann

Organisch-Chemisches Institut
der Technischen Universität Clausthal

D 3392 Clausthal-Zellerfeld, Leibnizstr. 6

In the first paper concerning luminol chemiluminescence published 1928 by H.O.ALBRECHT [1] a diazaquinone APD (1.4 dihydro-5-amino-phthalazine-1.4-dione) has been proposed as key intermediate:

APD was suggested to be hydrolyzed to give (in a dark reaction) 3-amino phthalate and diimine which was to react with another part of APD to yield electronically excited luminol, and nitrogen. Whereas this mechanism has been proved to be not valid in respect of very important parts, above all in respect of the emitting species, by E.H.WHITE and his coworkers[2] there remains the question whether diazaquinones are, nevertheless, intermediates in luminol type chemiluminescence. This problem could be investigated directly when the first diazaquinones had been synthesized by CLEMENT[3] and KEALY[4] in 1960 and 1962, respectively.

As KEALY's diazaquinones can be regarded as derivatives of non-fluorescent dicarboxylic acids (e.g. difluoromaleic or phthalic acid) no direct chemiluminescence is to be expected in their oxidation. E.H.WHITE, ROSWELL, and ZAFIRIOU [5] discussed phthalazinedione as a possible intermediate in the "anomalous" chemiluminescence reaction of phthalic hydrazide in aprotic medium but as diimine sources were found to have no effect on this chemiluminescence the role of phthalazinedione was regarded as very doubtful.

Before the first chemiluminescent diazaquinones were described in 1967 and 1968 RAUHUT, SEMSEL, and B.G. ROBERTS[6] postulated luminol diazaquinone as key product from the kinetics of the chemiluminescence reaction of luminol oxidized with peroxidisulfate/ hydrogen peroxide.

M.M. RAUHUT, A.M. SEMSEL, B.G. ROBERTS 1966

OMOTE, MIYAKE, and SUGIYAMA[7] were the first to isolate luminol diazaquinone - in the form of its Diels-Alder-adduct to cyclopentadiene - from solutions which exhibit luminol chemiluminescence normally i.e. containing luminol, potassium ferricyanide, and aqueous alkali; however, there is no chemiluminescence in the presence of cyclopentadiene.

E.H. WHITE and coworkers[8] obtained benzo(b)-phthalazine dione (BPD) by treatment of naphthalene-2.3-dicarboxylic hydrazide sodium salt with chlorine. The hydrazide is only weakly chemiluminescent as the fluorescence maximum of the corresponding acid is ca. 360 nm; the quantum yield of the hydrazide is about 5 % that of luminol[9]. On treatment of the relatively stable BPD with hydrogen peroxide and alkali chemiluminescence was observed the emission spectrum of

E.H. WHITE and COWORKERS 1968

which matches that of the corresponding hydrazide. A quantitative relation between the quantum yields of the diazaquinone and the hydrazide was difficult to obtain due to the very low solubility of the diazaquinone in aqueous media; moreover the diazaquinones are very liable to undergo hydrolysis and other "dark" reactions (see KEALY[4]). Qualitative observations, however, pointed on a possible 1:1 ratio.

We concentrated our investigations on diazaquinones derived from hydrazides the chemiluminescence of which is rather efficient and the emission spectra of which are in the blue range of the spectrum in respect of the important role of luminol itself in hydrazide chemiluminescence.

GUNDERMANN, FIEGE, KLOCKENBRING   1968, 1970

On treatment of luminol monosodium salt with tert. butylhypochlorite in dimethyl ether at -50° (a modified KEALY procedure) rather impure luminol diazaquinone could be obtained chemiluminescing on oxidation with aqueous alkaline hydrogen peroxide. In this context an observation of KAUTSKY and KAISER[11] should be mentioned who obtained violet solutions from luminol and aqueous calcium hypochlorite; these violet solutions chemiluminesce with excess alkali without additional hydrogen peroxide. In a control experiment we could confirm this finding of KAUTSKY and KAISER but the violet product surely is not a diazaquinone but perhaps a mauvein-like product. A diazaquinone is not obtainable under these reaction conditions. The weak chemiluminescence exhibited from the violet solutions is very probably due to traces of unreacted luminol oxidized by air.

4-Diethylamino phthalic hydrazide the chemiluminescence of which is even a bit more efficient than that of luminol could be transformed into the corresponding diazaquinone obtained in a rather pure crystalline form. This is certainly due to the higher oxidation stability of the tertiary amino group of these compounds.

An especially well defined rather stable diazaquinone is 1.4 dihydro(2.3g)naphtho-phthalazine-1.4-
dione(NPD) derived from anthracene 2.3 dicarboxylic
hydrazide (ADH). The hydrazide chemiluminesces in
hemin catalyzed oxidation with aqueous alkaline
hydrogen peroxide at 430 nm - very near the emission
of luminol; the quantum yield is about 25 % that of
the latter. NPD can be synthesized from ADH monosodium salt and t-butyl hypochlorite[10] yet better
results were obtained by treatment of ADH with the
hypochlorite without additional solvents, according
to a method developed by the WHITE group[12].

When a solution of NPD in acetone is rapidly mixed
with an aqueous-alkaline hydrogen peroxide solution
chemiluminescence occurs the emission spectrum of
which matches that of the hemin catalyzed oxidation
chemiluminescence of ADH. The light intensity shows
a first-order dependence from the hydrogen peroxide
concentration. The dependence of the light intensity
from alkali concentration is such that an optimal
concentration of $O_2H^{(-)}$ ions appears decisive.
However, one has to be aware of the rather great
extent of "dark" side reactions of a diazaquinone in
alkaline media. On the other hand there should be no
significant influence of the alkali concentration on
the fluorescence of the emitting species anthracene
2.3 dicarboxylate (ADS) - in contrast to the well
known situation concerning luminol chemiluminescence
(LEE and SELIGER[13], GORSUCH and HERCULES[14]). That
ADS is the emitting species as well as the main product
in NPD and ADH chemiluminescence in these aqueous-
acetonic systems was proved spectroscopically and by
isolation of ADS from the spent reaction mixtures[10].

The quantum yields of NPD were about four times those of the hemin catalyzed ADH chemiluminescence; it has to be mentioned, however, that acetone considerably quenches light emission in the latter case.

The duration of light emission of NPD chemiluminescence is very short compared with that of ADH chemiluminescence, e.g. 0.3 sec and 650 sec, respectively; therefore, the maximal intensities in NPD chemiluminescence are up to 9000 times that of ADH chemiluminescence.

To avoid dark side reactions as far as possible other solvents were searched appropriate for measuring NPD chemiluminescence at room temperature. Moreover stopped flow techniques were applied to ensure instananeous mixing of the reactants[15].

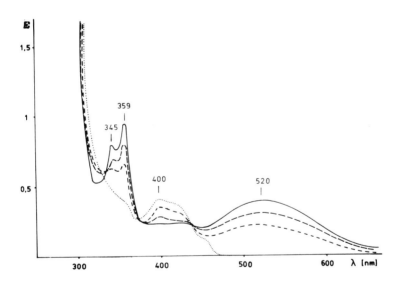

Fig. 1 : Extinction spectrum of Naphtho-phthal-azine-1.4-dione (NPD) in dimethyl phthalate (——— : freshly prepared solution; —————, ······ : decomposition products being formed within ca. 2 hours)

NPD stability was measured by means of its absorption spectrum (fig. 1), in the dry state, and in a series of solvents. Stored in a vacuum of 10^{-4} Torr about 5 % of the diazaquinone decomposed within the first 3 hours, at room temperature; after that time NPD decomposition continues only very slowly. Dimethylphthalate (DMP) proved to be the most appropriate solvent for quantitative chemiluminescence investigations, distinctly surpassing DMSO, DMF, or acetone, for example. Commercially available DMP had to be purified for this purpose the best method being repeated distillation of dried DMP over NPD. In such a

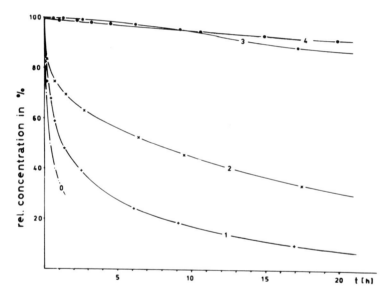

Fig. 2: Stability of NPD in Dimethyl phthalate (DMP)
(0: commercially available DMP, without purification
1: DMP dried on molecular sieve statically;
2: DMP dried on molecular sieve dynamically;
3: DMP treated like 2) then distilled over NPD;
4: same as 3), redistilled over NPD)

pretreated DMP NPD decomposition amounts to about 2 %
only, at room temperature under nitrogen atmosphere.
A disadvantage of DMP consists in the low solubility
of ADH, which is important for the comparative
measurements. The solubility of NPD is about 10^{-4} M,
in DMP (fig. 2).

The chemiluminescence of NPD in dimethylphthalate
was achieved by mixing NPD solutions with solutions of
anhydrous hydrogen peroxide and diethylamine in a
DURRUM-GIBSON stopped flow apparatus. Some characteristic decay curves obtained with constant NPD and
H_2O_2 - but varied base concentrations are shown in
fig. 3. As is seen maximal intensity is reached at the
mixing point whereas the oscillograms in fig. 4 display a different picture: in ADH chemiluminescence
(which only occurs in the presence of hemin) the
intensity maximum is reached only after a certain
"induction" period.

As to the dependence of NPD chemiluminescence from
NPD-, base- and hydrogen peroxide concentration a first
order dependence was stated for diethylamine concentrat-
ions $\leq$ 1×10^{-2}M and H_2O_2 concentrations up to 3×10^{-2}M,
with NPD concentrations of about 1×10^{-4}M. Within the
concentration range investigated also first order
dependence of chemiluminescence intensity from NPD
concentration was observed. Hemin has no influence
on NPD chemiluminescence up to concentrations of
about 10^{-6} M; above this value a decreasing effect
is exerted due to fluorescence inhibition on the
product ADS. As in aqueous acetone solutions the
chemiluminescence spectrum of NPD in DMP matches
the fluorescence of ADS and the chemiluminescence
of ADH (fig. 5).

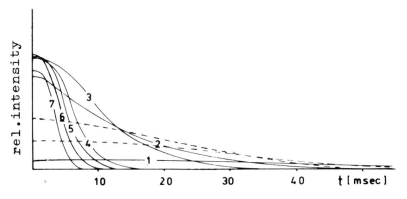

Fig. 3 : Oscillograms of NPD chemiluminescence as function of base (diethylamine)concentration.
(NPD: 0.96×10^{-4}M, H_2O_2: 1.6×10^{-2}M; diethylamine: 1 .... 7 = 0.482.... 482×10^{-3}M, e.g. 3): 9.64×10^{-3}M)

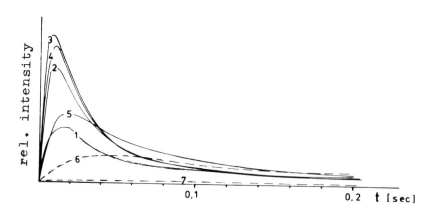

Fig. 4 : Oscillograms of Anthracene-2.3-dicarboxylic acid hydrazide (ADH) chemiluminescence as function of diethylamine concentration.
(ADH: 0.6×10^{-4} M; H_2O_2: $3,18 \times 10^{-2}$M, hemin 2.92×10^{-5} M; diethylamine: 1...7= 63.20... 0.11×10^{-2}M)

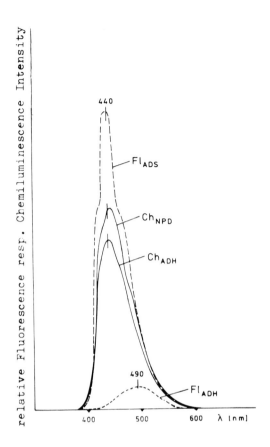

Fig. 5 : Chemiluminescence spectra of NPD
and ADH and Fluorescence spectrum
of anthracene 2.3 dicarboxylate (ADS)
in dimethyl phthalate systems.

Fl: Fluorescence spectrum of ADS; Ch_{NPD}, Ch_{ADH}: chemiluminescence spectra of NPD and ADH, respectively. The spectra are not normalized. Fl_{ADH} : fluorescence spectrum of ADH, being very weak in dimethyl phthalate.

Analogous relations between chemiluminescence intensity and reactant concentrations were found for ADH chemiluminescence in DMP. For comparative considerations, however, the role of hemin concentration has to be shortly discussed.

That the hemin concentrations necessary in DMP are about an order of magnitude higher than those used in aqueous systems may be due to the low polarity of DMP so that the concentration of hemin cation should be far lower than in aqueous solvents.

The base concentrations necessary for optimal ADH chemiluminescence are distinctly higher than those required for NPD chemiluminescence (figs. 6,7). This is probably a key relation concerning the role of the formation of hydrazide mono- or dianion: although it is not yet known whether the former or the latter is required for ADH chemiluminescence in DMP one can assume that a relatively high alkalinity is necessary for the hydrazide anion formation (see below).

<u>Optimal conditions provided concerning base and hydrogen peroxide concentrations the light intensity of NPD is 12 times, quantum yield about 4 times that of ADH in the dimethyl phthalate system.</u>

Summarizing these experimental facts one can state:
1) the chemiluminescence reactions of NPD and the corresponding hydrazide ADH yield the same emitting species anthracene 2.3 dicarboxylate, in aqueous acetone as well as in dimethyl phthalate.
2) The dependences of light intensity on NPD and ADH (respectively) concentrations, base-, and hydrogen peroxide concentrations are analogous (first order within certain ranges).

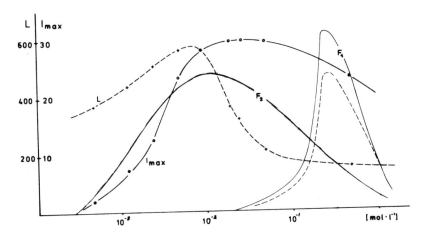

Fig. 6 : NPD chemiluminescence as function of base concentration
(—— maximal Intensity I_{max}, ---- L: quantum yield; reaction conditions as in fig.3; F_1: system NPD/acetone/water/air; F_2: same as F_1, instead of air H_2O_2

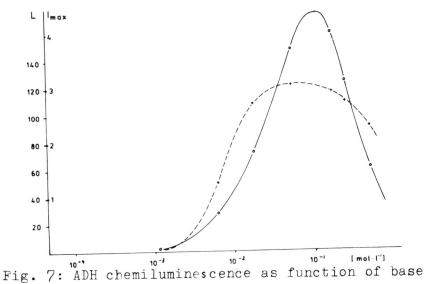

Fig. 7: ADH chemiluminescence as function of base concentration.
(—— maximal Intensity I_{max}; ---- :quantum yield; reaction conditions as in fig. 4)

3) The quantum yield of NPD chemiluminescence is about half an order of magnitude higher than that of ADH chemiluminescence.

All these facts are in accordance with those hydrazide chemiluminescence mechanisms which include a diazaquinone as intermediate. Some of the possible mechanisms are summarized in fig. 8 containing mainly the ideas of WHITE and ROSWELL[2].

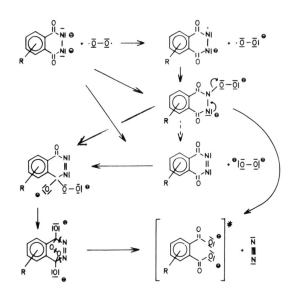

Fig. 8 : Possible pathways of cyclic hydrazide chemiluminescence mechanisms. The reaction of O_2 with hydrazide anion radical has been ommitted; it should yield a peroxy radical anion

Which form oxygen is attacking the hydrazide or the diazaquinone is evidently important. Molecular oxygen does not appear to react readily with luminol monoanion as is seen from the very weak luminol reaction in aqueous alkaline solution containing no catalyst: there must be conditions favoring radical formation as GORSUCH and HERCULES[14]) recently pointed out. In this case one should expect that not a "free" diazaquinone but hydrazide radical anion will recombine with oxygen to form a peroxide radical anion directly, which then decomposes, or, if oxygen radical anion $.O_2^-$ is present the corresponding peroxide dianion is formed. In DMSO/t-butylate/oxygen solutions where no catalyst is required oxygen can be present as $.O_2^{(-)}$ as MATHESON and LEE showed recently[16]); a diazaquinone intermediate would mean here the hydrazide first undergoing a two-electron oxidation followed by a one-electron reduction to give the radical anion, which seems unreasonable. It could be discussed that the two-electron oxidation of the hydrazide to form the diazaquinone is accompanied by a two electron-reduction of the oxygen producing $O_2^{2(-)}$.

As hydrogen peroxide was found to be by far the most efficient oxidant in NPD and in other diazaquinone chemiluminescence we think that in fact a diazaquinone if formed at all in hydrazide chemiluminescence will form the cyclic peroxide by a nucleophilic attack of $O_2^{(-)}$ or - more probably - $O_2H^{(-)}$ and subsequent ring closure on the diazaquinone carbonyl groups. Thus the question whether diazaquinones are in the main stream of cyclic hydrazide chemiluminescence, if they exhibit a special type of chemiluminescence mechanism very closely related to hydrazide chemiluminescence or if these two types of chemiluminescence are occurring simultaneously in different extent dependent on the reaction conditions cannot yet be fully answered; considerably more work will have to be done especially concerning the reacting oxidant.

Chemiluminescence of azodicarboxylates

That a cyclic peroxide suggested by RAUHUT and coworkers[6] and formulated in fig. 8 is an intermediate of hydrazide and diazaquinone chemiluminescence has a strong evidence by the experimental finding of the WHITE group[17] that on treatment of luminol with $^{18}O_2$ both carboxyl groups of the product aminophthalate contain one labelled oxygen atom - which also seems to exclude a dioxetane derivative (see [2]).
Δ1 -4.5 dioxa-1.2-diazine-3.6 dione (DDD) can be regarded as a model for the RAUHUT peroxide. DDD should be formed from hydrogen peroxide and reactive cis-azodicarboxylates:

$$RO_2C-N=N-CO_2R + H_2O_2 \xrightarrow{?} \underset{N=N}{O=C\overset{O-O}{\diagup\diagdown}C=O} \quad DDD$$

and its cleavage into two molecules of CO_2 and one N_2 molecule can be expected to be highly exergonic, leading to rather efficient chemiluminescence in the presence of appropriate fluorescers. As is well known dehydrogenation of azodicarboylates gives the trans isomers which cannot yield DDD for steric reasons. Photochemical trans-cis-isomerization was possible with certain aliphatic azodicarboxylates[18] which did not chemiluminescence on treatment with hydrogen peroxide in the presence of a fluorescer. We could not yet photoisomerize appropriate bis-aryl azodicarboxylates due to decomposition of the aryl groups on irradiation. However, some negatively substituted trans-

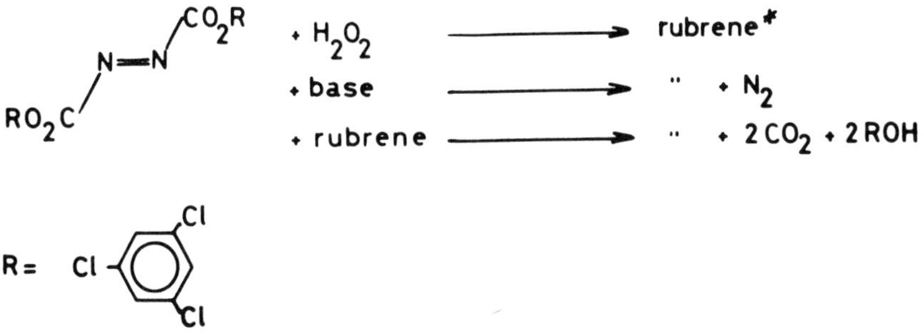

Fig. 9 : Chemiluminescence of trans-azodicarboxylates

diaryl azo dicarboxylates were found to be chemiluminescent. The most efficient reaction observed so far is that of trans bis(2.4.6-trichlorophenyl)azodicarboxylate with hydrogen peroxide/tert. butylate in ether, rubrene acting as fluorescer (fig. 9). The rather bright chemiluminescence is some msec in homogeneous phase, its duration being considerably longer when the azodicarboxylate is added in crystalline form. The emission spectrum matches that of rubrene fluorescence. Products are 2.4.6-trichlorophenol, CO_2, and N_2. Quantum yields could not yet be measured[19]. Azodicarboxyl acid monoperoxide is an intermediate perhaps as is suggested by the requirement of a base.

Sensitized phthaloylperoxide chemiluminescence

TOTTER and PHILBROOK[20] have proposed a luminol chemiluminescence mechanism (fig. 1o) involving a redox reaction of amino phthaloyl peroxide with hydrogen peroxide. We observed sensitized chemiluminescence on heating phthaloyl peroxide at about 7o°

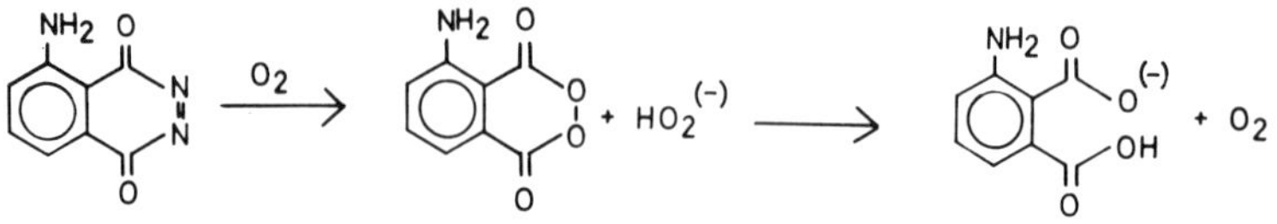

Fig. 10 : Luminol chemiluminescence mechanism of TOTTER and PHILBROOK[20]).

in dimethyl phthalate in the presence of 9.10 diphenyl anthracene[21] (fig. 11). The chemiluminescence quantum yield is about 7 % that of luminol the emission spectrum matching that of DPA. Phenylated dimethyl phthalate derivatives were isolated indicating a homolytic cleavage of phthaloyl peroxide; it is not yet known which of the products is the primary excited species. In the presence of base luminol chemiluminescence can also be induced by phthaloyl peroxide[21].

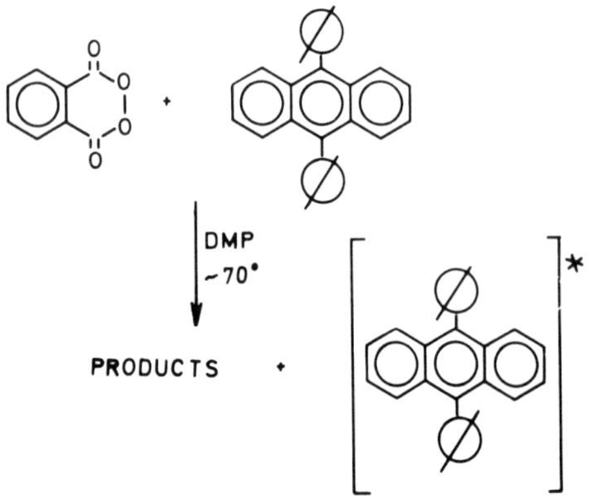

Fig. 11: Chemiluminescent phthaloperoxide decomposition. For luminol chemiluminescence initiation by phthaloyl peroxide a base is required.

References:

1) H.O. ALBRECHT, Z. physik. Chem. <u>136</u>, 321 (1928)
2) E.H. WHITE and D.F. ROSWELL, Accounts chem. Res. <u>2</u>, 80 (1969)
3) R.A. CLEMENT, J. org. Chemistry <u>25</u>, 1724 (1960)
4) T.J. KEALY, J.Amer. chem.Soc. <u>84</u>, 966 (1962)
5) E.H.WHITE, D.F.ROSWELL, and O.C.ZAFIRIOU, J. org. Chemistry <u>34</u>, 2462 (1969)
6) M.M.RAUHUT, A.M.SEMSEL, and B.G.ROBERTS, ibid. <u>34</u>, 2431
7) Y.OMOTE, T.MIYAKE, and N. SUGIYAMA, Bull. chem. Soc. Japan <u>4o</u>, 2446 (1967)-C.A.<u>68</u>, 49534 (1968)
8) E.H. WHITE et al., J.Amer.chem.Soc. <u>9o</u>,5932 (1968)
9) R.B. BRUNDRETT, D.F. ROSWELL, and E.H. WHITE, J.Amer.chem.Soc., in press
10) K.-D. GUNDERMANN, H. FIEGE, and G. KLOCKENBRING, Liebigs Ann. Chem. <u>738</u>, 14o(197o);<u>734</u>,2oo(1971); K.-D.GUNDERMANN, Angew. Chem. <u>8o</u>, 494 (1968)
11) H. KAUTSKY and K.H. KAISER, Z.Naturforsch. 5 b, 353 (195o)
12) D.F.ROSWELL, personal communication
13) J.LEE and H.H.SELIGER, Photochem.Photobiol. <u>11</u>, 247 (197o)
14) J.D.GORSUCH and D.M.HERCULES, ibid. <u>15</u>, 567(1972)
15) K.-D.GUNDERMANN and H.UNGER, unpublished results
16) I.B.MATHESON and J.LEE, Spectroscopy Letters <u>2</u>, 117 (1969)
17) E.H.WHITE et al., J.Amer.chem.Soc. <u>86</u>, 94o (1964)
18) G.O.SCHENCK et al., Z.Naturforsch. <u>2o b</u>, 637(1965)
19) K.-D.GUNDERMANN and K.H.SCHOLZ, unpublished results
20) J.R.TOTTER and G.E.PHILBROOK, Photochem. Photobiol. <u>5</u>, 177 (1966)
21) K.-D.GUNDERMANN, M.STEINFATT, and H.FIEGE, Angew. Chem. <u>83</u>, 43 (1971)

J. LEE: Baxendale (J. Chem. Soc. D (1971) 1489) recently reported on the chemiluminescence of luminol produced by pulse radiolysis of a basic aqueous solution. He reported on the lifetimes of various intermediate radical species. We have studied the effect of radical traps on luminol chemiluminescence in aqueous solution. We would expect them to react rapidly with the intermediates Baxendale has observed. Up to millimolar or more concentration of scavenger, the quantum yields are unaffected. I would like to ask whether you can rationalize the results if you propose such one-electron intermediates in the aqueous reaction?

GUNDERMANN: I also saw this Baxendale work and one can put it under one scheme as you remarked. On the question of radical scavenging of diazaquinones, this is complicated by the fact that if you add hydroquinone, just a reduction occurs. The hydroquinone gives quinone and the diazaquinone gives hydrazide, and so does pyrogallol and other compounds. So the question of what radical scavengers do to diazaquinone is a rather complicated one.

WHITE: You mentioned in an earlier paper that oxygen gas seemed to be required in the chemiluminescence of the diazaquinones; I wonder if you would comment on that?

GUNDERMANN: The first experiment we made was an attempt to try Albert Kavsky's method in which he claimed that only alkali was necessary for getting light from the diazaquinones. So we mixed isotonic solutions of the diazaquinone with sodium hydroxide solution-not excluding air, in which case, chemiluminescence was seen. We then purged the solutions with nitrogen to remove as much oxygen as possible. The light went down but there was still a residual emission, possibly due to the difficulty in removing the last traces of oxygen. If one than adds HOOH the light intensity increases enormously. We think, therefore, that the oxygen is just a source perhaps for HOOH.

ADAM: About the question concerning the scavenging of radicals. We have been trying very hard in our cyclic peroxide systems to trap, with normal kinds of scavengers, diradicals, but they are so short lived than an experiment designed to scavenge biradicals has, we have concluded to this point, an extremely small chance of succeeding. So, I wouldn't be too carried away with such a negative result. We haven't been able to trap and in 10 years.

J. LEE: The point is, however, that Baxendale has measured the reaction rates in pulse radiolysis and it seemed to us that we could then plug in those numbers. Of course, if these radicals are of a different sort of species than those in the normal aqueous reaction that would rule out that sort of calculation.

HERCULES: This was the point I was going to make. Another thing I think is interesting is that in our attempts to use luminol for analytical purposes the one thing we observed consistently was that those metals which catalyze the reaction are those which can undergo one electron transfer reactions to some unstable state. Now when exactly they do this I haven't the foggiest idea. What was interesting was that when iron (II) catalyzed the reaction, it catalyzed the reaction with molecular oxygen. It does not require peroxide. If you do thermodynamic calculations you can show that under the conditions of your solution, that what the iron is really doing is converting the oxygen to HOOH, and this is reacting subsequently.

GUNDERMANN: Yes, I think so too. We are just planning some experiments to measure electron transfer between luminol and radical anions. I think this should be quite interesting because in all luminescent systems you always have oxygen or HOOH present. And this is a crucial point, as I tried to point out; these two species, the hydrazide and the "oxygen salt" to use the expression, are always complementary to each other.

THE CHEMILUMINESCENCE OF ACYL HYDRAZIDES

Emil H. White and Robert B. Brundrett

Department of Chemistry, The Johns Hopkins University
Baltimore, Maryland 21218

Most hydrazides of fluorescent carboxylic acids are chemiluminescent on oxidation in basic media. Probably the best known hydrazide in this class of compounds is luminol (5-amino-2,3-dihydro-1,4-phthalazinedione, I), and much of our early work on hydrazides was carried out with this compound Since the early work on the chemiluminescence of luminol and other cyclic hydrazides has been thoroughly reviewed,[1] this article will cover only the more recent work.

The chemiluminescence of cyclic hydrazides such as luminol can be effected with a variety of oxidizing agents. In protic media, hydrogen peroxide and an oxidizing agent are required for efficient emission, whereas in aprotic solvents such as dimethyl sulfoxide (DMSO), only base and oxygen are required.[2] The reaction products and the stoichiometry in the latter system are indicated in equation 1.[3] By a comparison of the chemiluminescent spectrum

$$\text{I} + 1O_2 \xrightarrow[\substack{\text{NaOH} \\ O_2}]{\text{DMSO}} 1N_2 + \text{II}^* \rightarrow h\nu \qquad (1)$$

with the fluorescence spectrum of aminophthalate ion (II), it was shown that aminophthalate ion is the light emitter in the reaction, being formed in a singlet excited state.[4] A modest number of other cyclic hydrazides have been examined, and in all cases to date the corresponding carboxylic acid anion is the light emitter in the reaction[5] - <u>with one exception</u>. The chemiluminescence of phthalic hydrazide does yield phthalate ion as a product, but this particular ion is nonfluorescent; it undergoes a rapid intersystem

$$\text{(structures)} \quad (2)$$

crossing to the triplet state. The light emission seen in this case stems from the anion of phthalic hydrazide, which is fluorescent. An energy transfer from triplet phthalate ion to the anion of phthalic hydrazide followed by light emission (eq.2) has been proposed for the chemiluminescence observed.[6] A similar triplet-singlet transfer has been proposed as well for several related cases of intramolecular energy transfer.[7]

Luminol is not particularly efficient in light production, the quantum yield of emission being slightly over 1%.[8] More efficient compounds have been prepared using a rule implicit in the early work of Drew[9]-that electron releasing substituents increase the light emission. More highly conjugated systems also are more efficient than luminol, probably because the carboxylic acids they yield in the oxidation are highly fluorescent. In any event, several modestly efficient hydrazides have been tested lately, compounds III and IV in our laboratory,[10,11] and compounds V and VI by Professor K. D. Gundermann and his group at Clausthal;[12] the quantum yields are given below each compound. Compound IV is particularly interesting, since the fluorescence quantum yield of the emitting species, perylenedicarboxylic acid dianion is 0.14; thus, about one half of the hydrazide molecules oxidized lead to excited states![11]

We have recently turned to a more systematic approach to the study of how to improve quantum yields in hydrazide chemiluminescence. The overall efficiency of chemiluminescence, Φ_{ch}, is a product of efficiency of the component steps Φ_r, Φ_{es}, and Φ_{fl}, where Φ_r is the fraction of molecules that follow the "correct" chemistry, Φ_{es} is the fraction of molecules that cross over to the

$$\Phi_{ch} = \Phi_r \Phi_{es} \Phi_{fl}$$

singlet excited state (travelling the correct chemical path), and

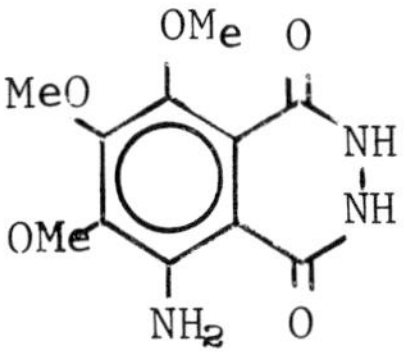

III (0.019) IV (0.07) V (0.03) VI (0.003)

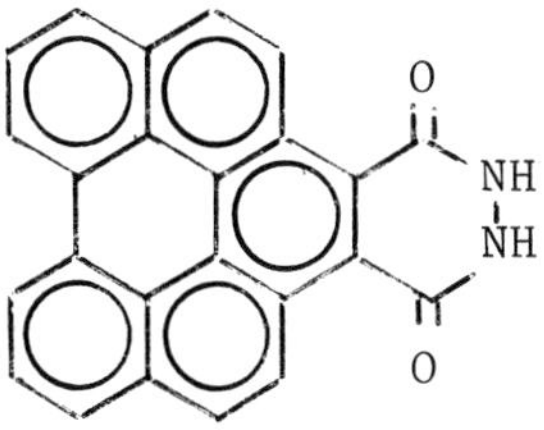

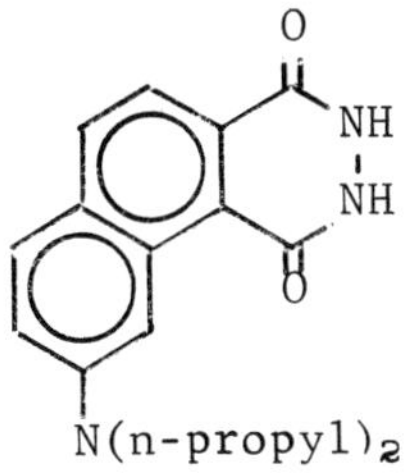

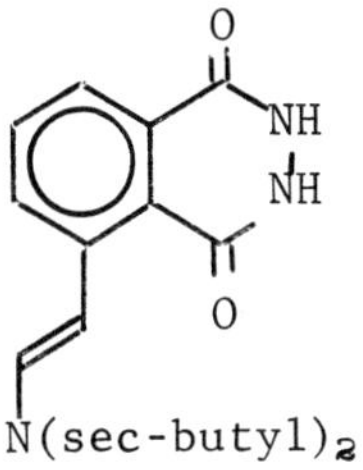

Φ_{fl} is the quantum efficiency of fluorescence of the emitter. We have measured Φ_{ch} and Φ_{fl} for a series of hydrazides and the corresponding carboxylates, and conclude that the effect of substituents occurs not only at the Φ_{fl} level but also at the $\Phi_r \Phi_{es}$ level. The factor Φ_r is unknown. However, since the chemiluminescent reactions lead to high yields of the corresponding carboxylates and since the reaction conditions used give nearly the maximum amount of light possible, we have assumed for this study that Φ_r has a common value of 1.

Using this general approach, the efficiency of the chemical production of excited states was measured for hydrazides VII-IX.

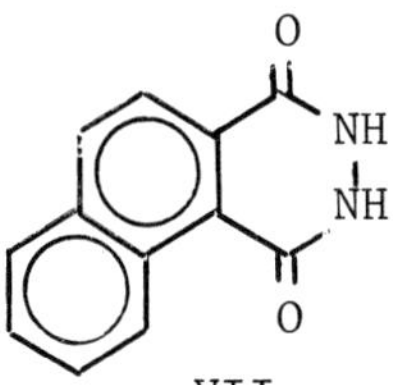

VII 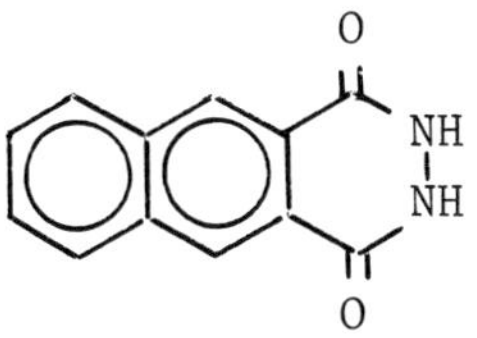   VIII 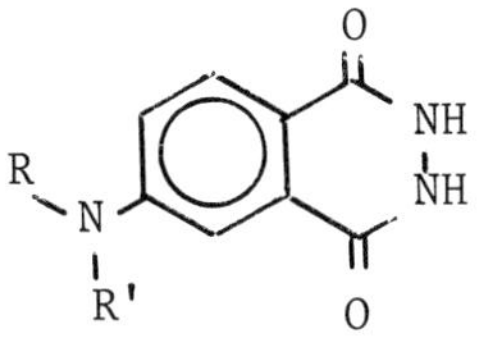   IX

a, R = R' = H
b, R = R' = CH₃
c, R = R' = C₂H₅

d, R = R' = $\underline{n}$C₄H₉
e, R = R' = $\underline{n}$C₇H₁₅
f, R = H, R' = $\underline{n}$C₄H₉

Table I. Chemiluminescence and Fluorescence Spectral Data[a]

	Hydrazides		Phthalic Acid Products		
Compound	Φ_{es}	Chemiluminescence λ_{max} (nm)[b]	Fluorescence λ_{max} (nm)[b]	Fluorescence ν_{max} (cm^{-1})[c]	Fluorescence ν_{min} (cm^{-1})[c,d]
VII	.0017	350	355	28,100	31,200
VIII	.0043	362	370	26,100	30,500
IX$_a$	.0092	416	419	23,400	28,200
IX$_b$	.030	437	438	22,200	25,700
IX$_c$	.030	439	439	22,200	25,700
IX$_d$	.046	442	442	21,900	25,500
IX$_e$	.015	442	442	21,900	25,500
IX$_f$	.021	435	435	22,400	26,400

[a]All in aqueous 0.1 M K$_2$CO$_3$ solution. [b]Determined from uncorrected spectra (error ~ ± 3 nm). [c]Determined from corrected spectra (error ~ ± 200 cm^{-1}). [d]Value is that obtained by extrapolating the straight portion of the high frequency side of the fluorescence curve to zero intensity.

The pertinent data are listed in Table I. A plot of the data[13] shows that there is a linear relationship between the energy of the excited state (E$_{fl}$, from ν_{min}, Table I) and Φ_{es} for compounds VII and VIII and series IX.

An interpretation of this result has been presented[13] based on the suggestion that excitation in the hydrazide chemiluminescence may stem from an electron transfer process.[13,14]

$$RCO_2 \cdot + 1e^{\ominus} \rightarrow RCO_2^{\ominus *} \qquad (3)$$

The explanation is derived from the theory of electron transfer excitations developed by Marcus.[15] Effectively, it was proposed that substituents will stabilize an excited state more than the ground state, a process which could lead to the selective lowering of the transition state energy for crossing to the excited state energy surface. A 30-fold effect in this direction was seen in the plot of the data of Table I. Clearly other factors are involved, however, since if the decrease in E$_{fl}$ were manifested entirely in a decrease in the transition state energy, an increase in Φ_{es} of about 10^{10} would result (E$_{fl}$[ν_{min}] for the phthalate ion from VII = 89 K$_{cal}$ and for the phthalate from IX$_d$ = 73 K$_{cal}$).

As might be expected in a multistep reaction, many variables are involved. Luminol (I) is more efficient than expected from the relationship between $\bar{\Phi}_{es}$ and the energy of the excited states discussed above.[13] Possibly, the ortho proton transfer this compound undergoes[16] is responsible. The methyl derivative X_a is even more efficient. In this case the overall quantum yield (0.028) is about twice that for luminol. The role of the methyl group is not known at the moment. The effect is presumably not steric in origin, judging from the effect of larger substituents (Φ_{ch} for compounds $X_{b,c}$, and d).[17] The quantum yields follow the

	Φ_{ch}
a, R = CH_3	0.028
b, R = nC_3H_7	0.016
c, R = $isoC_3H_7$	0.011
d, R = CH_2OCH_3	0.006

order of the hyperconjugative ability of the R substituents (X).[18]

Compound XI was prepared to see whether a bulky substituent might increase the efficiency of chemiluminescence.[17] It has been proposed, for example, that diphenylanthracene fluoresces with almost unit efficiency because the two phenyl rings, which are twisted out of the plane of the anthracene ring, block self quenching by virtue of their preventing near approaches of the molecule.[19] A similar effect was not apparent in the present case since Φ_{ch} for XI was 0.012 (compared to 0.015 for luminol).

The question of how excited states are produced in the chemiluminescence of the cyclic hydrazides remains an open one. Recent work on the chemiluminescence of monoacylhydrazides such as XII and XIII has led to the suggestion that the excitation stage is an electron transfer step (eq. 3).[14] In order to bridge the gap between the cyclic and monoacyl hydrazides, to see whether common intermediates or mechanisms might be involved, we have prepared compounds XIV-XVI. None of the compounds, however, were

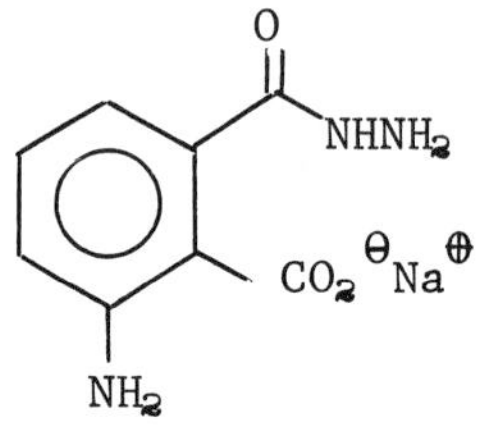

XIV 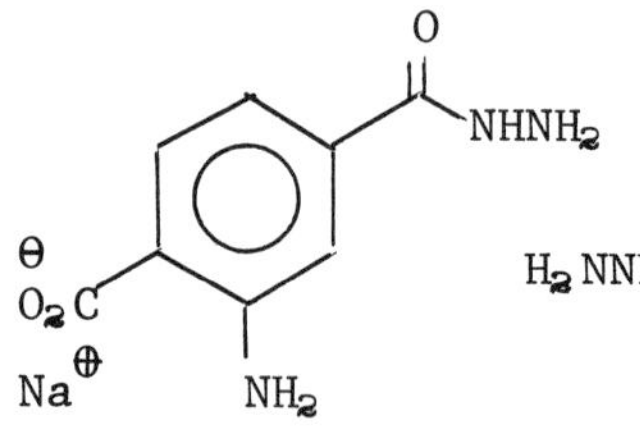   XV 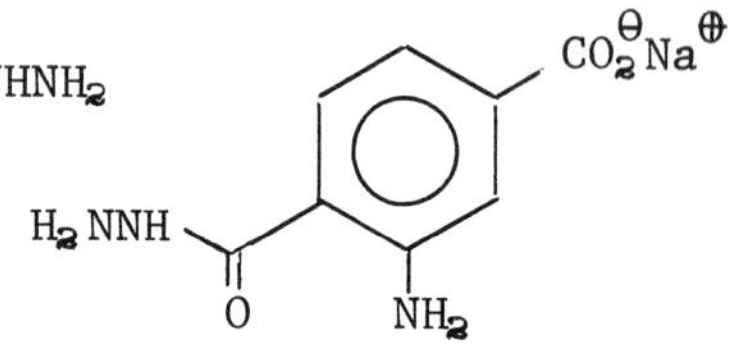   XVI

appreciably efficient in chemiluminescence (DMSO system).

In a second approach to the problem, the 2-electron oxidation product of the cyclic hydrazides, general formula XVII, has been investigated. A stable member of this case of diazaquinones, benzophthalazinedione (XVIII), has been prepared and shown to be

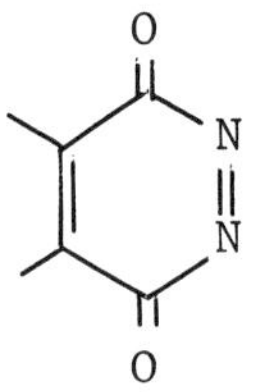

XVII 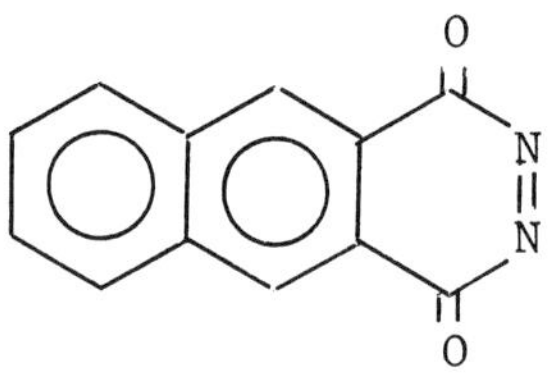   XVIII

chemiluminescent; only hydroperoxide ion is required for light emission.[20] The wavelength maximum of chemiluminescence is the same as that measured for the corresponding hydrazide (VIII), and this value is the same as that measured for the fluorescence of naphthalate ion.[20] Further, cyclopentadiene adducts of XVIII were isolated from chemiluminescing systems of VIII to which cyclopentadiene had been added. Thus it is attractive to consider the azaquinones as intermediates in the chemiluminescence of the cyclic hydrazides. Professor Gundermann, working with related compounds, has come to the same conclusion.[21]

How the azaquinones lead to excited states could, in principle, be revealed by an application of organic reaction mechanisms (eq. 4). Species C and E could lead to excited states by the electron transfer process (eq. 3 and ref. 14). Alternatively, various dioxetanes[1] (A, e.g.) could arise from the diazaquinone on reaction with peroxide ion. Tests of these possibilities have not been successful. Compound XIX on reaction with hydroxide ion should yield species B. No light was obtained in this reaction, however. This general approach is still

attractive, though, since aldehyde XX is produced in the chemiluminescence of hydrazide VIII. Formation of the bicyclic peroxide XXI (possibly from D) is often cited in the literature as a possible source of excited species.[22] The cleavage to nitrogen and the dicarboxylate ion is an allowed process[23] and should not yield excited states. A modification utilizing the back lobe of the O-O bonding orbital has been proposed recently[24] to account for the formation of excited states. The data available at the present time does not permit one to select the correct mechanism among the many possibilities, unfortunately.

SOME OBSERVATIONS ON LUMINOL CHEMILUMINESCENCE

Over the years, a number of unusual observations on the chemiluminescence of luminol have appeared in the literature. Behrens, Totter, and Philbrook[25] have reported, for example, that merely acidifying luminol in an alkali metal hydroxide solution yields light. In our hands, the light intensity, by eye, is on the borderline of detectibility. However, if hydrogen peroxide is added, the phenomenon is easily seen in a partially darkened room. In the original experiments of Behrens, et al., perhaps traces of hydrogen peroxide were formed by autooxidation of the luminol, e.g.

Far more light is produced if carbonate is used as the base instead of hydroxide. Alternatively, carbon dioxide gas can be used in place of the acid. Further, solutions of hydrogen peroxide in dilute sodium carbonate can be partially neutralized with acid to generate an active substance which leads to light emission as soon as luminol is added. Very little light is observed in any of the cases cited above unless the acid is added at some point.

Since free radicals in the presence of oxygen can lead to light emission from luminol and since benzoyl peroxide has been used as a radical source in this way,[1e] the formation of peroxycarbonates and their homolytic decomposition in the experiments cited above would appear to be a reasonable explanation of the

$$CO_2 + {}^{\ominus}OOH \rightarrow {}^{\ominus}O\text{-}\overset{O}{\underset{\|}{C}}\text{-}O\text{-}OH \rightarrow {}^{\ominus}O\text{-}\overset{O}{\underset{\|}{C}}\text{-}O\text{-}O\text{-}\overset{O}{\underset{\|}{C}}\text{-}O^{\ominus}, \text{ etc.} \rightarrow \text{radicals}$$

$$\qquad\qquad\qquad\quad XXI \qquad\qquad XXII$$

observations. In fact, salts of anions XXI and XXII have been reported by Partington and Fathallah,[26] and when prepared as directed, lead to chemiluminescence on addition to luminol solutions (in particular XXII). The efficiency of the system utilizing carbonate buffers or carbon dioxide approaches that of the hemin-hydrogen peroxide system, thus permitting a study of the chemiluminescence free of the difficulties attending the use of hemin.

Another interesting set of observations is due to Bersis and Nikokavouras.[27] They reported that if a rapid stream of carbon dioxide is passed through a basic solution of luminol containing sodium chloride, hydrogen peroxide, and manganous chloride, the intensity of the light emitted passes through four maxima before the reaction ceases; the reaction stops when enough of the base is neutralized to drop the pH of the mixture to ~ 8-9. We have been able to repeat this observation but feel that the various maxima are due to the formation successively of four catalytically active forms of the manganese, which differ in their co-ordination spheres. Thus, under the same reaction conditions but using hemin instead of manganous chloride, only a single maximum in light intensity (very near time zero) is seen for the chemiluminescence.

PREPARATIONS

4-Carbomethoxy-2-nitrobenzoic acid (mp 133-135°, lit.[28] mp 133.5-135°) was obtained by esterification of nitroterephthalic acid with hydrochloric acid in methanol; 4-carbomethoxy-3-nitrobenzoic acid (mp 170-172°, lit.[28] mp 174-175.5°) was obtained by half saponification with base of the dimethylester (mp 70-72°, lit.[28] mp 74-75°) obtained by esterification of nitroterephthalic acid with sulfuric acid in methanol.

4-Carbomethoxy-2-aminobenzoic acid and 4-carbomethoxy-3-aminobenzoic acid were obtained by reduction of the corresponding nitro compounds with 10% Pd/C in methanol under hydrogen (1 atm), filtering off the catalyst, evaporating the solvent under vacuum, and crystallizing the product from methanol-water: the former compound melted at 223-224° (lit.[29] mp 216-217°) and the latter at 213-215° (lit.[29] mp 216-217°).

4-Carboxy-3-aminobenzoic acid hydrazide and 4-carboxy-2-aminobenzoic acid hydrazide were obtained by heating the corresponding methyl esters in 95% hydrazine at 50° for 2 hr, adding 1 equivalent of sodium hydroxide in water, and drying under high vacuum. Each product showed one spot on tlc (8:1:1-ethanol, ammonium hydroxide, water on cellulose) which was more mobile than aminoterephthalic acid but less mobile than the monomethyl esters. Both compounds show ir bands (Nujol mull) at 1630 and 1580 cm^{-1} corresponding of the hydrazide and carboxylate groups.

3-Aminophthalic acid monohydrazide. 2-Aminophthalic anhydride was dissolved in THF and the solution was added dropwise to a solution of excess hydrazine in glyme at -30°. One equivalent of sodium hydroxide was added and the solvent was removed under high vacuum to give a white solid: uv (0.1 M K_2CO_3) 305 nm (3-aminophthalate gives 304 nm[8]); ir (KBr) 1650 (shoulder) and 1580 cm^{-1}.

N-Nitrosonaphthalimide. One equivalent of potassium hydroxide in 20% aqueous ethanol was added to a refluxing solution of 2,3-naphthalimide in ethanol. The resulting precipitate was collected, dried, and suspended in glyme at -40°. A solution of nitrosyl chloride (25% molar excess) in glyme was added dropwise and the mixture was stirred until all the solid had dissolved. Aliquots were removed for immediate use. On warming, nitrous fumes are produced.

ACKNOWLEDGEMENT

We thank the Public Health Service for its financial support (Research Grant No. 5 RO 1 NS-07868 from the National Institute of Neurological Diseases and Stroke), and Dr. W. Huang for carrying out the experiments on the isolation of carboxynaphthaldehyde.

REFERENCES

1. (a) K. D. Gundermann, "Chemilumineszenz Organischer Verbindungen," 1st ed, Springer-Verlag, New York, N.Y., 1968; (b) K. D. Gundermann, Angew. Chem. Intern. Ed. Engl., 4, 566 (1965); (c) F. McCapra, Quart. Rev. (London), 20, 485 (1966); (d) J. W. Haas, J. Chem. Educ., 44, 396 (1967); (e) E. H. White in "Light and Life," 1st ed, W.D.McElroy and B. Glass, Ed., Johns Hopkins Press, Baltimore, Md., 1961, p 183; (f) E. H. White and D. F. Roswell, Acc. of Chem. Res., 3, 54 (1970).

2. E. H. White, J. Chem. Educ., 34, 275 (1957).

3. E. H. White, O. C. Zafiriou, H. M. Kagi, and J. H. M. Hill, J. Amer. Chem. Soc., 86, 940 (1964).

4. E. H. White and M. M. Bursey, ibid., 86, 941 (1964).

5. M. M. Bursey, Ph.D. Thesis, The Johns Hopkins University, Baltimore, Maryland, 1963; see also references 1a, 4, 10, 11, 12a, and 13.

6. E. H. White, D. F. Roswell, and O. C. Zafiriou, J. Org. Chem., 34, 2462 (1969).

7. (a) D. R. Roberts and E. H. White, J. Amer. Chem. Soc., 92, 4861 (1970); (b) D. F. Roswell, V. Paul, and E. H. White, ibid., 92, 4855 (1970).

8. J. Lee and H. H. Seliger, Photochem. Photobiol., 4, 1015 (1965).

9. H. D. K. Drew and F. H. Pearman, J. Chem. Soc., 586 (1937); H. D. K. Drew and R. F. Garwood, ibid., 836 (1939).

10. E. H. White and M. M. Bursey, J. Org. Chem., 31, 1912 (1966).

11. C. C. Wei and E. H. White, Tetrahedron Lett., 39, 3559 (1971).

12. (a) K. D. Gundermann, W. Horstmann, and G. Bergman, Ann.,

684, 127 (1965); (b) K. D. Gundermann and D. Schedlitzki, Chem. Ber., 102, 3241 (1969).

13. R. B. Brundrett, D. F. Roswell, and E. H. White, J. Amer. Chem. Soc., in press.

14. E. Rapaport, M. W. Cass, and E. H. White, ibid., 94, 3153 (1972).

15. R. A. Marcus, J. Chem. Phys., 43, 2654 (1963).

16. E. H. White, D. F. Roswell, C. C. Wei, and P. D. Wildes, J. Amer. Chem. Soc., 94, 6223 (1972).

17. R. B. Brundrett, unpublished work.

18. J. W. Baker in "Conference on Hyperconjugation, Indiana University, 1958", Pergamon Press, New York, N. Y., 1959, p 135.

19. E. J. Bowen, Trans. Faraday Soc., 50, 97 (1954).

20. E. H. White, E. G. Nash, D. R. Roberts, and O. C. Zafiriou, J. Amer. Chem. Soc., 90, 5932 (1968).

21. K. D. Gundermann, H. Fiege, and G. Klockenbring, Liebigs Ann. Chem., 738, 140 (1970); K. D. Gundermann and H. Fiege, ibid., 743, 200 (1971).

22. (a) F. McCapra, Quart. Rev. Chem. Soc., 20, 485 (1969); (b) M. M. Rauhut, A. M. Semsel, and B. G. Roberts, J. Org. Chem., 31, 2431 (1966); (c) H. F. Eicke, H. Fiege, K. D. Gundermann, Z. Naturforsch, 25, 481 (1970).

23. R. B. Woodward and R. Hoffmann, "The Conservation of Orbital Symmetry," Academic Press, Inc., N. Y., 1970.

24. R. C. Dougherty, J. Amer. Chem. Soc., 93, 7187 (1971).

25. H. Behrens, J. R. Totter, and G. E. Philbrook, Nature, 199, 596 (1963).

26. J. R. Partington and A. H. Fathallah, J. Chem. Soc.,1934 (1950).

27. D. S. Bersis and J. Nikokavouras, Nature, 217, 451 (1968).

28. R. Wegscheider, Monatshefte fur Chemie, 23, 405 (1902).

29. R. Wegscheider, ibid., 28, 819 (1907).

SELIGER: I have a couple of questions I would like to ask about this. One, I think you expressed a very legitimate concern during the dioxetane discussion about the different quantum yields which might be obtained in the hands of different investigators. Were the quantum yields you reported for your compound the results of absolute measurements?

WHITE: They were all done relative to luminol and for that reason since they were done in one lab they are probably subject to an error of I wouldn't hesitate to say, 25 to 50%. For our pruposes that's sufficient.

SELIGER: No, I mean, did you establish a base line with an absolute quantum yield of one compound, or are all the numbers relative and, if so relative to what?

WHITE: They're all relative to luminol.

SELIGER: But what number for luminol?

WHITE: Oh, there is a slight difference, we differ from your value by about 10 or 15%, and I don't know whether that is due to the different pH we use of the buffer solution. We find an optimum value by using carbonate solution at a pH of about 11.4.

SELIGER: In order to differ by any number you have to have absolute measurements. Did you make an absolute calibration of the quantum yield for one of your compounds?

WHITE: We measure luminol under a number of conditions, took the optimum value, and assumed that the quantum yield was roughly 1.2, accepting your and John Lee's value luminol (Photochem. Photobiol., $\underline{4}$, 1015 (1965)). We then measured the chemiluminescence efficiency of all the other hydrazides under the same conditions relative to that.

SELIGER: I think that what you found in one of your recent papers (ref. 20), was an interesting effect -- that the fluorescence of the amino phthalate (APA) did not agree with the luminol chemiluminescence emission spectra, and you reported essentially two dashed curves; one at longer wave length of chemiluminescence emission and one at short wavelength from fluorescence emission of APA. Then there also is a summation curve, for photo excitation at a lower wavelength. I think that this is very interesting and very correct evidence that during chemiluminescence we are exciting one species only, while in the photoexcitation you have both species available to you and this sort of fits in with the amount of energy that is available during the chemiluminescent reaction which is only sufficient to populate the lower energy species in that mixture.

WHITE: Well, again, we may be saying the same thing in different ways. I think that the absorption spectrum of the sodium salt of the aminophthalate ion under our conditions is different from that of the cesium salt; e.g., there seems to be an influence of the ion pair; the sodium ion pair absorbs at slightly longer wave lengths than the free ion. Therefore, under these circumstances, by irradiating different parts of the envelope - we irradiate different fractions of the free ion and free ion pair - and, therefore, we get different emission spectra.

SELIGER: I think that comparison between these emission spectra is an excellent method of distinguishing between the formation of excited states by chemical and by photochemical methods.

WHITE: Yes, if other examples can be found.

SELIGER: I would like to propose a physical type of mechanism for the bubbling effect that has been observed ... we have always noted that we can obtain light from luminol solutions without the necessary addition of HOOH when we bubble air or oxygen through the solution. These are very low levels of light however, but they are significantly above the normal back-ground, and I would suspect that the shear forces involved in the bubbling (the physical bubbling mechanism itself) are sufficient to produce electrostatic fields which can give you rather small populations of the radicals that you need for the reaction itself. This might be some type of explanation for the CO_2 bubbling effect which was given earlier.

WHITE: You're saying that this is similar to the little tubes students make containing, say air at ten torr, and a drop of mercury and when these are sealed off and shaken they glow like a neon light bulb.

SELIGER: Yes, this is the triboluminescence.

HERCULES: I would like to suggest an alternate explanation for Seliger's effect, and that is that iron catalysis of the luminol reaction in aqueous solution is sensitive down to 10^{-10} molar. All you need is a small trace of iron in your solution and it works very well with oxygen or air.

McELROY: Isn't it true that when you initiate with CO_2 bubbling and then stop that you still continue to get the rhythmic emission of light in the reaction that you started?

WHITE: What do you mean by rhythmic, cyclical?

McELROY: Yes.

WHITE: No, I think it dies off montonically when one stops the oxygen flow; that's been our experience at least.

McELROY: Is this a result of decay of intermediates?

WHITE: Yes, definitely, that's because of the finite life time of these percarbonates. I did forget to mention--what's probably happening is that, just as in the case of the benzyl-peroxide, these are dissociating slowly to give radicals, which then initiate the chain.

RAUHUT: You mentioned that the excitation efficiency increases as the excitation energy of the fluorescer decreases?

WHITE: Yes, that would be in our case also the product of the reaction.

RAUHUT: I just thought I'd mention that we find this true for oxalic ester chemiluminescence as well, where ruberene is much more efficient, e.g., than 9,10-diphenylanthracene. These other fluorescers fall in between, and it is interesting I think that Eyring predicted this effect forty years ago.

GUNDERMANN: I have just a question on the scheme of the reaction mechanism. You have the dioxetane ring as I pointed out in my talk. The experiments with 18_{O_2} labeled oxygen don't seem to fit into this behavior if a dioxetane ring is formed. Now the question is was all the aminophthalate that you isolated in your reaction mixtures in 1964 fully or only partly labelled? If the label was on both carboxy groups then I think it would be impossible to have a dioxetane intermediate.

WHITE: Actually, mechanisms involving dioxetanes can be written that accommodate the O-18 observations (see equation 4 of the paper).

GUNDERMANN: Another little question. From the linear hydrazides--it was striking that on the left side where the chemiluminescent linear hydrazides were, there was a compound with both ends substituted with methyl. Whereas on the other side where the non-chemiluminescent compounds were, a terminal methylated compound was listed which is not chemiluminescent. Is there any reason why this compound does not give light?

WHITE: In this case ($RCONHNHCH_3$) one introduces a double bond here ($RCON=NCH_3$) and the nucleophile (RO^-) attacks the most logical place, the carbonyl group, and one simply goes to ester and that's a dead end. In the second case ($RCON(CH_3)NHCH_3$) our hypothesis is that we get a simple E-2 elimination reaction to form azomethane and the acyl anion, which is subsequently oxidized to give excited states. The azomethane then tautomerizes to the methyl hydrazone of formaldehyde or dissociates thermally to give methyl radicals which extract hydrogen to give methane. We see all those side products.

COMPARISON OF POTASSIUM IODIDE QUENCHING OF 3-AMINO-PHTHALATE FLUORESCENCE AND LUMINOL CHEMILUMINESCENCE IN AQUEOUS SOLUTION

John Lee and I.B.C. Matheson

Department of Biochemistry, University of Georgia

Athens, Georgia 30601

It has recently been suggested that the fluorescence of 3-aminophthalate in aqueous or dimethylsulfoxide (DMSO) solution can be described by a four level scheme[1]. The ground state species upon excitation undergoes considerable hydrogen bonding re-arrangements and in DMSO even complete intramolecular proton transfer occurs. In order to explain the differences in spectral and quantum yield behavior between the fluorescence of 3-aminophthalate and the chemiluminescence of luminol[1-4], we have suggested that the chemiluminescence process populates the lower of the two excited levels primarily[1].

In aqueous solution for instance the fluorescence is more effectively quenched by increasing base concentration than is the chemiluminescence[1-3]. A slight difference in oxygen quenching has also been observed[5].

We present here a comparison of the Stern-Volmer quenching constants $Q_0/Q = 1 + KC$ for potassium iodide quenching of the fluorescence and for the chemiluminescence in aqueous solution (pH 11.6, M/10 K_2CO_3, 23°C).

The lower line in the figure labelled "CL" shows that the chemiluminescence shows regular Stern-Volmer behavior up to half-saturating concentrations of potassium iodide. The K is essentially constant. All results have been corrected for the rather substantial refractive index changes incurred.

In contrast the upper line "F" for the quenching of the fluorescence of 3-aminophthalate shows that K increases approximately linearly with quencher concentration. This result is simply interpreted by the existence of two excited levels both of which can be quenched at different rates:

$$[AP^= \ldots H_2O] \xrightarrow{h\nu} {}^1[AP^= \ldots H_2O] \longrightarrow {}^1(AP^=) \ldots H_2O$$

$$k_q[KI] \downarrow \qquad\qquad k'_q[KI] \downarrow \qquad \searrow k_r$$

$$\text{quenched} \qquad\qquad \text{quenched} \quad (AP^=) \ldots H_2O + h\nu$$

The abbreviations are as used before for the different types of hydrogen-bonded species. Both excited species are quenchable but the first becomes more important as [KI] increases. Only the second $^1(AP^=)\ldots H_2O$ is populated by the chemiluminescence reaction.

The potassium iodide quenching of the fluorescence of fluorescein dianion also yields of Stern-Volmer constant dependent on [KI] but in a non-linear way. It is about ten fold greater than the 3-aminophthalate case.

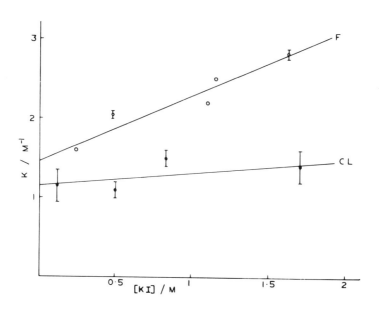

Figure 1. Stern-Volmer quenching constant (K) of 3-aminophthalate fluorescence (o) F and luminol chemiluminescence (●) Cl versus quencher (KI) concentration.

Since all species are negatively charged, stable ground state complexes would not be expected. There is no evidence of any sort of complex formation in the absorption spectra of fluorescein dianion or 3-aminophthalate in 4 M potassium iodide, except for a very slight red shift of the lowest energy absorption band maximum which is attributable to the change in solvent dielectric constant.

It is concluded that in the luminol chemiluminescence, the primary excited species find itself in a different environment from the primary species obtained by photophysically exciting the product molecule. This situation results in differences in behavior towards quenchers and differences in radiative and intersystem crossing rates would also not be unexpected.

Acknowledgement. This work was supported in part by a grant from the National Science Foundation.

1. J. Lee and H.H. Seliger, Photochem. Photobiol. 11 247 (1970).
2. E.H. White and M.M. Bursey, J. Am. Chem. Soc. 86 941 (1964).
3. J. Lee and H.H. Seliger, Photochem. Photobiol. 15 227 (1972).
4. E.H. White, D.F. Roswell, C.C. Wei and P.D. Wilder, J. Am. Chem. Soc. 94 6223 (1972).
5. J. Lee and H.H. Seliger, Photochem. Photobiol. 4 1015 (1965).

WHITE: This looks like it might be a powerful tool then for getting at the intimate details near the transition state in the chemical production of excited states.

HERCULES: I would like to ask, in terms of the figure for the sodium iodide quenching. I view this with some degree of caution in using sodium iodide, since iodine can cause light emission from luminol. It seems to me that if you are putting iodide in the highly basic solution with HOOH, you're going to get quite a variety of iodine species and we know in the catalysis of the luminol reaction by iodine that it is not the I_2 which catalyzes the reaction, but is, in fact, intermediate species in the oxidation of iodine in basic solution, and since it is extremely sensitive it seems that this could obscure the results.

J. LEE: We're not looking at the reaction rate but at the total quanta luminol is emitting. I don't think there's any possibility of getting more light from luminol using iodine rather than peroxide.

HERCULES: So what you're trying to do is to compare the effect of iodine on the intensity of the emission?

J. LEE: No, we're measuring the total quantum yield of luminol chemiluminescence and its quenching by iodide. There's no doubt that there are a lot of other effects on the reaction rate.

HERCULES: But if you're producing the light emission from iodine formed in your reactions then you would be producing a quenching effect you would not see; would you not? In other words you're comparing luminol reactions and the chemiluminescence which is simply a mixed reaction between the HOOH oxidation and the iodine oxidation.

J. LEE: No, I'm comparing the quenching of the quantum yield between the fluorescence of the product, 3-aminophthalate, and the total quantum yield of luminol chemiluminescence under basic conditions using the hemoglobin catalyzed reaction. Now the point, is that the luminol results are perfectly well behaved. In terms of absolute quenching rate constants, this corresponds to about 10^8 M^{-1} sec^{-1} or so, and it looks like the expected fluorescence quenching rate, for instance, for the quenching by iodide of any fluorescent dianion. It's reasonably slow; it's not diffusion controlled anyway. I think that the luminol results are quite well behaved, and that is the point. The Stern-Vomer constant should be constant and shouldn't depend on concentration.

WHITE: Is it possible that the luminol results are the combination of two variables, and that what Hercules is saying is correct, that sodium hypoiodite might be oxidizing the luminol in a completely independent pathway?

J. LEE: And leading to a dark reaction?

WHITE: Yes.

J. LEE: That's certainly possible, but I think one would not see a regular behavior like I see here -- the chemiluminescence behavior is perfectly as expected.

WHITE: But you're changing the KI in a regular fashion might also be regular.

THE CHEMILUMINESCENT AUTOXIDATION OF REDUCED BIISOQUINOLINIUM DICATIONS

Carl A. Heller and Ronald A. Henry

Naval Weapons Center (Code 605), China Lake, CA 93555

John M. Fritsch

Washington University, St. Louis, MO 63110

INTRODUCTION

Numerous chemiluminescent reactions involve molecular oxygen or hydrogen peroxide reacting with an organic compound. In many cases the formation of an intermediate 1,2-dioxetane can be demonstrated or is implicated as one of the steps in the reaction mechanism. The excitation step then involves splitting the dioxetane ring to form two carbonyl groups. The excitation energy is sometimes found in the carbonyl compounds and sometimes in a convenient fluorescer.

The reaction by which the dioxetane ring is formed for any class of compounds is of interest. This is particularly true when the reaction proceeds from ground state O_2. We have studied two classes of electron-rich olefins which oxyluminesce and which are catalyzed by alcohols.[1,2] The initiating reaction is proposed[1] to be of the type:

$$C=C + O_2 + ROH \rightleftharpoons (+)C-C(+) + HOO^- + RO^- \qquad (1)$$

The equilibrium position of reaction 1 will depend upon the reduction potential of the olefinic compound and upon the concentration of ROH. Once peroxy anion is formed, the reaction with dication to form dioxetanes should proceed as with base-catalyzed H_2O_2 reactions.

Tetraaminoethylenes, which are air reactive and some of which chemiluminesce, have large reduction potentials; for example, that of tetrakisdimethylaminoethylene (TMAE) is comparable to that of zinc. Mason and Roberts[3] suggested that biisoquinolinium (BIQ) salts, which chemiluminesce when made basic in air, first react to form olefins. We have previously described the reduction and subsequent autoxidation of one BIQ salt.[2] In this paper we want to describe the reactions of 8,9-dihydropyrazino-[1,2-a:3,4-a'][1,1']-biisoquinolinium dibromide monohydrate (BIQ I++), whose chemiluminescence was the brightest of the several salts studied by Mason and Roberts. In the next section we will describe how BIQ I++ is reduced to a mixture of olefins. The subsequent section describes the chemiluminescent autoxidation reaction.

FORMATION AND PROPERTIES OF THE OLEFINS

The reduction by base of BIQ I++ would be expected to be similar to that of the more frequently studied bipyridinium dications such as Diquat and Paraquat. Two very different mechanisms[4,5a,5b,6] have been advanced for these reductions. Our results support a proton abstraction mechanism as opposed to reduction by the methoxide-formate anion redox couple.

Reduction of the BIQ I++ at a Pt electrode gives two clean, reversible, one-electron additions. The expected products[6,7,8,9] would be the following:

The product I·+ was red.

The reduction potentials were measured versus an s.c.e. in acetonitrile using tetraethylammonium perchlorate as supporting electrolyte. The results mean that I is a somewhat weaker reducing agent than 1,1'-dimethyl-1,1'-dihydro-4,4'-bipyridyl.[6,7] We can use the reduction potentials to calculate an equilibrium constant for the reaction

$$I^{++} + I \rightleftharpoons 2I^{\cdot +} \quad (2)$$

of K = 1.43 × 10^6. This large constant suggested that the radical should be formed quite readily if I⁺⁺ or I were present with trace amounts of reducing or oxidizing materials.

It has indeed proven simple to obtain I·⁺. Methanol solutions of I⁺⁺ in Pyrex or soft glass vessels under N_2 slowly change from yellow to wine-red, and an EPR signal appears. Figure 1 shows the visible absorption spectrum, and Fig. 2 shows the EPR spectrum. The same spectra are obtained from methanol or water solutions of I⁺⁺ in contact with Zn powder. The color and EPR signal disappear if O_2 is added, but the amount of I⁺⁺ is little affected.

Solutions in methanol under N_2 do not form I·⁺ if in N51A neutral glass vials, nor do aqueous solutions in Pyrex turn wine-red. This probably indicates that basic properties of the glass, as well as trace reducing materials, will give the radical. The following experiments show the mechanism of reduction by strong base.

If sodium methoxide is added in less than stoichiometric amounts to a methanol solution of BIQ I⁺⁺ under N_2, the solution changes to orange, but the typical I·⁺ spectrum remains under the new absorbance. If sodium methoxide is added in excess to the wine-red or yellow solution under N_2, it turns orange-red and the EPR signal disappears. The visible absorbance spectrum is shown in Fig. 1. This orange-red solution reacts with O_2 to give chemiluminescence and the bis-isoquinolones shown in Fig. 3; A and C have been isolated and their structures determined (see appendix). These two products would be expected from the autoxidation of olefins I and III, shown in Fig. 4.

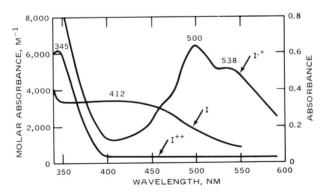

FIG. 1. Absorbance Spectra of Three Solutions. I⁺⁺ is BIQ I⁺⁺ in methanol under air (left scale); yellow solution. I is same solution made basic with NaOMe (left scale); orange-red solution. I·⁺ is first solution after standing under N_2 (right scale); wind-red solution.

FIG. 2. EPR Spectrum of I·+ in Methanol, With g = 2.0025. Spectrum run by D. W. Moore and W. Thun.

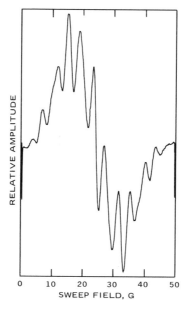

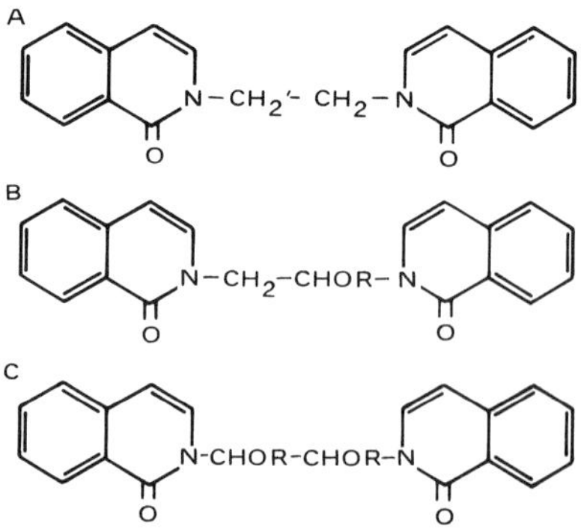

FIG. 3. Bis-isoquinolones Found or Expected From Autoxidation of BIQ I+.

The mass spectrometric data in Tables 1 and 2 show evidence for all the olefins of Fig. 4, as well as for the bis-isoquinolones of Fig. 3. Traces of O_2 account for these oxidation products. These mass spectrometric samples were prepared under N_2 in the small cavity of the quartz solid-sample holder of the Hitachi RMU-6E mass spectrometer. A small sample of BIQ I+ was placed in the cavity. Zinc powder was added to two samples and then either water or methanol. After the samples were thoroughly dried by evacuation, they were transferred to the Hitachi with a short exposure to air. Similar samples were

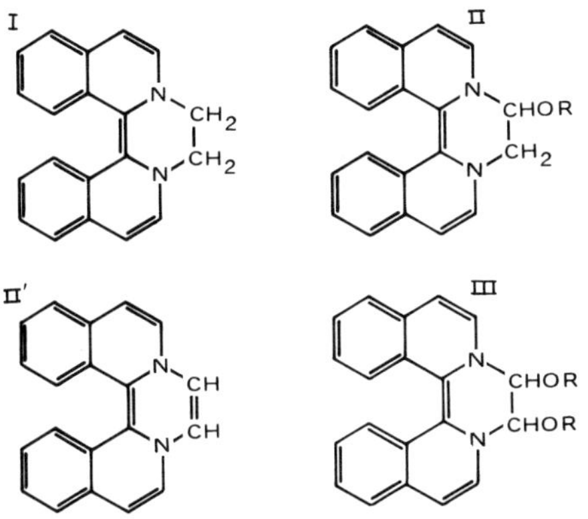

FIG. 4. Reduced Forms of BIQ I Dication.

TABLE 1. Mass Spectra of Species Formed in Methanol Under N_2.

Mass	Intensity[a]	Intensity[b]	Species[c]
255	137.1	144	...
269	8.3	19.7	...
282	2.2	13.4	II'
283	6.6	16.3	...
284	51.1	52.2	I
299	...	5.52	...
314	...	48.1	II
315	...	12.9	...
316	0.56	9.4	A
329	...	13.3	...
330	...	3.9	...
331	...	36.9	...
344	...	25.0	III
345	...	7.2	...
346	...	12.5	B
361	...	1.1	...
362	...	0.6	...
376	...	0.97	C
408	...	0.15	C·CH_3OH or III·2CH_3OH

[a] BIQ I^{++} + Zn + CH_3OH. CH_3OH quickly evaporated from sample.
[b] BIQ I^{++} + $NaOCH_3$ (1.0 M). CH_3OH quickly evaporated from sample.
[c] R is CH_3 in formulas of Fig. 3 and 4.

TABLE 2. Mass Spectra of Species Formed in Water.

Mass	Intensity[a]	Species[b]
255	138.0	...
256	72.9	...
269	2.0	I - 15
270	0.73	...
271	0.80	...
272	0.74	...
282	0.5	Fragment of I, no II'
283	2.0	...
284	6.4	I
285	2.2	...
286	3.5	...
300	0.14	Trace of II
302	3.82	Fragment of III - 14
303	0.80	...
316	5.10	III (A unlikely)
317	1.12	...

[a] BIQ I^{++} + NaOH (1.0 M). Water quickly evaporated from sample.
[b] R is H in formulas of Fig. 3 and 4.

prepared by adding sodium methoxide or sodium hydroxide solutions to the salts.

The solid samples were run at 220, 240, and 250°C. The zinc-treated samples gave nearly identical spectra whether wet with water or methanol. The parent peaks indicate I and a trace of A. The basic methanol solutions produced I, II', II, and III with some A, B, and C (with R = CH_3). Aqueous NaOH, in which the olefins precipitate rapidly, gave mainly I and III (with R = H). This reduction in aqueous solutions seems to rule out the methoxide ion reduction mechanism for these compounds.[5a,5b]

Other mass spectrometric samples were made by evaporating basic solutions which had stood for hours or days. The spectra were more complex than those in Tables 1 and 2. It appears that the olefins react further in basic solution, giving species with still higher masses and with intermediate masses. These results are consistent with the slow color changes and with a changing quantum yield of the oxyluminescent reaction as the solutions age. We have not investigated these complex, slow changes.

The complete sequence of reactions to produce compounds like I, II, II', and III has been discussed elsewhere.[2] A simplified sequence would be

$$I^{++} + 2RO^- \rightleftharpoons II + ROH \tag{3}$$

$$I^{++} + 2RO^- \rightleftharpoons II' + 2ROH \tag{4}$$

$$II + I^{++} \rightleftharpoons II^{++} + I \tag{5}$$

$$II^{++} + 2RO^- \rightleftharpoons III + ROH \tag{6}$$

This reaction scheme would produce the reduced species found in our mass spectral analysis.

It seems reasonable that the bipyridinium and biquinolinium compounds will react analogously. As mentioned above, our mechanism agrees generally with that of Corwin and others[4] for the 1,1'-dibenzyl-4,4'-bipyridinium dication. In presenting the evidence for reduction by the methoxide-formaldehyde redox couple, the appearance of formaldehyde from dimethyl bipyridinium dication was important.[5a] However, Urry and Sheeto[10] have shown that formaldehyde can arise from a methylamino group in octamethyloxamidinium dication following proton abstraction. Furthermore, the methoxide-formaldehyde couple has too low a redox potential for the observed reductions to olefins.[6]

Our evidence shows clearly that proton abstraction is the important initial mechanism for reduction of BIQ I⁺⁺ to the mixed olefins. Along with the earlier work,[4] it suggests that a similar mechanism should be reconsidered for dimethyl bipyridinium dications. Perhaps the bridged bipyridinium dications recently studied by Black and Summers[11] would be useful to test the mechanistic points.

We have observed that when aqueous NaOH solution is added to an aqueous solution of BIQ I⁺⁺, a red precipitate of I and III is formed, even in air. The mass spectra of the precipitate were similar to those of Table 2. Compounds II or II' are nearly absent, although our mechanism shows them as primary products. The reasonable explanation for their absence is that II is a stronger reducing agent than I, so that the equilibrium of reaction (5) is to the right. Reaction (5), being ionic, should be rapid—more rapid than the autoxidation, which explains why we see no light from the aqueous system. Probably III would be a still stronger reducing agent.

We could not isolate II, II', or III for a direct measurement of their oxidation or reduction potentials. We did prepare a tetramethoxy BIQ according to the method of Mason and Roberts.[3,12]

It has reduction potentials of -0.43 and -0.70 V versus an s.c.e. in acetonitrile.[9] The presence of electron-donating CH_3O groups on the aromatic ring would be expected to increase the reduction potentials. Apparently, OH or OR groups on a carbon atom alpha to the nitrogen in the pyrazino ring have a similar effect.

Olefin I has recently been synthesized by an alternate route by W. H. Urry of the University of Chicago; its solution shows the EPR spectrum of I·⁺ and oxyluminesces to form bis-isoquinolone A (m.p. 195 to 197°C).[13]

CHEMILUMINESCENT AUTOXIDATION

General

The oxidation of BIQ I⁺⁺ in basic methanol is fast (about 90% of the light appears in 10 sec) and therefore bright, even though the violet light has a poor photopic factor. As an air-reactive material it seems to have possible applications in automatic emergency light systems if the emission could be red-shifted by suitable changes. Since it was not possible to separate the mixed olefins prepared under nitrogen by reduction of BIQ I⁺⁺ by either base or zinc powder (previous section), most of the studies reported here are for the equilibrium mixture of olefins obtained from BIQ I⁺⁺ and NaOCH$_3$ in air-saturated methanol.

Fluorescence Spectra and Quantum Yields of Products

The major products formed when NaOCH$_3$ solution is added to BIQ I⁺⁺ in air are the bis-isoquinolones A and C. The separated compounds have been studied in a Turner Model 210 spectrofluorometer, which produces corrected spectra.[14] The fluorescence emission spectra of A and C are shown in Fig. 5. For A the peaks are at 370 and 385 nm, with a height ratio of 1.00/0.96. For C the peaks are at 374 and 388 nm, with a height ratio of 0.97/1.00.

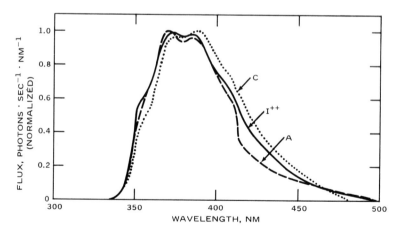

FIG. 5. Spectra for Comparison. I shows chemiluminescence of BIQ I⁺⁺. [I⁺⁺] ≅ 10^{-4} M in methanol. A and C show fluorescence of bis-isoquinolones A and C in methanol. [A] = 2.17 × 10^{-4} M. [C] = 1.17 × 10^{-4} M.

Quantum yields were measured versus quinine sulfate under air at 700 torr. The measured yields for A and C are 0.04 and 0.069, while a product mixture in basic methanol gave 0.042. The product mixture was also measured under N_2 where we obtained 0.047, or a Stern-Volmer constant of $K = 0.91 \times 10^{-3}$ torr^{-1}, if we considered the quenching to be linear. The bis-isoquinolone spectra change somewhat with aging in solution.

Chemiluminescent Emission Spectra

A flow system was built to produce a constant luminescence in a flow cell in the fluorometer. BIQ I$^+$ and NaOCH$_3$ solutions were mixed just outside the cell. The time interval of the reaction irradiating the monochromator was determined by the flow rate and cell size.

At the shortest possible time, the chemiluminescent emission was as shown in Fig. 5. This would appear to be due to A*, with some C* possibly included. The difficulty of a positive identification was due to the spectral shift shown in Fig. 6. If the flow rate was slowed, a different spectrum appeared, with a maximum around 418 nm. This was dim enough to require a wide slit, so we got little resolution. This 418-nm peak was not found in fluorescence of the reaction product.

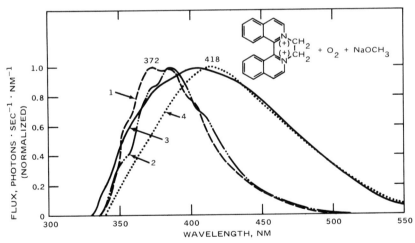

FIG. 6. BIQ I$^+$ Chemiluminescence Showing Spectral Change During Autoxidation. Spectra were obtained from flow cells in a spectrofluorometer (without excitation). Reaction was initiated by flowing together solutions of BIQ I$^+$ (2×10^{-4} M) and NaOCH$_3$ (10^{-2} M) in about 1:1 ratio. Spectrum 1 taken from 0.14-ml cell using 60-ml · min^{-1} flow and 10-nm slit. Spectrum 2 taken from 2-ml cell using 60-ml · min^{-1} flow and 2.5-nm slit. Spectra 3 and 4 taken from 2-ml cell using slow but undetermined flow rates and 25-nm slit.

Kinetic Measurements

It was possible to stop the flow through the fluorometer cell and to monitor the decay of light at any one wavelength. Figure 7 shows one such run plotted as first-order decay curves. Monitoring at 375 nm would give only one curve. All runs gave similar first-order decay kinetics with good straight lines. However, the decay slopes were poorly reproducible, which may have been due to the effects of moisture or other impurities.

At 25°C the rate constants were $k' = 0.17 \pm 0.11$ sec^{-1} and $k'' \cong 0.01$ sec^{-1}. An Arrhenius plot of runs at 25° and 2 to 3°C gave an activation energy of $E' \cong 10$ kcal mole^{-1} for k'. The value of E' was somewhat larger. Thus there are two reactions occurring to give the two types of emission. This would seem to involve two intermediates as well as two products. If either of these decay rates are for dioxetanes, they are clearly faster than those for the more stable alkyl or alkoxy dioxetanes. For those the room-temperature rate constants are about 10^{-4} sec^{-1} and the activation energies about 20.[15,16]

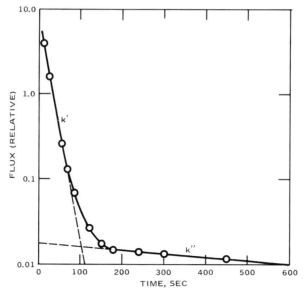

FIG. 7. Decay Kinetics of Light Monitored at 430 nm With 25-nm Slit. Conditions: [BIQ I$^+$] = 1.05 × 10^{-4} M; [NaOCH$_3$] = 50 × 10^{-4} M; T = 2.3 to 2.5°C. Results: $k' = 60 \times 10^{-3}$ sec^{-1}; $k'' = 4 \times 10^{-3}$ sec^{-1}.

Since the plot in Fig. 7 is semilog, it is clear that most of the light comes from the fast reaction; that is, the area under k" is only a few percent of that under k'.

Quantum Yield Effects

Earlier we reported the absolute chemiluminescent quantum yields of BIQ I[+] under carefully controlled conditions.[17] Using the same apparatus, we have measured approximate quantum yields to study the effect of temperature, O_2 pressure, and initial concentration of BIQ I[+]. These are overall quantum yields defined as $\bar{q} = \int F dt/n_i$, where F is flux, $\int F dt$ is the graphically integrated light output in einsteins, and n_i is initial moles of BIQ I[+].

Although temperature affects the rate of light output, it has little effect on $\bar{q}$ (Table 3). This probably means that there is no dark reaction path in competition with the light-producing reaction path. Only if both paths fortuitously had equal activation energies would $\bar{q}$ remain constant over a 78°C range.

TABLE 3. Effect of Temperature on $\bar{q}$.

[BIQ I[+]] $\cong 10^{-4}$ M; [NaOCH$_3$] $\cong$ 0.1 M.

T, °C	$\bar{q}$, einstein · mole^{-1}
0	0.0114
−30	0.0110
−78	0.0132

The O_2 results (Table 4) show some effect and perhaps a balancing of kinetics and quenching. However there is no strong, single effect of O_2 pressure on $\bar{q}$.

The effect of initial concentration of BIQ I[+] is perhaps the most revealing (Table 5). It shows that the differential quantum yield defined as $q = (dF/dt)/(dn/dt)$ must equal the overall yield $\bar{q}$. This in turn means that q must be constant over the run and not depend upon [I[+]], [O_2], or temperature. This implies a mechanism without dark reactions or strongly competing light paths. This is very different from TMAE, where both [TMAE] and [O_2] affect q strongly.

TABLE 4. Effect of O_2 Pressure on $\bar{q}$.

$[BIQ\ I^+] \cong 10^{-4}$ M; $[NaOCH_3] \cong 0.1$ M; T = 0°C.

P_{O_2}, torr	$\bar{q}$, einstein · mole^{-1}
66.3	0.0068
121.6	0.0079
125.7	0.0079
243.3	0.0065
614	0.0061

TABLE 5. Effect of Initial BIQ I^+ Concentration on $\bar{q}$.

$[NaOCH_3] \cong 0.1$ M; T = −78°C.

$[BIQ\ I^+]$, M	$\bar{q}$, einstein · mole^{-1}
10^{-5}	0.0122
10^{-4}	0.0122
5×10^{-4}	0.0117

SUMMARY

The reaction of BIQ I^+ with base produces two olefins as the main products. These olefins react rapidly with O_2 or basic H_2O_2.[2,3] The major products are two bis-isoquinolones corresponding to O_2 addition across the olefinic double bond and subsequent splitting. Most of the light emitted comes from the excited bis-isoquinolones, but there is a slow reaction producing light at 418 nm.

The reaction is simple in the sense that $\bar{q} = q$ and there seem to be no competing reactions with different kinetic rates or activation energies. Any intermediates are unstable, and in particular any dioxetane intermediate decomposes with higher rates and lower activation energies than are common. These same high rates also hold for TMAE, suggesting that the electron-donating groups are responsible for the high reactivity of the dioxetane.

Since olefins can now be synthesized directly as single species,[13] the chemiluminescent autoxidation of pure compounds can now be studied.

[This work was supported in part by the Office of Naval Research.]

APPENDIX

Isolation and Characterization of Oxidation Products

8,9-Dihydropyrazino-[1,2-a: 3,4-a'][1,1']-biisoquinolinium dibromide (0.5 g) was dissolved in 400 ml of absolute methanol and treated with 1.0 g of sodium methoxide (approximately 10-fold excess) in 100 ml of methanol. The red-orange colored solution first became deep blue, then rapidly changed to red-brown. Dry air was aspirated through the solution for 4 hours; the color faded to yellow in about 1 hour. The solution was evaporated to dryness under reduced pressure at ambient temperature, and the solid residue extracted several times with diethyl ether. Evaporation of the extracts left 0.3 g of off-white solid, m.p. 100 to 150°C. (An additional 0.1 g was recovered by slurrying the ether-insoluble residue with water and filtering.)

The crude product was dissolved in 10 ml of hot 50% aqueous ethanol and cooled slowly to 25°C; two different types of crystals formed, which were easily separated mechanically:

1. Coarse prisms, m.p. 200 to 202°C, after a second recrystallization from 50% ethanol; reported[12] for 1,2-di-(N-isoquinolon-1-yl)ethane, 203 to 205°C: mass spectrum m/e (rel. intensity) 316 (parent, 10), 188 (19), 171 (2-vinylisoquinolone-1, a McLafferty rearrangement product, 100), 170 (66), 158 (M/2, 6), 145 (1-hydroxyisoquinoline, the other McLafferty rearrangement product, 13), 128 (63); nmr (CDCl$_3$) τ 5.60 (s, 2, C$\underline{H}_2$ in ethylene bridge); 3.65 (d, 1, $\underline{J}$ = 7 Hz, H$_4$); 3.12 (d, 1, $\underline{J}$ = 7 Hz, H$_3$); 2.28-2.60 (m, 3, H$_5$, H$_6$, H$_7$); 1.45 (m, 1, H$_8$).

2. Felted needles, m.p. 190 to 191°C, after two more recrystallizations from 50% ethanol. The data are consistent with those required for 1,2-dimethoxy-1,2-di-(N-isoquinolon-1-yl)ethane: mass spectrum m/e (rel. intensity) 376 (parent, 9), 232 (53), 231 (2-dimethoxyvinylisoquinolone-1, a McLafferty rearrangement product, 7), 198 (M/2, symmetrical bond cleavage between carbon atoms bearing the methoxyl groups, 100), 185 (15), 171 (22), 170 (M/2−CO, 21), 160 (12), 145 (1-hydroxyisoquinoline, the other McLafferty rearrangement product, 7), 128 (98). None of the processes lead to fragments with methoxy on the ring. Nmr (CDCl$_3$) τ 6.52 (s, 3, C$\underline{H}_3$O-), 3.43 (s, 1, CH$_3$OC$\underline{H}$-), 3.38 (d, 1, $\underline{J}$ = 6.5 Hz, H$_4$), 2.22-2.60 (m, 4, H$_3$, H$_5$, H$_6$, H$_7$,), 1.55 (m, 1, H$_8$).
Anal. Calcd. for C$_{22}$H$_{20}$N$_2$O$_4$: C, 70.20; H, 5.36; N, 7.44. Found: C, 70.97; H, 5.39; N, 7.34.

REFERENCES

1. C. A. Heller, "Oxidation of Organic Compounds", Vol. I, Advances in Chemistry Series (#75), 225-244, American Chemical Society, Washington, D.C. (1968).

2. R. A. Henry and C. A. Heller. *Journal of Luminescence, 4,* 105 (1971).

3. S. F. Mason and D. R. Roberts. *Chem. Comm., 1967,* 476.

4. A. H. Corwin, R. R. Arellano, and A. B. Chivvis. *Biochim. Biophys. Acta.,* 162, 533-538 (1968).

5a. J. A. Farrington, A. Ledwith, and M. F. Stam. *Chem. Comm., 1969,* 259.

5b. A. Ledwith. *Accounts of Chemical Research, 5,* 133 (1972).

6. J. G. Carey, J. F. Cairns, and J. E. Colchester. *Chem. Comm., 1969,* 1280.

7. J. E. Dickeson and L. A. Summers. *J. of Heterocyclic Chem., 7* (2), 401 (1970).

8. K. Kuwata and D. H. Geske. *J. Am. Chem. Soc., 86,* 2101 (1964).

9. J. M. Fritsch, H. Weingarten, and J. D. Wilson. *J. Am. Chem. Soc., 92,* 4038 (1970).

10. W. H. Urry and J. Sheeto. *Photochem. and Photobiol., 4,* 1067 (1965).

11. A. L. Black and L. A. Summers. *J. Chem. Soc.* (C), *1969,* 610-611.

12. Preprint of thesis of D. R. Roberts, University of East Anglia.

13. W. H. Urry. Private communication about recent work.

14. G. K. Turner. *Science, 146,* 183 (1964).

15. T. Wilson and A. P. Schaap. *J. Am. Chem. Soc., 93,* 4126 (1971).

16. W. H. Richardson and V. F. Hodge. *J. Am. Chem. Soc., 93,* 3996 (1971).

17. C. A. Heller, D. T. Carlisle, and R. A. Henry. *J. Luminescence, 4,* 81-88 (1971).

SCHAAP: I wouldn't necessarily discount a dioxetane structure for your intermediate, just because it doesn't decompose at the same temperature as some of those we've worked with. We've just observed recently an amino substituted dioxetane, by the addition of singlet oxygen to an eneamine, and found that these dioxetanes decompose below zero degrees to the expected products. This is a big difference in stability; it is not surprising, of course, in light of Dr. Wilson's work on the catalytic decomposition by amines. We've got an internal catalyst in the dioxetanes.

HELLER: Well that's certainly possible. The amine catalysis effect could make our dioxetane decompose very rapidly, as we see with the TMAE. I might mention something about Emil White's paper, that chemiluminescence and photoluminescence may form species in different states. In the case TMAE, the intermediate forming the TMAE was ionic. If the dioxetane was th final intermediate it would be non-ionic, and yet the quenching of that state formed in the chemiluminescence shows that it was formed in an ionic environment.

HERCULES: These carbonyl species formed have accessible excited states such that you might be able to form an excited carbonyl and have a subsequent energy transfer from that as your excitation mechanism.

HELLER: Let's see, the biisoquinoline which I have been talking about primarily has an accessible state, and that's what we see as the excited product.

HERCULES: Yes, in other words you see the carbonyl derived from that as the excited.....

HELLER: That is, we see the fluorescence emission of the biisoquinone product...

HERCULES: No, I was asking about the TMAE.

HELLER: In the TMAE case we see the TMAE itself, and I believe that the excited states of the ureas which are formed, tetramethyl urea, the major product, are probably too high to be accessible.

HERCULES: Even the triplet?

HELLER: The triplet might be accessible. While we look for that in trying to show that there was an intermediate excited state. We looked for second order quenching. If you have formed one excited state and it transfers its energy to a second molecule, to a second excited state, then both of these states should be quenched. We knew that in the case of the TMAE that the second one was quenched; the triplets would be quenched by oxygen, but we found no second order quenching as we changed the oxygen pressure. So, we said that the primary excited species was TMAE, and it must enter into the intermediate, there must be two TMAE's in the intermediate, one of which is oxidized, and one of which is 'picked' off in the excited state.

OXYGEN IN CHEMILUMINESCENCE. A COMPETITIVE PATHWAY OF DIOXETANE DECOMPOSITION CATALYZED BY ELECTRON DONORS

Daniel Chia-Sen Lee and Thérèse Wilson

The Biological Laboratories, Harvard University

Cambridge, Massachusetts 02138

I. INTRODUCTION

It is striking and well known that most of the familiar cases of chemiluminescence in solution involve oxygen, bioluminescence being the foremost example (1). Yet it is clear that not one particular state of the oxygen molecule, nor one single mechanism can be given credit for this prevalence. Ground state triplet oxygen is involved in the many free-radical autoxidation processes which emit low yield luminescence. The step responsible here for the generation of electronically excited products is probably the disproportionation of two peroxy radicals, in what may be a concerted Russell chain termination (2). Singlet oxygen is itself directly the origin of the remarkable red chemiluminescence attending the heterolytic decomposition of hydrogen peroxide. Here the light emitting step is a unique energy pooling process combining the energies of two $O_2(^1\Delta_g)$ into one double-size quantum (3). Examples of such a pooling process followed by energy transfer to a fluorescer have been observed, but it is evident that such processes will have low quantum yields and can hardly be a general mechanism of chemiluminescence, as envisioned by Khan and Kasha (4). However, singlet oxygen is efficient at forming peroxides (3c,d), which are potential sources of luminescence. The superoxide ion is the likely intermediate in another class of peroxidation and in the decomposition of hydrogen peroxide catalyzed by some metal ions or by peroxidases (5). The dismutation of $O_2^{\cdot-}$ may generate singlet oxygen (6), but here again such a step would give only very low yields of luminescence.

Therefore what these three species appear to have in common is their role in the formation or breakdown of peroxides. One

$$\underset{\text{Dioxetane}}{\begin{array}{c}\text{O}\text{---}\text{O}\\ R_1\text{---}\!\!\!\text{---}\!\!\!\text{---}R_4\\ R_2\quad R_3\end{array}}\xrightarrow{\Delta\atop k_1}\begin{array}{c}\text{O}\\ R_1\text{---}\!\!\!\!\text{---}\\ R_2\end{array}+\begin{array}{c}\text{O}^*\\ \text{---}\!\!\!\!\text{---}R_4\\ R_3\end{array}\quad(1)$$

$$\downarrow + \text{Fluorescer}$$
$$\longrightarrow \text{Fluorescer}^*$$

particular class of cyclic peroxides, the dioxetanes, has been looked at as a possible common precursor of many chemiluminescences (7). Its qualifications are several: a) cleavage products with two new carbonyl groups, as is often the case in chemi- and bioluminescence; b) very exergonic cleavage; c) a bent $>\!\!C\!\!<^{\text{O}}_{}$ configuration, as in singlet and triplet carbonyl in contrast with planar ground state $>\!\!C\!\!=\!\!O$; d) the possibility that the cleavage is concerted, along an orbital symmetry forbidden path, which would provide a way of exciting efficiently or necessarily one of the cleavage fragments (7a,8).

Several simple dioxetanes have now been successfully synthetized.* This paper will first summarize the results of their study, from the standpoint of yields and spin-multiplicities of excited products. Some complications inherent to these investigations will be discussed; recent results on the catalysis of a dioxetane decomposition by electron donors will then be presented.

II. RESULTS WITH DIOXETANES

It is interesting to note that all but the last two dioxetanes (11 and 12) of Table I appear to have very similar lifetimes: from a few minutes to a few hours at 60°. Increasing substitution by methyl groups stabilizes the dioxetane ring. Replacing a methyl by a phenyl group has hardly any effect. Exchanging four methyl groups for four methoxy groups increases the rate, but only by a factor of two or three. cis-Diethoxydioxetane 5 appears to be the least stable of the dioxetanes 1 to 9; but replacing one of the ethoxy groups by a phenoxy group seems to have little effect. The lifetimes of the two bicyclic dioxetanes 8 and 9 are about the same as that of 2.

---

* by one or a variant of three methods: a) treating the corresponding bromohydroperoxide with base (9), b) 1,2-cycloaddition of $^1O_2$ to the olefin (10) or c) ozonolysis of the olefin in certain aldehydes and ketones as solvent (11).

Table I. Relatively stable, isolated dioxetanes.

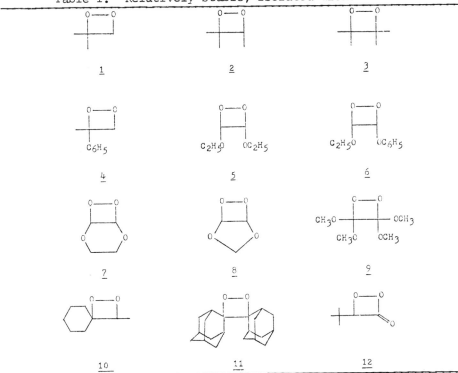

1 $k_1 = 1.1 \times 10^{-3}$ sec^{-1} at 60°, E_a = 23.0 kcal/mole, log A = 12.2 in CCl$_4$ (12)

2 $k_1 = 5.5 \times 10^{-4}$ sec^{-1} at 60° in benzene (9); E_a = 23.7, log A = 12.2 in CCl$_4$ (13)

3 $k_1 = 4.0 \times 10^{-4}$ sec^{-1} at 70° in benzene (15)

4 $k_1 = 1.1 \times 10^{-3}$ sec^{-1} at 60°, E_a = 22.9 kcal/mole, log A = 12.1 in CCl$_4$ (12)

5 $k_1 = 3.2 \times 10^{-3}$ sec^{-1} at 60°, E_a = 23.6 kcal/mole, log A = 13.1 in benzene (16)

6 similar to 5 (17)

7 and 8 lifetimes of the order of 20 to 30 min. at 60° in acetone (18)

9 $k_1 = 1.2 \times 10^{-4}$ sec^{-1} at 56° in benzene (10b)

10 no kinetic data (11)

11 melting point 163-164° (19)

12 lifetimes 5 to 8 min. at room temp. in CCl$_4$ (20)

(References in text)

On the other hand, the last two entries of the table have quite different stabilities. The dioxetane of adamantylidene adamantane 11 has to be heated to 200° to emit visibly (it melts at 163°), whereas the α-peroxylactone 12, of particular interest considering that such structures have been suggested as intermediates in bioluminescence, lasts only 5-8 minutes at room temperature.

Yields and spin states of excited products have been reported for only three dioxetanes (Table II). These results are intriguing. The apparent yields of excited singlet carbonyl fragments are very low, whereas the yields of triplet carbonyls are high, perhaps quantitative. Unusually fast and efficient intersystem crossing from the singlet to the triplet of the carbonyl products is not ruled out in some cases (for example in ethyl formate from 5), but

Table II. Yields of carbonyl cleavage products.

Dioxetanes	Methods	Yields[a]
2	Kopecky and Mumford (9):	
	direct emission 430-440 nm	
	(acetone excimer ? acetaldehyde ?)	
	energy transfer to anthracene, pyrene, biacetyl	
	White et al. (21):	
	isomerization of stilbene, acenaphthylene, 4,4-diphenyl-2,5-hexadienone	highest $\phi_3 \simeq 0.04$
	energy transfer to and fluorescence from a europium chelate	$\phi_3 \simeq 0.14$
5	Wilson and Schaap (16):	
	energy transfer to and fluorescence from 9,10-diphenyl- and 9,10-dibromoanthracene	$\phi_1 \simeq 0$ $\phi_3 \simeq 1$
4	Turro and Lechten (15):	
	reaction of acetone singlets and triplets with trans-dicyanoethylene	$\phi_1 \simeq 0.01$
	also direct emission (max 404 nm); energy transfer to and fluorescence from biacetyl	$\phi_3 \simeq 0.5$

[a] ϕ_1 and ϕ_3 are moles of excited singlet or triplet carbonyl products per mole of dioxetane decomposed.

this is not a likely explanation of the results with 3, in the
light of the known properties of singlet excited acetone (15). In
any case, enough energy is available for exciting one carbonyl
product to either its singlet or its triplet state, but not for
generating two triplet excited products (14,16).

A concerted, symmetry-forbidden pathway could explain high,
if not quantitative, yields of excited products, but it seems more
difficult to predict a high yield of triplet molecules. A spin
conversion during decomposition should, it seems, slow down the
rates, which are fast (log A = 13.1 for 5, 12.2 for 1 and 2). On
the other hand, a diradical pathway, compatible with available
thermochemical data (14), is not immediately expected to give high
yields of electronically excited fragments, on which the reaction
exothermicity must somehow be concentrated. One can perhaps
speculate that quick intersystem-crossing of the diradical occurs,
from the original singlet to a statistically favored triplet, hence
generation of a triplet cleavage fragment (Fig. 1). When the
oxygen atoms have sufficiently separated, the energies of the
singlet and triplet states of the diradical should indeed be very
close.

New and more reliable data on the identities and yields of
excited products from different dioxetanes are very much needed in
order to distinguish between these alternative mechanisms. Bicyclic dioxetanes, such as 7 and 8, where rotation around the C-C
bond would be prevented, would be particularly interesting. Syn-
chronous versus sequential bond cleavage and formation are, however,
two extreme concepts, bracketing in fact a wide range of "diradical"
lifetimes.

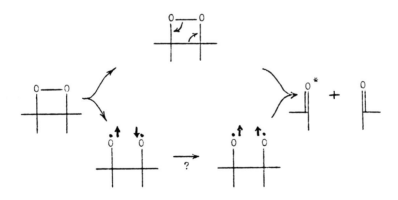

Fig. 1. Concerted versus sequential bond cleavage.

III. ALTERNATE PATHWAYS OF DIOXETANE DECOMPOSITION

Different types of problems complicate yield studies. For example, when "photochemistry without light" is used to count the triplet products, the decomposition of the dioxetane is usually carried out at higher temperature than the photochemical reaction to which its yields are compared. On the other hand, when a spin-forbidden energy transfer to a fluorescer is the key step on which primary yields are based, the experimental uncertainty with which the rate of this transfer is known affects the results.

A more interesting type of complication, however, is inherent to the dioxetanes themselves, and concerned with alternate, "dark" modes of decomposition.

As reported earlier (16), the routine degassing of a solution of 5 and 9,10-dibromoanthracene (DBA) carried out with the hope of increasing the lifetime of the triplet carbonyl donor, hence the yield of energy transfer, brought about an unexpected acceleration of the rate of dioxetane decomposition as well as a lower yield of chemiluminescence. This was attributed to a "photo"-decomposition of the dioxetane, sensitized by triplet DBA which had a longer life as a result of the degassing. Turro and Lechten (15) observed the same effect with dioxetane 3 and interpreted it similarly, by the decomposition of 3 self-sensitized by triplet acetone. These sensitized dioxetane decompositions seem to generate no excited products. Therefore when yields of excited products are measured, it is advisable to keep the dioxetane concentration low in order to minimize this competitive pathway.

Another intriguing and unexpected complication turned up in an attempt to capitalize on this oxygen effect. If indeed oxygen is important as a quencher of triplet excited DBA (or triplet acetone), this quenching process must generate singlet oxygen, which could then react with a suitable acceptor. In the case of dioxetane 5, the logical candidate for the acceptor is cis-diethoxyethylene (DEE), since 5 is synthetized in the first place by 1,2-cycloaddition of $^1O_2$ to DEE (10a,16). Thus, we hoped, singlet oxygen formed by quenching of ^{3}DBA* should react with added DEE to reform 5, a "recycling" process which should result in slower apparent rates of decomposition of the dioxetane and of chemiluminescence decay. To our surprise, exactly the opposite was observed. The addition of DEE markedly shortens and weakens the chemiluminescence. We wish to report here our results with this electron-rich olefin and with other electron donors.

IV. EXPERIMENTAL

cis-Diethoxydioxetane $\underline{5}$ was prepared as in (10a) and (16). 9,10-Dibromoanthracene (Aldrich) was recrystallized from xylene. Triethylamine and diethylamine (Eastman) were distilled and dried before use. A sample cis-diethoxyethylene, purified by vpc, was kindly given to us by Dr. A. P. Schaap. Benzene, either Reagent Grade (Fisher) or Spectrograde (Eastman), dried over molecular sieves, was used; no difference was observed. Spectrograde acetonitrile (Eastman) was used without further purification. Pyridine (Fisher) was freshly distilled. Tetramethylethylene (Aldrich) and 1,4-diazabicyclooctane (Aldrich) were used without further purification. All experimental methods were as in (16).

V. CATALYSIS OF THE DECOMPOSITION OF CIS-DIETHOXYDIOXETANE BY ELECTRON DONORS

We found that, like DEE, aliphatic amines also accelerate the rate of chemiluminescence decay at the expense of the intensity, even when present only in catalytic amounts. Diazabicyclooctane (DABCO), triethylamine (Et_3N), diethylamine (Et_2NH)* are all very effective, with benzene as solvent and DBA as fluorescer, whereas pyridine is much less efficient. These faster rates of chemiluminescence decay reflect faster decomposition of the dioxetane, monitored by nmr or by iodometric titration: the rates obtained by these two methods agree within experimental errors with the chemiluminescence decay rates. With or without Et_3N or DEE, plots of the chemiluminescence intensity, or concentration of $\underline{5}$, are first order as a function of time and remain linear down to 1 or 2%** of the initial dioxetane concentration. Nmr shows that the reaction product is the same, i.e., ethyl formate only, and that Et_3N or DEE appear unchanged. This is supported by the lack of curvature in the plots log I versus time: a bimolecular reaction removing the catalyst would introduce an upward curvature. Typically, at 43.5° the observed rate constant, k_{obs}, with 0.015 M Et_3N (in equimolar initial concentration with $\underline{5}$) is 35 times larger than the uncatalyzed rate, k_{un}.

The rate of the catalyzed reaction, defined as $k_{cat}=k_{obs}-k_{un}$, increases with the concentration of Et_3N although not linearly (Fig. 2).*** DEE gives similar results.

---

* The only aliphatic amines tried so far.

** Where measurements become inaccurate.

*** Plots of $1/k_{cat}$ versus catalysts concentration are not linear, either.

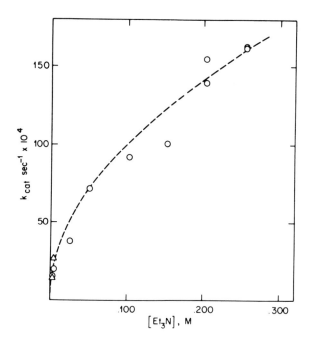

Fig. 2. Rate of cis-diethoxydioxetane decomposition catalyzed by Et$_3$N as a function of Et$_3$N concentration. O, chemiluminescence decay rates (with DBA as a fluorescer); Δ, from iodometric measurements of dioxetane concentration. (Solvent, benzene; temperature, 43.5°; initial concn. of 5, 0.015 M).

Figure 3 shows the temperature dependence of the catalyzed rates for three catalysts. Surprisingly, the slopes of the Arrhenius plots are very much the same for the olefin and the two amines. They correspond to an activation energy of 11.9 kcal, roughly half that of the uncatalyzed reaction (E_a = 23.6 kcal).

In order to find out if the catalyzed portion of the overall decomposition generates chemiluminescence, we carried out experiments in which the intensity of chemiluminescence from a dioxetane solution was measured immediately before (I_1) and after (I_2) a quick drop in solution temperature (from T_1 to T_2). In the case of a reaction such as (1), the activation energy for chemiluminescence

$$E_{chl} = \frac{R \ln I_1/I_2}{\frac{1}{T_2} - \frac{1}{T_1}}$$

is related to the activation energy for the unimolecular decomposition of the dioxetane (E_a, from the Arrhenius plot) by

$$E_a = E_{chl} - E_{\phi_F}$$

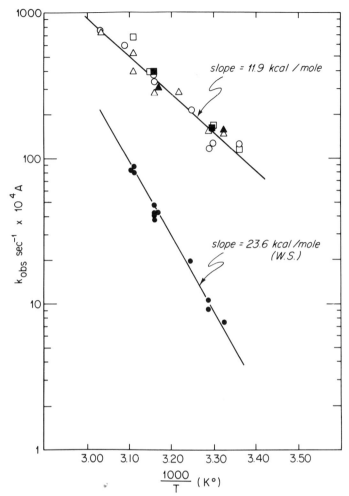

Fig. 3. Temperature dependence of the first-order rate constant k_{obs} of chemiluminescence decay, with or without "catalyst." ●, with no catalyst; ○, with Et$_3$N, 0.025 M (A=1); ◻, with Et$_2$NH, 0.039 M (A=.74); ◼, with Et$_2$NH, 0.034 M (A=1); △, with DEE, 0.052 M (A=1); ▲, with DEE, 0.021 M (A=2.6). (Solvent, benzene; fluorescer, DBA; initial concn. of 5, 0.015 M).

(where E_{ϕ_F} is the activation energy for the fluorescence of DBA $\simeq$ -4kcal) (16).

For example, in the case of the uncatalyzed decomposition of 5, the luminescence <u>intensity</u> is directly proportional to the <u>rate</u>

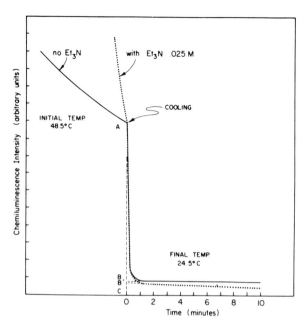

Fig. 4. Effect of a temperature drop on the chemiluminescence of a solution of cis-diethoxydioxetane, with or without Et₃N. Ratio of light intensities before and after cooling: AC/BC (no Et₃N); AC/B'C (with Et₃N). (Solvent, benzene; fluorescer, DBA; initial concn. of 5, 0.015 M).

of the light emitting reaction, so that the ratio of intensities before and after cooling gives the ratio of rates, hence the activation energy E_{chl} (for an example, see Fig. 4). The agreement between E_{chl}, after correcting for E_{ϕ_F}, and the activation energy E_a obtained from the slopes of intensity decay is excellent (Table III, first row, and ref. 16). It establishes that there is only one observable path in the uncatalyzed reaction. The situation is very different for the catalyzed reaction. Although the reaction rates at the two temperatures, measured by the rates of decay of the chemiluminescence, are now much faster (Fig. 4), the drop in intensity, i.e., the ratio I_1/I_2 and therefore E_{chl}, are the same for the catalyzed and uncatalyzed reactions. This means that the light emitting reaction is the same, even though the ratio of the two slopes (rates of decay by all paths) now corresponds to a very different activation energy (E_a = 11.9 kcal). Similar results were obtained with other catalysts. Hence we conclude that there are now two paths leading to the same products and that only the uncatalyzed path emits chemiluminescence. A specific example illustrates this conclusion as well as the sensitivity of the method. Cooling a dioxetane solution with Et₃N 0.256 M from 43.6° to 30.2° gave I_1/I_2 = 4.26 as in the control run without Et₃N, whereas the ratio of the catalyzed rates before and after cooling

Table III. Activation energy E_{chl} (from temperature drop).[a]

Catalyst	E_{chl}, kcal M^{-1}
-	18.5 ± 0.5 (6 expts.)[b]
DEE	18.7 ± 1.2 (5 expts.)
Et$_3$N	18.6 ± 2.0 (10 expts.)
Et$_2$NH	19.4 ± 0.4 (3 expts.)

[a] Solvent, benzene; fluorescer, DBA; initial concn. of 5, 0.015 M.
[b] Previous value: 18.9 ± 0.2 kcal M^{-1} (16).

was only 1.7. At the lower temperature, the rate of the catalyzed reaction was 100 times the uncatalyzed rate, yet evidently the chemiluminescence emission monitored only that hundredth part of the dioxetane reaction which took place non-catalytically. In fact, if as little as 1% of the catalyzed reaction had generated chemiluminescence, the ratio I_1/I_2 would have been reduced by 30%, far outside the limits of error of the measurements.

There are two ready hypotheses to explain these observations. The simplest is that the catalyzed path is truly a dark one, not forming excited products. This is in quantitative agreement with the observed quantum yields, after the obvious corrections for the quenching of ^{1}DBA* and ^{3}F* by the catalyst in the bulk of the solution (Table IV). The quenching of the fluorescence of aromatic hydrocarbons by amines is indeed well established (22) and quenching by electron-rich olefins has recently been reported (23). Charge-transfer interactions are implicated in both cases. We found here, in separate experiments, that the fluorescence of DBA is indeed quenched at diffusion controlled rate by DEE and by Et$_3$N. A Stern-Volmer analysis gave $k_q\tau_s$ = 24.0 M^{-1} for DEE and 18.5 M^{-1} for Et$_3$N.* Unlike the rates of quenching of ^{1}DBA*, the k_q values for ^{3}F* could not be checked experimentally. But the average value, k_q = 7 ± 3 x 10^8 M^{-1} sec^{-1} which fits the data (third row, Table IV) does not seem out of line with the literature values for quenching of triplet ketones by amines and olefins (25).

An alternative explanation of our results cannot be ruled out. The catalyzed decomposition presumably requires the immediate

---

* In aerated benzene solutions, at room temperature, measured with an Aminco-Bowman spectrofluorimeter. In degassed solutions $\tau_{^1DBA^*}$ = 1.8 nsec (24).

proximity of the amine and the dioxetane. Hence the catalyst may always be present in the solvent cage of the excited carbonyl product. Under such conditions, it is conceivable that the quenching efficiency of the catalyst would be much greater than that calculated from even diffusion controlled bulk quenching. At this point, we cannot decide between those two alternatives, although we feel that our results would call for such complete quenching in the cage that we favor the view that the catalyzed pathway is intrinsically dark.

Amines are well known to induce the decomposition of different classes of peroxidic compounds, such as benzoyl peroxide (26), ozonides (27), triphenylphosphite-ozone adduct (28), etc. A charge-transfer interaction between the amines as the donors and the

Table IV. Effects of catalysts on relative quantum yields of chemiluminescence.[a]

	DEE concentration			Et_3N concentration		
	0	0.020 M	0.052 M	0	0.025 M	0.256 M
Φ_{obs}[b]	1.00[c]	0.27	0.047	1.00	0.007	0.002
Φ_{calc}[d]		0.90	0.66		1.02	0.47
k_q calc[e]		6×10^8	1×10^9		[f]	4×10^8

[a] Temperature, 43.5°; solvent, benzene; fluorescer, DBA; initial concn. of 5, 0.015 M.

[b] Relative chemiluminescence yield, defined as $\Phi_{obs} = I_{obs}/k_{obs}$, where I_{obs} is the peak chemiluminescence intensity emitted immediately after the solution reaches the constant temperature; in arbitrary units.

[c] Relative chemiluminescence yield in absence of catalyst, defined in text as Φ_{un}.

[d] Φ_{calc} is a relative quantum yield calculated on the assumption that only the uncatalyzed reaction is chemiluminescent and that $^1DBA^*$ is quenched by the catalysts at diffusion controlled-rate.

[e] Calculated values of k_q (rate of quenching of $^3F^*$ by catalysts) which would account for remaining discrepancy between Φ_{calc} and Φ_{un}; in $M^{-1} sec^{-1}$.

[f] No apparent quenching of $^3F^*$ here; but with $k_q = 7 \times 10^8$ $M^{-1} sec^{-1}$ the expected quenching would amount to less than 20% of the yield, within the limits of errors of these measurements.

peroxides as acceptors (in the vacant antibonding orbital of the O-O bond) has been invoked for the first step of these processes (29). In the dioxetane case, there is only a rough correlation between a catalyst's ionization potential and its efficiency as a catalyst (Table V), probably because steric effects are important in the catalysis. The correlation between catalytic efficiency and rate of quenching of DBA fluorescence--a likely donor-acceptor interaction also--is much better.

Increasing the solvent polarity seems to have only a very small effect on the rate of the catalyzed reaction. In one pair of experiments with mixed acetonitrile:benzene (6:1, v:v) instead of benzene as solvent, with and without Et_3N, there was a 3-fold increase in k_{cat} compared to a 1.4-fold increase in k_{un}. This small solvent effect (30) is consistent with a transition state having a larger dipolar character in the catalyzed reaction than in the uncatalyzed reaction, but it seems far too small to be compatible with total charge separation along the catalyzed path.

Table V. Catalytic effectiveness, quenching of DBA fluorescence and ionization potential of the catalysts.

Catalyst	First Adiabatic Ionization Potential (eV)	$\dfrac{k_{cat}}{k_{un}}$ [a]	$k_q \times 10^{-9}$ [b] $(M^{-1} sec^{-1})$
Triethylamine (Et_3N)	7.50[c]	11	11
Diethylamine (Et_2NH)	8.01[c]	7	4.4
cis-Diethoxyethylene (DEE)	7.38[d]	5.5	14
Tetramethylethylene (TME)	8.05[e]	0.02	0.06
Diazabicyclooctane (DABCO)	7.2[f]	67	23

[a] Catalytic effectiveness at 50°; catalyst concn., 0.05 M.

[b] Rate of quenching of DBA fluorescence, in aerated benzene at room temperature; from Stern-Volmer plots with τ_{1DBA*} = 1.8 nsec.

[c] From D. W. Turner, Adv. Phys. Org. Chem. 4, 31 (1966).

[d] G. N. Taylor, personal communication.

[e] Reference 23 in text.

[f] A. M. Halpern, J. L. Roebler and K. Weiss, J. Chem. Phys. 49, 1348 (1968).

VI. CONCLUSIONS

The powerful effect of cis-diethoxyethylene on the decomposition of dioxetane 5 was not anticipated, since this dioxetane is synthetized by 1,2-photoaddition of singlet oxygen to DEE, at low temperature (10a). Evidently it is mainly because the activation energy of $^1O_2$ addition is much smaller than that of the DEE induced cleavage of D (E_a = 11.9 kcal) that the near-quantitative synthesis of D is possible. The use of a solvent (Freon 11) where D is largely insoluble is certainly of help too.

This dual behavior of DEE is only an apparent paradox. It is its low ionization potential which is responsible for 1,2- instead of 1,3- ("ene") addition of electrophilic $^1O_2$ to the double bond (31). In contrast, tetramethylethylene with its appreciably higher ionization potential undergoes an "ene" type reaction with singlet oxygen, while being 2 or 3 orders of magnitude less efficient than DEE at catalyzing the decomposition of D or at quenching the fluorescence of DBA (Table V). There is therefore no hope of "recycling" a dioxetane: any olefin which would react with singlet oxygen to give a dioxetane would tend to decompose it at the temperature necessary for the dioxetane cleavage.

The striking feature of the catalyzed reaction is the independence of its activation energy on the type of catalyst. It suggests that in the presence of any good donor, the dioxetane decomposition can proceed through a new transition state of more polar character, located about 12 kcal below the transition state of the uncatalyzed reaction. The catalyzed path is not attended by observable luminescence, either because it generates no excited products, or because these are quenched in the primary solvent cage, an unresolved question. One can only speculate at this point on the mechanism of the decomposition. The role of the catalyst is probably to allow an easier cleavage of the O-O bond, thereby forming a diradical with more or less ionic character. The effect of electron donors on other dioxetanes should be investigated.

Our results suggest caution in the interpretation of triplet titrating experiments, where yields of products but not reaction rates are measured in order to estimate the yield of excited products from the decomposition of a dioxetane. The possibility that a reagent such as a diene may induce a dark decomposition should be kept in mind.

Acknowledgements

We are grateful to Professor J. W. Hastings for his interest in this work, which was supported in part by National Science

Foundation Grants GB-16512 and GB-31977X to his laboratory, and also to Dr. A. P. Schaap for helpful suggestions and a sample of cis-diethoxyethylene.

References

1a. F. McCapra, Quart. Rev. Chem. Soc. 20, 485 (1966).

b. K. D. Gunderman, Angew. Chem. Int. Ed. Engl. 4, 566 (1965).

c. J. W. Hastings, Ann. Rev. Biochem. 37, 597 (1968).

2a. V. A. Belyakov and R. F. Vassil'ev, Photochem. Photobiol. 11, 179 (1970).

b. R. E. Kellogg, J. Am. Chem. Soc. 91, 5433 (1969).

c. G. A. Russell, Ibid. 79, 3871 (1957).

3a. A. U. Khan and M. Kasha, J. Am. Chem. Soc. 92, 3293 (1970).

b. E. A. Ogryzlo, in Photobiology (A. C. Giese, ed.), Vol. V, pp. 35-47 (1970).

c. T. Wilson and J. W. Hastings, Ibid., pp. 49-95.

d. D. R. Kearns, Chem. Rev. 71, 395 (1971).

4. A. U. Khan and M. Kasha, J. Am. Chem. Soc. 88, 1574 (1966).

5a. P. F. Knowles, J. F. Gibson, F. M. Pick and R. C. Bray, Biochem. J. 111, 53 (1969).

b. I. Fridovich, Accounts Chem. Res. 5, 321 (1972).

c. I. Yamazaki, L. H. Yokota and R. Nakajuma, in Oxidases and Related Redox Systems (T. E. King, H. S. Mason and M. Morrison, eds.) Wiley, N. Y. (1965), p. 485.

6. R. M. Arneson, Arch. Biochem. Biophys. 136, 352 (1970).

7a. F. McCapra, Chem. Comm. 155 (1968).

b. M. M. Rauhut, Accounts Chem. Res. 2, 80 (1969).

8. F. McCapra, Pure Applied Chem. 24, 611 (1970).

9a. K. R. Kopecky, J. H. van de Sande and C. Mumford, Can. J. Chem. 46, 25 (1968).

b. K. R. Kopecky and C. Mumford, Ibid. 47, 709 (1969).

c. W. H. Richardson and V. F. Hodge, J. Am. Chem. Soc. 93, 3996 (1971).

10a. P. D. Bartlett and A. P. Schaap, J. Am. Chem. Soc. 92, 3223 (1970).

b. S. Mazur and C. S. Foote, Ibid. 92, 3225 (1970).

11. P. R. Story, E. A. Whited and J. A. Alford, Ibid. 94, 2142 (1972).

12. W. H. Richardson, M. B. Yelvington and H. E. O'Neal, Ibid. 94, 1619 (1972).

13. K. R. Kopecky, quoted by O'Neal and Richardson, ref. 14.

14. H. E. O'Neal and W. H. Richardson, J. Am. Chem. Soc. 92, 6553 (1970).

15. N. J. Turro and P. Lechten, Ibid. 94, 2886 (1972).

16. T. Wilson and A. P. Schaap, Ibid. 93, 4126 (1971).

17a. A.P. Schaap and N. Tontapanish, "Symposium on Oxidation by Singlet Oxygen," Am. Chem. Soc., Washington, D. C., Vol. 16, No. 4, 1971, p. A78.

b. A. P. Schaap, personal communication.

18. A. P. Schaap, Tetrahedron Lett., 1757 (1971).

19. J. H. Heringa, J. Strating, H. Wynberg and W. Adam, Tetrahedron Lett., 169 (1972).

20. W. Adam and J. C. Liu, J. Am. Chem. Soc. 94, 2894 (1972).

21a. E. H. White, J. Wiecko and D. R. Roswell, Ibid. 91, 5194 (1969)

b. E. H. White, J. Wiecko and D. R. Roswell, Ibid. 91, 5194 (1969).

c. P. D. Wildes and E. H. White, Ibid. 93, 6286 (1971).

22. A. Weller, Pure Appl. Chem. 16, 115 (1968), and references therein.

23. G. N. Taylor, Chem. Phys. Lett. 10, 355 (1971).

24. R. S. H. Liu and D. M. Gale, J. Am. Chem. Soc. 90, 1897 (1968).

25. See for example: (a) G. A. Davis and S. G. Cohen, Chem. Comm. 622 (1970); (b) I. E. Kochevar and P. J. Wagner,

J. Am. Chem. Soc. 94, 3859 (1972); (c) J. B. Guttenplan and S. G. Cohen, Ibid. 94, 4040 (1972), and references therein.

26. See for example: (a) W. A. Pryor, Free Radicals (McGraw-Hill Book Co., New York, 1966), pp. 98-99, and references therein. (b) K. Tokumaru and O. Simamura, Bull. Chem. Soc. Japan 36, 333 (1963).

27. R. M. Ellam and J. M. Padbury, Chem. Comm., 1094 (1971).

28. P. D. Bartlett and G. D. Mendenhall, unpublished results.

29. K. Tokumaru and O. Simamura, Bull. Chem. Soc. Japan 36, 333 (1963).

30. Comparable to the solvent effects on the rate constants for quenching of naphthalene fluorescence by conjugated dienes, observed by G. N. Taylor and G. S. Hammond, J. Am. Chem. Soc. 94, 3684 (1972).

31. D. R. Kearns, J. Am. Chem. Soc. 91, 6554 (1969).

ADAM: Do you think you could determine the difference between the diradical and the concerted mechanism by this attack on a double bond.

WILSON: I think that diradicals in general are very very difficult to study in a way that you can be sure that it doesn't go by one path or another.

J. LEE: In order to point out the difficulty of making quantitative measurements of the yield of triplet formation, one of the things that I think must be considered is the possibility of complex formation in the ground state between the quencher and the quenchee and this possibility would be very difficult to test. Does the system obey Beer's law for instance? Would this offer an explanation for the apparent catalysis? Is it merely that the amine is forming a complex with the dye in the ground state, so you no longer have the normal path of decomposition to give light?

WILSON: You don't need to have any fluorescer in the system because by NMR you can follow the oxygen reaction catalyzed by amines.

HELLER: Just a comment on compounds that we think go through dioxetanes as judged from the products. The molecules that we have been working with all have a nitrogen right next to the carbon which will be forming the dioxetane. These decompose very rapidly if they are dioxetane with the nitrogen there apparently catalyzing by a factor of at least 10^6-10^8. These also apparently go to singlets rather than to triplets. I wonder whether the amines might have the same effect as a catalyst. In catalyzing your dioxetanes, the nitrogen might tend to give a singlet product rather than a triplet state as they seem to do in the case in which you have the nitrogen right in the compound

WILSON: Have you been able to isolate these dioxetanes in any way?

HELLER: No, the impression we get from the studies we've done is that the dioxetane may only last on the order of 10^{-12} sec. It's rather hard to isolate them. So we think that the evidence points to them we haven't been able to isolate them. The rates are very much higher than the rates you have mentioned.

RAUHUT: I wonder if we can really exclude the possibility of an interaction between the fluorescer and the dioxetane, which leads to a singlet excitation of the fluorescer which is more a question than it is an argument. But what bothers me is the low quantum yields that one gets even when a fluoreser is added, which could indicate that you are accellerating or adding a new reaction into the overall scheme to an extent of perhaps only 1% of the overall process and you probably wouldn't see that 1% kinetically.

WILSON: Yes. The only thing that I can say is that fluorescers that have been tried do not change the rate of decomposition of the dioxetane decomposition, let's say, by NMR or by iodometric titration. Without fluorescer we get rates which are identical to those that we get with the fluorescer. So, I think that the fluorescer doesn't seem to accelerate the decomposition or modify it in any way.

RAUHUT: Yes, but you see what bothers me is that the kinetics are probably not adequate to detect a rate change if the total change is only 1% or so of the total process.

WILSON: Yes, but I think that the rates--the yields that you obtain with dibromoanthracene are extremely high.

RAUHUT: Are the quantum yields in the presence of something like diphenylanthracene actually very high?

WILSON: No, not with diphenylanthracene. Our interpretation here is that diphenylanthracene cannot accept energy from a triplet donor, only from a singlet donor and that as soon as you put something in that, because of its heavy atom is able to make the conversion with any kind of efficiency, which is also the case I guess, of the europium chelates of Dr. White, then you get a much more intense chemiluminescence which of course is the product of the fluorescence efficiency, and the efficiency of production of carbonyls in the first place. You have to multiply that by the efficiency of this energy transfer from carbonyl to the fluorescer and multiply this again by the quantum yield of the fluorescer.

McCAPRA: I don't want to anticipate what I am going to say later about that. I would just like to mention that calculations on the atoms that involve the symmetry induced barrier seem to suggest that you will find that the route via the triplet surface is actually the lower energy route. In other words it is not only that the barrier imposed by Woodward-Hoffman appears (this is a thermodynamic question and not a symmetry problem) so much as just simply that you have this high barrier. You have a trigger holding something that is really quite explosive on the molecular level.

WILSON: I had intended to say this, but I got lost.

McCAPRA: I wonder whether your catalysis by amines and olefins is not due to the fact that you don't know actually the thermodynamics of the situation and the fact is that you have lost sufficient activation energy to find yourself thermodynamically incapable of populating the high levels that you have to .

WILSON: I don't believe so because, as I said, the triplet state of the formate is apparently quite low (maybe only 70 Kcal), and I think we still have plenty of energy to produce that.

McCAPRA: The other thing is that olefins and amines do in fact catalyze the decomposition of simple dioxide peroxides. This has to involve formation of a diradical and therefore supports your findings.

WILSON: Yes, that's what I would like to say, that the noncatalytic reaction is perhaps concerted whereas the catalyzed reaction goes through a diradical.

McCAPRA: I have reservations about accepting this, but I think I've taken enough time.

ADAM: This is more of a comment regarding the difficulty in accepting a diradical mechanism. If you look at the thermal decomposition of dioxetanes at room temperature as a function of solvent you see that there is an interplay between the energy factor and the entropy factor. Now if it is a diradical mechanism it is very difficult to understand this vast change in the entropy factor. Therefore, the idea that comes out of Goulds and Lessen's work is to retain the concerted mechanism with some vibronic interactions of the peroxide which converts a singlet into a triplet by appropriate vibrational changes coupled with electronic motion. There is a possibility that this would be more or less half of a Woodward-Hoffman path.

CHEMICAL AND ENZYMATIC MECHANISMS OF FIREFLY LUMINESCENCE

W. D. McElroy and Marlene DeLuca

University of California, San Diego

Introduction

The chemistry of firefly luminescence is gradually being resolved. After a number of years of intensive work by a large number of investigators it is now possible to propose a reasonable organic mechanism for the overall reaction leading to light emission (1,2,3,4). The most recent findings which are of considerable interest to all investigators in the field of bioluminescence concern the mechanism of the chemical processes leading to the excited state and the identification and synthesis of the product emitter (4,5).

Several recent papers have reviewed our extensive knowledge about the firefly systems; therefore the present paper will be limited to a capsule presentation of the facts which are essential to our understanding of the chemical and enzymatic mechanisms which have been proposed.

It should be kept in mind that most of the detailed chemistry and enzymology has been done primarily with the lymprid beetle, _Photinus pyralis_; however enough work has been done on over twenty other species of true fireflies to suggest that the same proposed mechanisms are valid. In addition, the present evidence indicates that the luciferin structure and the action of ATP serve the same function in the _Elateridae_, the _Phengodidae_ and possibly the _Drilidae_ as they do in the _Lampyridae_ (6).

Structure of the Product Emitter, (Oxyluciferin), Luciferin (LH_2) and Dehydroluciferin (L)

Before discussing the detailed mechanisms which have been proposed for firefly bioluminescence we would like to review briefly the structure of luciferin and point out some of the key features that are important for the light emitting process. The structure of $D(-)LH_2$ and L are shown in Figure 1. (7,8).

1. The carboxyl group of LH_2 is the important site for the formation of the anhydride with adenylic acid from ATP and as discussed later is the source of the CO_2 which is released in the chemical reaction leading to light emission.

2. The hydrogens at the 4 and 5 carbon atoms are of great importance. The data indicate that the enzyme must abstract a proton from the 4 position prior to the addition of oxygen at that point. The fact that the substitution of deuterium at that position inhibits the rate of the light reaction by almost fifty percent supports this conclusion (9).

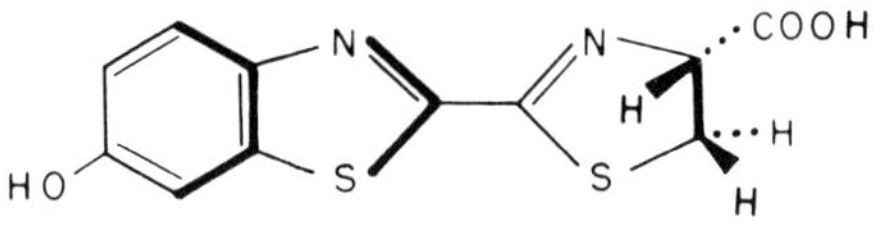

L (+) LUCIFERIN

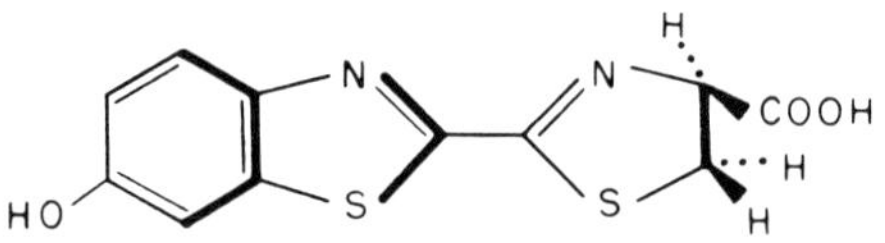

D (−) LUCIFERIN

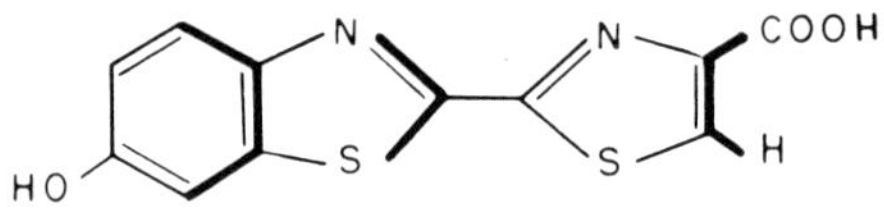

DEHYDROLUCIFERIN

Fig. 1. Structure of firefly luciferin and dehydroluciferin.

Proton abstraction at the 5 carbon atom is essential in determining the color of the light emitted, and will be discussed later.

3. The state of the hydroxyl group at the 6' position of the benzothiazole ring is important for both the luminescent and fluorescent properties. All evidence indicates however that it is the phenolate ion that is essential for both red and yellow-green light emission.

   The titration of LH_2 both potentiometrically and spectrophotometrically shows a single ionization between pH 4 and 11.5. This reflects the dissociation of the 6'-hydroxyl group to the phenolate ion with pK_a = 8.7. The ionization of this group has a large effect on the absorption spectrum. This shift in the spectrum on ionization is both necessary and sufficient to conclude that the pK_a of ionization of the excited state is different from the ground state. From the calculated pK_a, one would predict that, if equilibrium were established in the excited state, the predominant form between pH 2 and 12 would be the phenolate ion. The fluorescence emission spectrum of LH_2 at pH 4.5 shows that if the phenol form is excited, proton transfer to the solvent occurs and emission is from the excited state of the phenolate ion (LH_2-O^{-*}). The quantum yields of fluorescence due to absorption by the two ground-state forms are not the same; the phenolate ion is higher by a factor of 2.5, suggesting only partial equilibrium in the excited state is obtained. We will discuss this proton transfer to the enzyme later as well as the effect of substituting other groups at the 6' position on the color of light (10).

4. The binding of dehydroluciferin to the enzyme brings out a blue fluorescence property which is characteristic of the phenol excited state. This property can be used to study proton transfer and suggests the presence of a proton acceptor located in a highly hydrophobic environment of the enzyme (See below). Dehydrolucerfin, when it reacts with ATP, can also form the adenylate anhydride. The high fluorescence of L essentially disappears when it forms the adenylate, a property that is very useful in studying the activation reaction.

5. The strength of binding of luciferin to luciferase has been determined indirectly by studying the binding of competitive analogues and calculating the corresponding K_i values.

The fact that the K_i values of all the luciferin analogues are at 10^{-5} M level suggest that these ring structures alone are responsible for most of the binding forces. Using the average K_i values, the free energies of binding of the two ring systems were calculated. These values are given below, together with those of the thiazoline ring system and the carboxylate group which were calculated from the difference in K_i values of the compounds with and without such groups (11, 12).

```
2-(2-benzothiazolyl)- Δ²-thiazoline . . .-7.5 kcal
benzothiazole . . . . . . . . . . . . . .-6.0 kcal
thiazoline . . . . . . . . . . . . . . .-1.5 kcal
carboxylate group . . . . . . . . . . . .+1.1 kcal
```

The difference in the effect of a methyl group at the 5'- and 6'- position of the benzothiazole ring suggests that a benzothiazole derivative or a luciferin analogue is bound to its site only in one fixed position. If it could also be bound upside down, the 5'-position would be equivalent to the 6'-position. We do not have any evidence at this moment concerning the group or groups on the enzyme which are responsible for this binding. As to the groups on the benzothiazole molecule, the data are also suggestive that the OH group at the 6'-position is probably not needed for this specific binding. Replacement of the OH group with other groups did not impair the binding to any significant extent. It is likely, therefore, that the hetero-atoms in the ring, either N or S or both, are responsible in directing the compound to the set position.

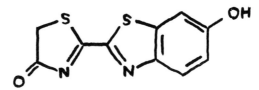

Fig. 2   Structure of Product-Emitter (Suzuki and Goto 1971)

6. Hydrophobic Nature of the Active Site

The use of dyes as probes for hydrophobic sites on proteins is now well documented (13, 14). The interaction between dye and protein may be followed by monitoring the change in the intensity or spectrum of fluorescence when the dye is bound in a hydrophobic environment. The binding of dyes to luciferase has been used in an attempt to obtain more information about the nature of the "active site" of this enzyme.

A surprising finding is that 2,6-TNS (toluidinonaphthalene sulfonate) binds much better than 1,5-ANS or the corresponding isomer, 2,6-ANS (anilinonaphthalene sulfonate). The only difference in the structure of these two classes of dyes is a methyl group, yet the K_A for TNS is tenfold greater than that for ANS. It is not obvious why such a small change in the dye should result in such an increase in affinity for the protein.

Calculations of the ΔF of binding of the dyes from the equilibrium constants shows for 1,5-ANS, $\Delta F = -6.3$ kcal/mole while for 2,6-TNS, $\Delta F = -8.2$ kcal/mole. Therefore, the addition of a methyl group to ANS results in a change of ΔF of binding of 1.9 kcal/mole. The large difference in ΔF observed between the binding of 1,5-ANS and 2,6-TNS to luciferase cannot be attributed entirely to the increased hydrophobic character of the latter molecule. The position of substituents on the naphthalene ring does not seem important for binding since all of the ANS isomers tested have similar binding constants. The results may mean a larger change in the conformation of the enzyme when 2,6-TNS combines with luciferase which results in an apparent tighter complex than observed for 1,5-ANS (14).

7. Structure of the Emitter Product

The emitter product of the luminescent reaction has been identified and shown to be identical for both the chemiluminescent and the enzyme catalyzed light reaction (5). In addition, Suzuki et al. have been able to synthesize the product and to demonstrate fluorescent properties identical to the natural products. The structure of "oxyluciferin" is shown in Fig. 2.

The Enzyme Catalyzed Light Reaction

When one starts with free luciferin and luciferase, it is necessary to add ATP and magnesium or manganese ions in order to obtain light emission. In the initial reaction, there is an adenyl transfer from ATP to the carboxyl group of luciferin with the elimination of inorganic pyrophosphate (15). The reaction is analogous to the fatty-acid- and amino-acid -activating reaction. The luciferyl-adenylate (LH_2-AMP) remains tightly bound to the enzyme and subsequently reacts with molecular oxygen to give light emission as indicated in the following reactions:

$$E + LH_2 + ATP \xrightarrow{Mg^{++}} E \cdot LH_2\text{-AMP} + PP \qquad (1)$$

$$E \cdot LH_2\text{-AMP} + O_2 \longrightarrow \text{Oxyluciferin} + CO_2 + AMP + \text{light} \qquad (2)$$

It is possible to eliminate the necessity of ATP for light emission if one synthetically makes LH_2-AMP from LH_2 and AMP. The addition of LH_2-AMP to an enzyme solution leads to a rapid production of light indicating that the activation step is the rate limiting reaction for the overall process.

In addition to reactions 1 and 2, luciferase will catalyze the formation of dehydroluciferyl-adenylate (L-AMP) as shown in the following reaction:

$$E + L + ATP \xrightarrow{Mg^{++}} E \cdot L\text{-AMP} + PP \qquad (3)$$

The quantitative utilization of substrate and product production has been studied in great detail during light emission. It has been established that for each LH_2 molecule used one quantum of light is emitted (16). In addition one mole of oxygen is consumed per mole of LH_2 utilized (17), and one mole of CO_2 is released (18). In the activation of LH_2 to form LH_2-AMP one ATP is used and one PP is formed.

These facts eliminate a large number of possible mechanisms and suggest that at least one atom of oxygen is incorporated into the product emitter.

Oxidative Mechanism for Light Emission

Plant et al. using ^{14}C- carboxyl labeled luciferin in the presence of excess enzyme, ATP and O_2 demonstrated the quantitative liberation of $^{14}CO_2$ from the luciferin during the light reaction. Based on studies from chemiluminescent reactions by McCapra and associates (19,20,21) and White and associates (4) a mechanism of the organic reaction was proposed. The suggested scheme is presented in Fig. 3. Following the loss of a proton at carbon 4 oxygen addition occurs at that position which eventually led to the formation of a four-membered peroxide ring after the loss of AMP. Decarboxylation would lead to the expected products and theoretically calculations indicate that the final reaction yields more than enough energy to give rise to the excited state.

Although this mechanism was very attractive for a number of reasons subsequent experiments indicate that an alternate pathway for oxygen utilization must be proposed. In addition the necessity for water ($^-$OH) is also suggested.

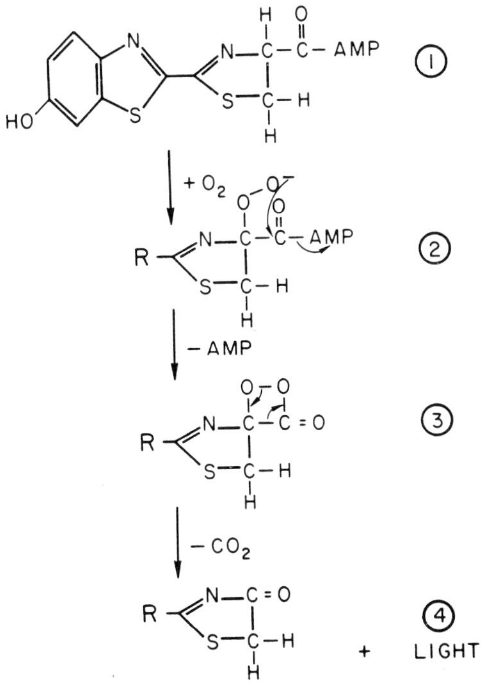

Fig. 3 Proposed Oxidative Mechanism for Light Emission (Plant, White and McElroy 1968)

For example, the proposed scheme indicates that one of the oxygens in the CO_2 must come from molecular oxygen. Using $^{18}O_2$ and $H^{18}OH$ DeLuca and Dempsey (22) have shown that one of the oxygens of the released CO_2 originates from water and that neither of the oxygens of CO_2 is derived from molecular oxygen. Similar results have been obtained for the chemiluminescence of LH_2-AMP in alkaline DMSO (23); furthermore it has been shown recently that the addition of water to the DMSO solution greatly accelerates light emission (23).

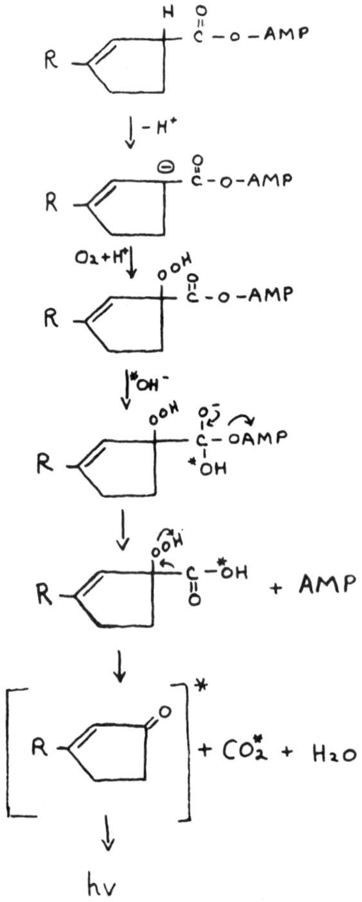

Fig. 4 Modified Oxidative Mechanism for Light Emission Based on the Results of DeLuca and Dempsey (1970).

A possible explanation of these results is presented in Fig. 4. Starting with luciferyl-adenylate, the first step is removal of a proton from the number 4 carbon atom of luciferin. The fact that the rate of the light emission starting with LH_2-AMP is slower if deuterium is substituted for hydrogen at the number 4 carbon supports this conclusion. The carbanion then adds oxygen at carbon number 4. The peroxide does not cyclize but OH is added at the carbonyl carbon. If the reaction medium is $H_2{}^{18}O$, this is the step where ^{18}OH is incorporated. In the next step AMP is assumed to be removed leaving a linear peroxide of luciferin bound to the enzyme. This is followed by a rapid dehydration and decarboxylation leading to the excited product.

If the reaction was carried out in the presence of $^{18}O_2$, it can be seen one of the oxygen atoms would appear in the keto group of the product, while the other oxygen would be released to the medium water and no ^{18}O would be incorporated into CO_2.

Factors Affecting the Color of Light Emission

The peak emission for bioluminescence of Photinus pyralis is 562 rμ (2). As an absolute minimum the energy requirement for the light reaction is estimated to be 57 Kcal/mole. Since the color of light emitted in in vitro reactions can be altered and the fact that other fireflies show different peak emissions studies of the chemiluminescent reaction under various conditions are useful in the interpretation of color changes.

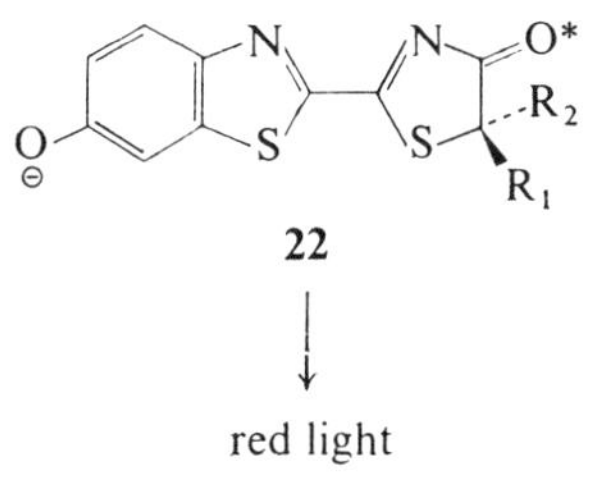

Fig. 5 Proposed structure of the Red Light Emitter (White et al. 1971)

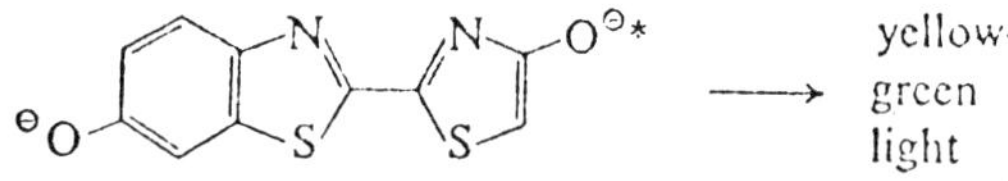

Fig. 6 Proposed Structure of the
Yellow-Green Emitter
(White et al. 1971)

a. <u>Chemiluminescence</u>

Recent data from studies of White et al., indicates that the product emitter in the red chemiluminescence of luciferyl-adenylate in organic solvents is the monoanion of the decarboxylated, 4-keto derivative of LH_2 (4). The nature of the reaction is shown in Fig. 5. The chemiluminescence of LH_2 and its derivatives in base-organic solvent was first studied by Seliger and McElroy (24). Since then, a number of hypotheses have been suggested for the mechanism of the enzyme-catalyzed emission. The fact that one can obtain red emission from enzyme-catalyzed reactions (acid pH, high temperature, etc.) suggests that the red chemiluminescence in organic solvents may be of some biological significance.

In the presence of excess base, the red chemiluminescence of esters or anhydrides of LH_2 shifts to a yellow-green emission. These results suggest that proton abstraction at carbon 5 is essential for obtaining the yellow-green emitter. The nature of the reaction proposed is shown in Fig. 6. By measuring fluorescence of spent reaction mixtures, it was possible to obtain data to support the hypothesis that the dianion is the light emitter in the yellow-green chemiluminescence.

b. <u>Bioluminescence</u>

It has been shown that the variations in color of light emitted by various species of fireflies is due to a difference in the structure of the luciferase (1). Since the structure of LH_2 and the product emitter are identical for all species the shifts in color must be attributed to a change in the polarity or relative hydrophobicity of the binding site of luciferase. A charge change on the enzyme can affect the binding as well as abstraction of protons at both carbon 6 and carbon 5; in addition the relative hydrophobicity of the solvent is known to affect the fluorescence properties of LH_2 and the product.

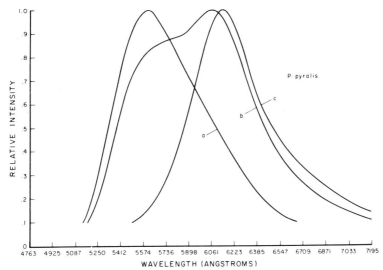

Fig. 7 Effect of pH on the *in vitro* emission spectrum. P. pyralis luciferase. a: pH 7.6; b: pH 6.5; c: pH 5.0. (Seliger and McElroy 1960)

(1) Temperature, pH and metal ions

As the pH of the P. pyralis extract is lowered, it can be observed that the intensity of the yellow-green bioluminescence decreases, leaving a dull brick-orange glow (25). This variation in bioluminescence emission with pH is shown in Fig. 7. As can be seen at neutral (and alkaline) pH, there is a single emission band in the yellow-green region. At intermediate pH, a red emission band appears at 616 mμ, and at pH value below 5.5, the yellow-green emission is completely suppressed and only the red band is evident: At acid pH, the number of light quanta emitted per luciferin molecule oxidized is markedly lower than 1 and indicates a predominantly dark reaction. However, at alkaline pH, although the rate of light emission is reduced to a fraction of the rate of pH 7.6, the quantum yield is essentially unity. Since the red emission is known to be due to the monoanion form of the product emitter this suggests that the pH change must be affecting a group on the enzyme concerned with the abstraction of a proton from carbon 5. The pK for the appearance of red light is approximately 6.8. This suggests the strong possibility that a histidine residue in the enzyme is the active group concerned with proton abstraction.

Except for the partial denaturation of the enzyme in acidic buffer, the pH effect on the emission spectrum shift is completely reversible. A reversible red shifts in emission spectra can be observed by increasing and the decreasing the temperature of the reaction, by carrying out the reaction in 0.2 M urea at normal pH values (7.6) in glycylglycine buffer, or by adding small concentrations of Zn^{++}, Cd^{++} cations, as chlorides.

(2) <u>Effect of substrate structures</u>

The above facts support the idea that the color of the emitted light depends upon the nature of the binding of the intermediate to the enzyme. It seemed likely, therefore, that a change in the structure of the substrate molecules (luciferin or ATP) may alter the binding and in turn affect the color of the light. Unfortunately, it is not possible to change greatly the luciferin structure and still obtain an active light-emitting substrate. It turns out that the 6'-aminobenzthiazole compound is an active substrate, and in this case, a red emission instead of the yellow-green is observed even at neutral pH. The emission at neutral and alkaline pH is red, peaking at 605 m_μ, very close to the bioluminescence emission of firefly luciferin at acid pH. More significantly, the color of the 6'-aminoluciferin bioluminescence is entirely independent of pH, from below 6 to above 10 in exactly the range where native firefly luciferin shows the remarkable color shifts outlined above. Since phenols are stronger acids than anilines, this observation supports the idea that it is the phenolate ion of firefly luciferin that is involved in the normal yellow-green bioluminescence. The results also suggest that the amino group has a strong effect on the ability of the enzyme to abstract a proton from carbon 5.

Until recently, only adenosine triphosphate (ATP) was shown to be active for the enzymatic reaction leading to light emission. UTP, CTP, GTP, and ADP, and other pyrophosphate-containing nucleotides were inactive. Recently, Leonard and associates prepared an ATP with the ribose attached to the 3 position of the adenine ring (3-isoATP) and made a sample available to us. This compound appears to be about 10 - 15% as effective in the light reaction as normal ATP. The additional interesting observation, however, is that at pH 7.5, a significant fraction of the light emitted is red when 3-isoATP is used. Thus, the nature (stereochemistry) of the nucleotide attachment to the enzyme is also of importance in determining the color of the light.

A second modified ATP has been prepared by Leonard and associates known as ε-ATP (26). The structural alterations are due to the addition of an aldehyde group which couples the 6-amino group of adenine to the number 2 nitrogen thus making a four membered ring at that position on the purine moiety. The ε ATP is completely inactive for initiating the light reaction. However, when ε AMP is used as a substrate to make LH_2-εAMP the latter is active for light emission and the emission is red instead of the usual yellow-green. The results indicate that the 6-amino group on the purine ring is essential for the activation reaction; a result in agreement with the observation that ITP is ineffective as a substitute for ATP in the light reaction. In addition, the nature of the binding of the 6-amino group to the enzyme also influences the structure of the excited product in a manner which determines whether red or yellow-green light is emitted.

The results from the ε ATP, ε-AMP and iso-ATP indicates that the binding of the adenylate to the enzyme induces changes in the enzyme structure that must be sustained during the subsequent decarboxylation that leads to the enzyme-product complex excited state. If this were not true then the structure of the AMP should not affect the color of light since it must be removed from LH_2 before the final creation of the excited state according to all proposed mechanisms.

Mechanism of Enzyme Action

a. Number of Binding Sites and the Activation of L

Studies on the substrate binding properties of firefly luciferase (11, 12) have shown that there are two binding sites each for luciferin and ATP per 100,000 molecular weight of enzyme. These results are consistent with the model that luciferase is a dimer of identical 50,000 molecular weight subunits (27), each with one binding site for each substrate. However, the enzymatically active substrate, the MgATP complex, is bound to only one site per 100,000 molecular weight of enzyme. Similarly, only one site is found for dehydroluciferyl adenylate (L-AMP) formed from the following reaction:

$$E + L + MgATP \rightleftharpoons E \cdot L\text{-}AMP + MgPPi$$

Dehydroluciferin (L) is activated by the enzyme to
form dehydroluciferyl adenylate which remains tightly
bound to the enzyme with no production of light. In
order to have only one binding site per dimer of
identical subunits, there must be asymmetry in the
system. Therefore, the physical properties of firefly
luciferase were examined in greater detail. The
results indicate that the minimum molecular weight of
enzymatically active protein is 50,000 and that only
one of the subunits in the 100,000 molecular weight
aggregate is enzymatically active (see below).

b. Binding Sites of Dehydroluciferyl Adenylate

By adding small amounts of L to luciferase in the
presence of MgATP it is possible to determine the
number of L-AMP formed per enzyme molecule by following the decrease in fluorescence. Such experiments
demonstrate that one L-AMP is formed per 100,000 molecular
weight of protein. The enzyme concentration was varied
from 0.4 mg/ml to 2.0 mg/ml in different experiments.
Under all these conditions, there was only one L-AMP
site per 98,000 ± 4000 molecular weight of enzyme.

Isolation of the E·L-AMP complex by Sephadex G-25
chromatography produced similar results. Measurements
by fluorescence of the material bound to the protein
gave 1.01 ± 0.05 molecules of L-AMP per 100,000
molecular weight. Counts of the radioactivity incorporated into the adenylate from ^{14}C ATP yielded 1.1 ± 0.1
molecules of L-AMP per 100,000 molecular weight.

c. Binding of Dehydroluciferin to E·L-AMP

Denburg et al. (12) has shown that there are two
binding sites for dehydroluciferin per 100,000 molecular
weight. Since only one L-AMP is bound to the enzyme
an attempt was made to demonstrate further binding of
dehydroluciferin to the E·L-AMP complex. The E·L-AMP
complex was isolated by chromatography on a G-25 column
in 0.1 M Tris, pH 8.0. The increase in fluorescence
at 440 mμ when dehydroluciferin is bound to the enzyme
was used to measure binding. When dehydroluciferin
was titrated into a solution of the E·L-AMP complex no
increase in the 440 mμ fluorescence was observed. This
technique is sensitive enough to detect a tenfold
increase in the affinity of the enzyme for dehydroluciferin. Therefore, when one L-AMP molecule was bound
to the enzyme two dehydroluciferin sites were no longer
accessible. The possibility that dehydroluciferin can

still bind to E·L-AMP but without the increase in
fluorescence at 440 mµ seemed unlikely since this
fluorescence change arises from putting the molecule
in a hydrophobic environment. Most of the energy
for binding comes from such hydrophobic interactions
and without them binding could not take place.

d. Equilibrium Dialyses

Using equilibrium dialysis, the average number
of molecules of dehydroluciferin bound per mole of
enzyme (100,000 m.w.) was calculated. Using these
parameters, a Scatchard plot was made, from which
was obtained n, the number of binding sites per enzyme
molecule and K_A, the association constant for this
binding. In 0.05 M phosphate buffer at pH 7.8 at
$25°$, n = 1.8 and K_A = 1.7 x 10^5. Equilibrium dialyses
could not be performed with luciferin because of its
instability under the conditions in which the enzyme
maintains its activity.

Other evidence which indicates two sites for L
and LH_2 come from luminescence studies. Hopkins and
Seliger (unpublished) have demonstrated that starting
with LH_2-AMP, two molecules are consumed per 100,000
molecular weight before the luciferase is completely
product inhibited. Recently, Denburg and McElroy (27)
have shown that there is a strong tendency of this
protein to aggregate under certain conditions which
was probably responsible for the previously reported
value of 100,000 for the molecular weight. The
aggregation was observed in solvents of relatively low
ionic strength as the solubility limit of the protein
was reached. The data were not sufficient to decide
whether there was a monomer-dimer or monomer-dimer-
trimer equilibrium occurring or to calculate the
equilibrium constants. The single, symmetrical peak
observed in the schlieren patterns of experiments in
which the sedimentation coefficient was increasing
as the protein concentration was increased suggested
a rapidly reversible monomer-dimer system. However,
the asymmetry observed in the schlieren pattern at
saturating concentrations of enzyme indicated the
presence of polymers greater than dimer. Since no
change in the specific activity of the protein was
observed as the molecular weight of the enzyme increased,
this aggregation plays no physiological role in the
regulation of the enzymatic activity and may be
fairly nonspecific.

The tendency to aggregate and the poor solubility properties of luciferase may be expected in light of the very high percentage of nonpolar amino acids found in the enzyme. The average hydrophobicity of luciferase was calculated to be 1240 cal/residue from the amino acid composition as reported by DeLuca et al. (28). This makes luciferase one of the most hydrophobic proteins ever reported. The high percentage of nonpolar amino acids necessitates that some of them be on the external surface of luciferase. The possibility of hydrophobic intermolecular interactions in luciferase helps to explain its physical properties.

Heterogeneity in the purified luciferase preparations of maximal specific activity was indicated by the results of the binding studies. One L-AMP binding site per 100,000 molecular weight of protein was observed under conditions in which only a single size species of 50,000 molecular weight was present. In addition, the active substrate, MgATP, had only one binding site per 100,000 molecular weight, while there were two sites for ATP.

e. ATP Binding Site

Kinetically, the MgATP complex has been shown to be the substrate in the light reaction catalyzed by the firefly luciferase (29). Kinetic as well as inhibition studies showed that uncomplexed ATP is also bound to the luciferase, and is a competitive inhibitor with respect to MgATP. Similar studies for Mg^{2+} indicated that it is not bound to the luciferase by itself.

Kinetic inhibition studies with ATP analogs showed that the 6-amino group of adenine is important for the binding of bases and nucleosides at the MgATP site. Energetically, adenine and negative charges on phosphate groups contribute 57% and 43%, respectively, toward binding (total binding energy = 4.8 kcal).

dATP can also serve as a substrate in the formation of adenylate of both luciferin and dehydroluciferin. It is a poor substrate, having maximal velocity of light reaction only 5% that of ATP and equilibrium constant of dehydroluciferyladenylate formation five times smaller than that with ATP.

f. Effects of AMP

Under the normal assay conditions (at pH 8.0), where only yellow-green light is emitted, 5'-adenylic acid (AMP) is a competitive inhibitor of luciferase with respect to one of its substrates, MgATP (30). However, AMP serves as a competitive activator of the yellow-green light emission and a noncompetitive inhibitor of the red light emission at pH 6.5 or below. The activation appears to be specific for AMP. Other nucleotides tested were all ineffective.

Luciferase has one MgATP site and one ATP site per molecular weight of 100,000. The MgATP site can also bind ATP, but the ATP site cannot bind MgATP. This latter site, the ATP site, seems to be the site involved in the AMP activation.

The competitive nature of activation suggests that a conformational change occurs in luciferase at pH 6.5 in the presence of AMP. This was demonstrated by comparing optical rotatory dispersion curves of luciferase in the presence and absence of AMP at pH 6.5. Since luciferase is in the monomeric state (molecular weight 50,000) under these experimental conditions, it is suggested that both the MgATP site and the ATP site are located on one of the monomers.

The differences in the subunits' composition must be very small. A single N-terminal serine was found, and the number of peptides from a tryptic digest agreed with the theoretical number predicted on the basis of identical subunits of 50,000 molecular weight (31). However, recent evidence for heterogeneity was the observation of two different C-terminal amino acids, leucine and serine (32).

From the facts cited above, one concludes that only one of the 50,000 molecular weight species is catalytically active. Furthermore, it appears that both the L binding sites and the Mg-ATP and ATP (AMP) binding sites are all on this subunit.

The only data that do not completely support this conclusion is the appearance of two active bands on electrofocusing (27). However, these experiments should be interpreted with caution because of the close similarity of the two species and their strong tendency to aggregate.

g. Role and Reactivity of Sulfhydryl Groups

The number of sulfhydryl groups of firefly luciferase has been determined by spectrophotometric titration with p-mercuribenzoate in the presence and absence of a competitive inhibitor (28). Between six and seven sufhydryls are titrated with p-mercuribenzoate in the native enzyme. In the presence of the inhibitor only four to five sulfhydryls will react with the p-mercuribenzoate. Four or five moles of p-mercuribenzoate can be reacted with the enzyme-inhibitor complex and subsequent removal of the inhibitor results in recovery of 90% of the original enzymatic activity. Addition of 4 moles of p-mercuribenzoate to the enzyme in the absence of inhibitor results in complete loss of activity. The enzyme is also inhibited by dithiol reagents such as arsenite-2.3-dimercaptopropanol, $CdCl_2$, and γ-(p-arsenosophenyl)-n-butyric acid. The data show that four or five of the enzyme sulfhydryls have no effect on the catalytic activity, but the two sulfhydryl groups which are "covered" by the inhibitor are essential in some way for the enzymatic reactions leading to light emission.

Recently, it has been found that TPCK (chloromethyl ketone derivative of N-tosyl-L-phenylalanine), an aromatic inhibitor of chymotrypsin, also inhibits luciferase activity (33). The inhibition of enzymatic activity is accompanied by a loss of approximately two sulfhydryl groups. TPCK is a competitive inhibitor for luciferin and a non-competitive inhibitor with respect to ATP. The aromatic character of TPCK appears to be the major factor for its binding to the active site of luciferase. The observation that N-tosyl-L-phenylalanine alone is a competitive inhibitor also with a K_i of the same order of magnitude of TPCK supports this conclusion. Inactivation by TPCK is pH dependent and it is of interest to note that the inactivation - pH curve corresponds very closely to the luciferase activity - pH curve.

The fact that TPCK competes with LH_2 and not with ATP suggests that the inhibitor is most likely reacting at the catalytic site. If the inhibitor were reacting at a different site thus preventing conformational changes, then one might expect TPCK to be competitive with both LH_2 and ATP rather than just for the one (LH_2).

In the studies with luciferase, TPCK also appears to react only with SH groups. This conclusion is supported by (i) the stoichiometry between number of SH groups lost and the number of TPCK groups incorporated into luciferase; (ii) decrease in cysteic acid content in oxidized TPCK-luciferase equivalent to the amount of SH groups lost; and (iii) absence of either 1-carboxymethyl or 3-carboxymethyl histidine in the hydrolysate of performic acid-oxidized TPCK-luciferase.

Sulfhydryl groups are apparently reactive enough so that almost all the SH groups of luciferase react with TPCK in an excess of TPCK. However, the two essential SH groups of luciferase react more rapidly than do the other SH groups.

A comparison of the inactivation of luciferase by several reagents of different reactivity shows that the ability of TPCK to inactivate luciferase far exceeds that expected from its chemical reactivity, suggesting that something more than chemical reactivity is operating in the TPCK inactivation of luciferase. Since N-tosylphenylalanine itself can inhibit luciferase competitively, and TLCK does not inhibit luciferase under the comparable conditions used for TPCK, it appears that the hydrophobic nature of TPCK brings this compound to the binding site of LH_2 (or L), thus achieving an effect of affinity labeling. As might be expected, the tight binding of L-AMP to the active site protects the enzyme from TPCK inactivation. Luciferin alone in high concentrations also retards the TPCK inactivation rate whereas ATP-Mg^{2+} had no effect. Since neither of the substrates alone has any measurable effect on the conformational changes in luciferase, the above results must mean that LH_2 (or L) interferes with TPCK inactivation by competing for the same site.

Since the carboxylic acid group of LH_2 (or L) must react with the AMP-PP bond of ATP, the two substrates of luciferase must be in close proximity to each other on the luciferase surface. However, since TPCK is strictly noncompetitive with respect to ATP, it suggests that the inhibitor is specifically bound to the LH_2 (or L) site.

Because the titration of all of the sulfhydryl groups in luciferase results in total inhibition, it was of interest to determine the amino acid sequence in the vicinity of the protected sulfhydryl groups.

Labeling of the essential sulfhydryl groups with
(1 - ^{14}C)N-ethyl-maleimide followed by tryptic digestion
resulted in the isolation of a single radioactive
decapeptide whose sequence was determined.

The amino acid analysis and partial acid hydrolysis
of the radioactive peptides proves, unambiguously,
that the two SH groups covered by dehydroluciferyl
adenylate and the reactive pair in the native enzyme
are the same. This is further substantiated by the
inhibition studies performed on the native enzyme
with NEM where the addition of only 1 mole of NEM
results in approximately 50% inactivation.

Luciferase binds approximately 2 moles of dye
per mole of enzyme suggesting that there are two catalytic
sites on luciferase. This is consistent not only with
the dye-binding stoichiometry but also with the
fact that 2 moles of L-AMP are required to remove all
of the bound dye. If two molecules of dye were bound
at a single 'active site' then 1 mole of L-AMP should
completely remove the dye. The observation that two
SH groups are essential for catalytic activity and that
one each of these appear in identical peptides is also
consistent with two active sites per molecule.

Since 2,6-TNS, unlike TPCK, is competitive with
both LH_2 and ATP, it suggests that the dye is binding
at or near the normal substrate binding sites. However, the complete lack of pH effect on binding over
the range pH 6-9 indicates that the groups which
ionize in this region: imidazole; lysine; sulfhydryl;
must not alter the binding site for the dye. There
is a large change of enzymatic activity in this pH
range with an optimum pH of 7.8. Functional groups
essential for catalysis, however, appear to have no
effect on binding of the dye.

Dehydroluciferin when bound to luciferase shows a
slow rate of proton transfer, suggesting that the
binding site is hydrophobic (34). Fluorescence
lifetime measurements have been used to obtain nano-
second time-resolved emission spectra of dehydroluciferin in various solvents and when bound to luciferase.
The blue fluorescence caused by the phenol decreases
with decay time relative to the green emission caused
by the phenolate. The time course of excited state
ionization may thus be measured directly. The rate of
proton transfer is very fast in aqueous solution but
slower in 80% ethanol. Addition of imidazole increases
the rate of proton transfer.

h. Effect of NEM and Its Analogues

It was previously shown that NEM preferentially attacks the two SH-groups that are essential for luciferase activity (31). Incubation of luciferase with a concentration of NEM twice that of the enzyme in 0.05 M phosphate, pH 7.8 at $0°$, results in essentially complete loss of activity. Luciferase inactivated in such a manner by NEM did not have the ability to bind dehydroluciferin as measured by fluorescence and equilibrium dialysis. The apparent competition of NEM and luciferin for the same site on the enzyme is seen by the protection given by luciferin against NEM inactivation. The half-life of inactivation is three (3) hours without and eleven (11) hours with luciferin. These experiments implicate the position of the essential SH group to be near the luciferin-binding site. Detailed studies have been made using various analogues of luciferin and K_i values were determined (12).

i. Peptide Derived from Luciferin-Binding Site

2-Cyano-6-chlorobenzothiazole (CCB), a substrate analog of firefly luciferase, inactivates this enzyme slowly at pH 8 to about 20% of original activity without affecting free sulfhydryl groups (32). The inactivated luciferase contained 1.5 - 2.0 moles of CCB per 100,000 daltons of luciferase. It is believed that the benzothiazole derivatives are incorporated at the luciferin-binding sites. Tryptic digest of the inactivated luciferase and subsequent electrophoresis yielded a fluorescent peptide containing benzothiazole derivative. The peptide was found to contain pyroglutamic acid at the N-terminal end. The sequence of the peptide was established tentatively to be: pyroglu-X-gly-ala-val-(asp)-ile-leu where X is an amino acid (possibly tyr) to which the benzothiazole derivative is attached. There is little resemblance in the composition of the peptide isolated here and that of the peptide containing the reactive SH groups. This is not unreasonable, however, when one considers the distance between these two binding points. It is interesting to note that while the SH-peptide contained many hydrophilic amino acids (composition: 2 ser, 2 gly, 1/2-cys, glu, gln, asn, ala, lys), CCB-peptide has a high proportion of hydrophobic amino acids. This is in accord with the earlier observations that the SH groups are not located in a very hydrophobic environment, perhaps near the entrance of the LH_2

binding site, while the interior of the LH_2 binding site, including the 6'-position of the benzothiazole ring, is in a very hydrophobic environment.

Since the C-terminal residue of CCB-peptide is leucine, it is thought that this might actually be the C-terminal peptide of the luciferase. Preliminary studies of the carboxypeptidase digestion of the luciferase in native form and in 6 M urea indicated that there are one mole each of leucine and serine per 100,000 daltons. Isoleucine was not produced during such digestions (unpublished results). Therefore, it seems that CCB-peptide is not derived from the C-terminal end.

j. <u>Concluding Summary on Luciferase Action</u>

From the results presented in the last section we can make the following conclusions concerning the properties of luciferase:

1. The enzyme at relative low concentrations of protein readily associates into a dimer of 100,000 molecular weight. However, the catalytically active unit is 50,000 m.w.

2. There are two binding sites for LH_2 and L per 100,000 molecular weight while there is only one Mg_2ATP binding site. There is a second site which binds ATP as well as AMP.

3. There are two binding sites for L-AMP per 100,000 molecular weight while only one L-AMP is formed from L and Mg_2ATP.

4. Dye binding indicates a very hydrophobic site for LH_2 binding. Two dye molecules are bound per 100,000 m.w. and two L-AMP molecules are required to displace the two dye molecules.

5. Starting with LH_2-AMP it is possible to show that only two molecules of the substrate are used for light production at which time the enzyme is 100 percent inhibited.

From this data we conclude that the 100,000 molecular weight is composed of two subunits of which only one is enzymatically active. If this is correct, the active subunit must contain two sites for LH_2 (LH_2-AMP) binding but only one Mg_2 ATP binding site. The second binding site for free ATP (AMP) appears to be concerned with regulatory activity. Unfortunately it has not been possible to separate the two different subunits which must be almost identical in amino acid composition.

Acknowledgement: The research summarized in this paper has been supported by the National Science Foundation, the National Institutes of Health and the Atomic Energy Commission.

References

1. McElroy, W.D., Seliger, H.H., and DeLuca, M. (1965), In Evolving Genes and Proteins (V.Bryson and H. Vogel, eds.), p. 319. Academic Press, New York, New York.
2. McElroy, W.D. and Seliger, H.H. (1966), In Molecular Architecture in Cell Physiology (Hayashi and Szent-Gyorgi, eds.) p. 63, Prentice-Hall, Inc., Englewood Cliffs, New Jersey.
3. McElroy, W.D., Seliger,H.H., and White, E. (1969), Photochem. and Photobiol. 10:153.
4. White,E.H., Rapaport, E., Seliger, H.H., and Hopkins, T.A. (1971), Biorganic Chemistry 1:92(1971).
5. Suzuki, N., and Goto, T., Tetrahedron Lett. 2021 (1971).
6. McElroy, W.D., Seliger, H.H., and Deluca, M., Insect Bioluminescence In Insect Physiology (Rockstein, M. ed.) Academic Press 1973.
7. White, E.H., McCapra, F., Field, G.M., and McElroy, W.D. (1961), J. Am. Chem. Soc. 83:2402.
8. Bitler, B. and McElroy, W.D. (1957), Arch. Biochem. Biophys. 72:358.
9. DeLuca, M. and White, E.H., unpublished.
10. Morton, R.A., Hopkins, T.A., and Seliger, H.H. (1969), Biochemistry 8:1598.

11. Denburg, J. and McElroy, W.D. (1970), Arch. Biochem. Biophys. 141:668.
12. Denburg, J., Lee, R.T., and McElroy, W.D. (1969), Arch, Biochem. Biophys. 134:381.
13. Turner, D.C. and Brand, L. (1968), Biochemistry 7:3381.
14. DeLuca, M. (1969), Biochemistry 8:160.
15. Rhodes, W.C. and McElroy, W.D. (1958), J. Biol. Chem. 233:1528.
16. Seliger, H.H. and McElroy, W.D. (1960A), Arch. Biochem, Biophys. 88:136.
17. Seliger, H.H. and Morton, R.A. (1968), In Photophysiology Vol. IV (A.C. Giese, Ed.) p. 253 Academic Press, Inc., New York.
18. Plant, P.J., White, E.H., and McElroy, W.D. (1968), Biochem. Biophys. Res. Comm. 31:98.
19. McCapra, F. and Richardson, D.G. (1964), Tetrahedron Letters, 3167.
20. McCapra, F., Chang, Y.C. (1967), Chem. Comm. 1011.
21. McCapra, F., Chang, Y.C., and Franciose, V.P. (1968), Chem. Comm. 22.
22. DeLuca, M. and Dempsey, M. (1970), Biochem. Biophys. Res. Comm. 40:117.
23. DeLuca, M., and Dempsey, M.E., manuscript in preparation.
24. Seliger, H.H. and McElroy, W.D. (1964), Proc. Nat. Acad. Sci. U.S. 52:75.
25. McElroy, W.D. and Seliger, H.H. (1962), Federation Proc. 21:1006.
26. Secrist, J.A., Barrio, J.R., Leonard, N.J., Villar-Palase, C; and Gilman, A.G.; Science 177, 279 (1972).
27. Denburg, J. and McElroy, W.D. (1970), Biochemistry 9:4619.
28. DeLuca, M., Wirtz, G., and McElroy, W.D. (1964), Biochemistry 3:935.
29. Lee, R.T., Denburg, J., and McElroy, W.D. (1970), Arch. Biochem. Biophys. 141:38.
30. Lee, R.T. and McElroy, W.D. (1971A), Arch. Biochem. Biophys. 146:551.
31. Travis, J. and McElroy, W.D. (1966), Biochemistry 5:2170.
32. Lee, R.T. and McElroy, W.D. (1971B), Arch. Biochem. Biophys. 145:78.
33. Lee, R.T. and McElroy, W.D. (1969), Biochemistry 8:130.
34. DeLuca, M., Brand, L., Cebula, T.A., Seliger, H.H., and MaKula, A.F. (1971), J. Biol. Chem. 246:6702.

J. LEE: I would like to discuss the proposal made by Dr. Seliger and Dr. McElroy to account for the difference in spectral emission maximum obtained for the different luciferases. There are several perturbing effects which can effect the position of a spectral emission maximum. One is the dielectric constant of the medium which many believe to be equivalent to hydrophobicity in an enzyme active site. Another is solvent polarizability and yet a third is viscosity. It occurs to me that to postulate the site as being very hydrophobic and at the same time that the spectral maximum be shifted by ionic effects is contradictory.

McELROY: I don't see any contradiction when we know there is both a hydrophobic site (luciferin) and a hydrophilic site (AMP). We know that charge (structure) on both substrates can alter the color, therefore we conclude that the nature of the binding as well as the conformational change in enzyme are important in determining the color. The factors above could be in some cases equated to dielectric constant, solvent polarizability, or viscosity.

J. LEE: I think that the contradiction is that an ionic environment cannot exist in a solution of low dielectric constant.

McELROY: Don't forget here too that we have to have a very high hydrophilic area where the ATP attaches. The two sites must be close to one another.

J. LEE: What is difficult to believe is that one part of the luciferin molecule lies in the region of very low dielectric constant and that the biochemistry takes place in a highly polar environment.

McELROY: This depends on the nature of the tertiary structure of the enzyme. Distant amino acids in a sequence could be brought very close to one another under appropriate conditions.

WHITE: I have two points to make on that question. The first point is that Frank McCapra came up independently with the mechanism of the red chemiluminescence of luciferin. The second point concerns the mechanism you outlined for the bioluminescence. I am reluctant to let the old mechanism go, but will wait for the later paper by DeLuca for further comment. The question I have now pertains to your mention that you had an analogy for the reaction mechanism you proposed. What evidence exists that peroxycarboxylates can lead to light emission?

McELROY: No I don't have any analogy except for the work that has been done by Adam in Puerto Rico. The work he has done is leading in that direction.

WHITE: Maybe he could comment on that. I believe you have to close the ring to form a dioxetanone before you get light emission.

McELROY: If the mechanism involves a dioxetane ring then by definition it must close; however it is not necessary from an energetic point of view. Certainly the isotopic data eliminates the earlier data.

WHITE: Lastly I would like to point out that a negative charge at the sixth position of luciferin may not be necessary. We recently synthesized an analogue with the dimethylamino group in that position and this compound is active. Red light is emitted although it is impossible to put a negative charge there. I think that with an OH at C-6 it is probably true, but the charge is not really necessary.

McELROY: You don't have to have a negative charge at the sixth position to get luminescence; the evidence indicates that when you use luciferin the OH group must exist as O^- in order to obtain a yellow green emission.

WHITE: Right, with OH, that's true, but not with an amino function substituted with two methyls or ethyls.

HASTINGS: I missed what the evidence was that the 50,000 unit is the active unit. Is this in solution?

McELROY: Yes this is in solution. When you use very dilute enzyme there is no question you have a 50,000 molecular weight active catalyst.

HASTINGS: Then there are conditions where association doesn't take place and yet you have full activity.

McELROY: That's right.

HASTINGS: And the quantum yield data per unit of protein is that per monomer or per dimer?

McELROY: The enzyme is still active even in the dimer form. There is no change in specific activity in going from monomer to dimer.

HASTINGS: So in the dimer form you get two photons per protein molecule?

McELROY: One obtains one photon per 50,000 molecular weight unit; only half of the units appear to be active.

HASTINGS: You get one per 100,000 then?

McELROY: You get one per 100,000 but we don't know which subunit is inactive. You only know about that statistically.

HASTINGS: The measurement must state it just in terms of average and the average is in terms of one photon per 100,000.

McELROY: That's right.

HASTINGS: There is another thing I wanted to ask about. We have also tried Leonard's ε-ATP and we did get activity - an appreciable amount.

McELROY: You've got to get rid of the contaminating ATP. There is a small amount of ATP in the ε-ATP and that's what is giving you light.

HOPKINS: Is it possible using two photons per 100,000 and one photon per 50,000 to give an alternate reaction that one of the subunits may be an activating subunit and the other one a light emitting subunit? That is to say that the bifunctionality of the enzyme is really carried in two different subunits.

McELROY: That was an attractive idea and we did everything we knew how to test it. We noted no interaction as judged by dilution experiments. Seliger did those experiments some time ago and we repeated them again. If one subunit was the activating and the other one was the one doing the oxidizing then one should see a non-linear fall-off in rate upon dilution. This was not observed.

HOPKINS: Was molecular weight followed at the same time?

McELROY: It has been followed over the same range of dilution and it stayed at 50,000 all the way. We have observed light intensity at protein concentrations where you start getting dimers. When one dilutes out such a mixture you see no effect on rate of light emission when the dimer goes over to primarily monomers.

TOTTER: Could I ask if the quantum yield has been run under these different circumstances?

McELROY: Seliger originally did the quantum yield. He used very low luciferin and excess enzyme. Under the conditions used the enzyme was probably in the form of dimer. It was very high in protein. It has not really been done with very dilute enzyme because that is a very difficult experiment to do because the enzyme does not like to turn over.

MODEL COMPOUNDS IN THE STUDY OF BIOLUMINESCENCE

Frank McCapra, M. Roth, D. Hysert and K.A. Zaklika

School of Molecular Sciences

University of Sussex

Introduction

The problems presented by the phenomenon of bioluminescence span a uniquely fascinating range. Some of the more obvious concern the evolution of light production in organisms from both chemical and biological points of view, the nature of the molecules involved, the chemical and physical mechanisms of light emission and the role of the enzyme. We believe that although the discovery of bioluminescence is understandably easy, the difficulty of finding sufficient material and the novelty of the reactions taking place, make a thorough study of model reactions absolutely essential to a proper investigation. Almost all the questions above can at least be partially answered by such an approach, and we feel that some of our work briefly described here, demonstrates its usefulness.

Structure and Mechanism

Naturally enough, model compounds can only be useful where adequate information concerning the structures of the chemical entities involved is available. At the present time model studies may be applied to four organisms, and the relevant structures are shown:

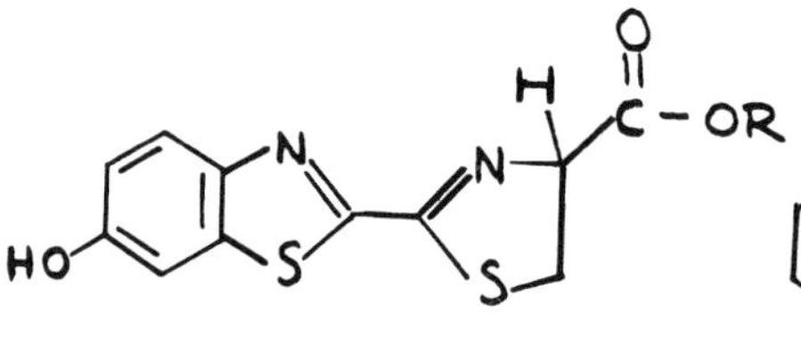

Firefly

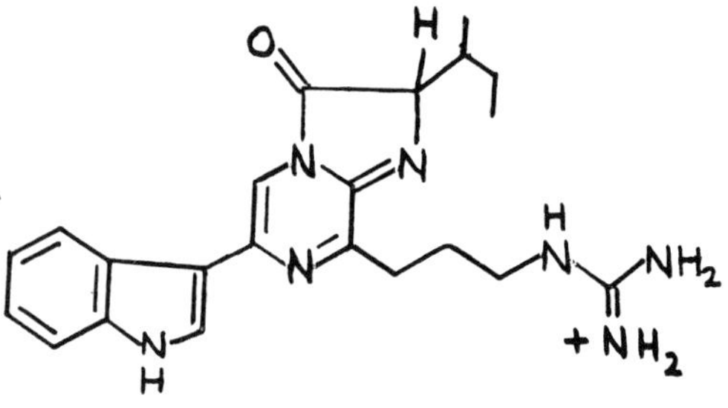

Cypridina

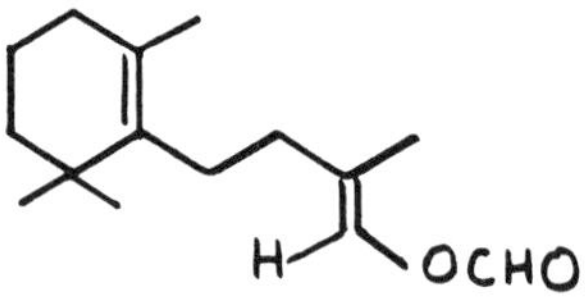

Latia

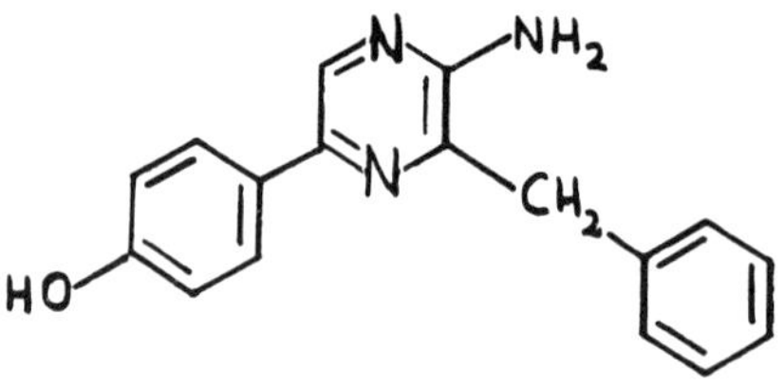

Aequorea (chromophore)

The construction of model compounds for 1 and 2 follows naturally from their common properties. Each has an acidic hydrogen atom, with the resulting anion so conjugated as to stabilise any radical formed. The neighbouring carbonyl function belongs to the class of active esters in both cases (in the case of 2 there are few properties of an amide, solvolysis occurring with formation of the acid or ester merely by heating in water or methanol at neutral pH). For 3 and 4 certain assumptions must be made in our present state of knowledge. A suggestion for the reaction of 3 has already been made,[1] and we will return to 4 later.

It seems clear that the oxidation represented by equation (1) will be chemiluminescent provided only that, in addition to the features already mentioned, the carbonyl product is fluorescent or that energy transfer to another fluorescent molecule is possible.

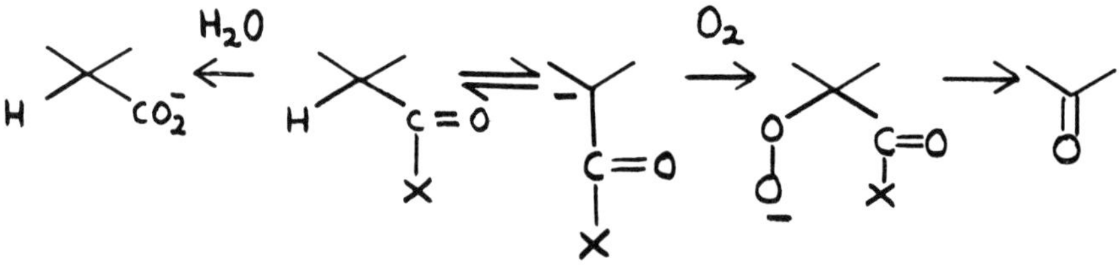

(1)

That the reaction is general is confirmed by more than twenty
examples discovered in our own laboratories, some of which are
shown below:

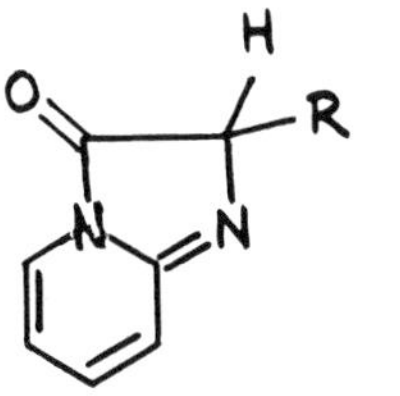

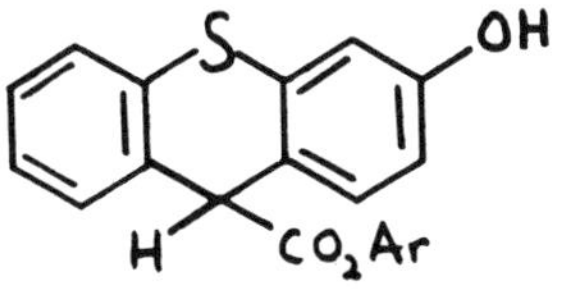

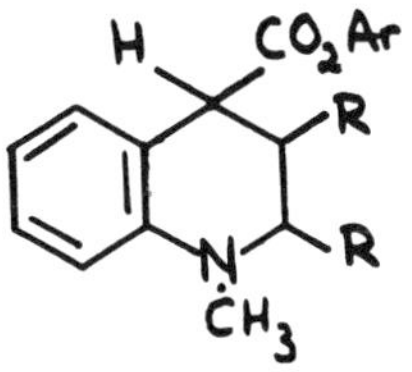

We have selected the series of acridans 5 for detailed study.
They show very similar behaviour to that of the firefly model
compound 1 (R = Ph) with certain important advantages. Most
significant of these is the ability to isolate the intermediate
peroxides 9 and 10 in pure crystalline form.

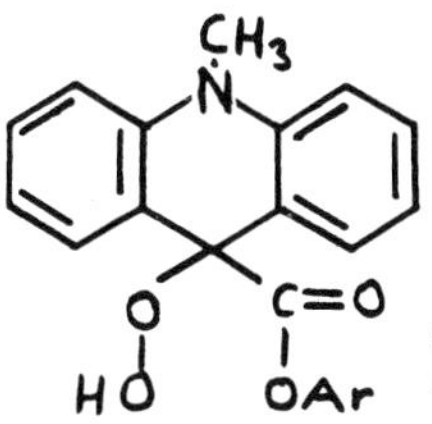

Some of our results in this series have already been reported and
these will be summarised together with more recent work. Since
what we may call a linear route of decomposition seems indicated
in the enzymic firefly reaction[2], as opposed to the proven intermediacy of the cyclic dioxetanone in chemiluminescence[3,4], these
alternatives, shown in equations (2) and (3) will receive most
attention.

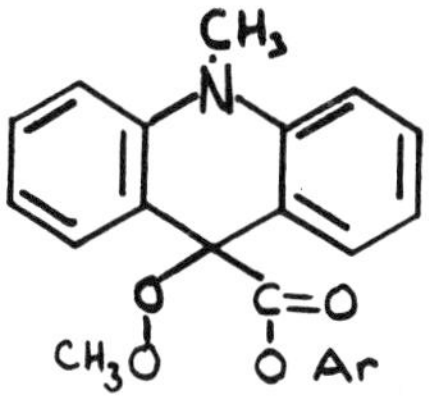

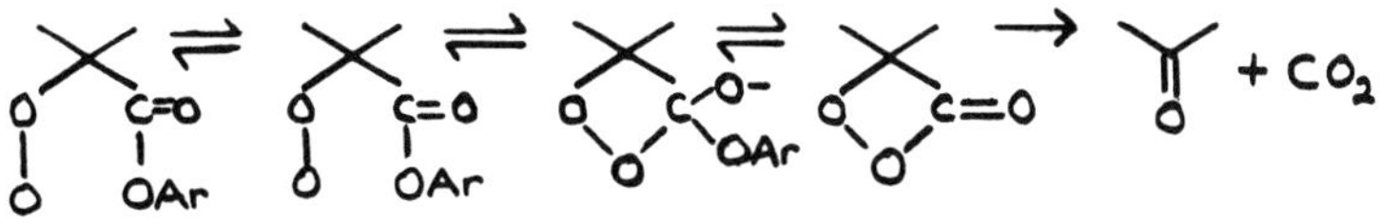

$$ + CO_2 \qquad (2)$$

$$\text{[structure]} \underset{H_2O}{\rightleftharpoons} \text{[structure]} \rightarrow \text{[structure]} + CO_2 + ArO^- \qquad (3)$$

Acridan Ester Chemiluminescence

There is an isotope effect K_H/K_D for oxidation of the acridan in DMF (potassium phthalimide as base) of 3.8, and it is clear that removal of the proton at C-9 is partially rate determining. The Hammett reaction constant for the series of substituted esters is +1.7, differing greatly from that of the intermediate peroxide. It follows that the study of the excitation step is only accessible by direct examination of the peroxides. Relevant observations and their relationship to equations (2) and (3) are set out as follows:

1) The reaction constant for the series of peroxide esters 9 is +4.6 in aqueous ethanol.

    This is very much higher than that for addition of water (or hydroxyl ion) as in equation (3) which should be that for hydrolysis, i.e. +2.1. Intramolecular reactions show such enhanced values of ρ, such enhancement also being expected of the formation of a strained ring.

2) External nucleophiles (alcohols, NH_2OH, H_2O_2, amines, etc.) do not affect the rate or quantum yield significantly.

    Competition with hydroxyl ion (in equation (3)) is known to be exceedingly effective for these nucleophiles. The products which should result, e.g. alkyl esters and amides, are not chemiluminescent. Therefore hydrolytic mechanisms of whatever kind are not involved. It also follows that spontaneous decomposition of the dioxetane in (2) must occur more rapidly than does trapping.

3) Only 'active' esters are effectively chemiluminescent.

    By raising the pH alkyl esters can be made to react at the same rate as phenyl esters[5], yet do not emit light during the reaction. That is, only displacement by peroxide leads to chemiluminescence. (Alcohols are known not to be so displaced).

CHEMICAL MECHANISM IN BIOLUMINESCENCE

4) The hydrolysis product (carboxylic acid) derived from either 9 or 10 decomposes in base quantitatively to N-methylacridone without light emission.

5) The rate of formation of N-methylacridone from $\underline{9}$ is $1.54 \pm 0.1 \times 10^{-1}$ sec^{-1} and from $\underline{10}$ $5.9 \pm 0.2 \times 10^{-6}$ sec^{-1}, both at pH 11.1, under pseudo first order conditions.

The difference in rate of about 10^4 can be ascribed to the expected 'α-effect,'[6] the lower pKa of peroxides and the intramolecular nature of the reaction of $\underline{9}$.

6) The base catalysed formation of N-methylacridone from $\underline{10}$ gives about **15,000** times less light than from $\underline{9}$.

Strong base, 0.1N NaOH is used to allow the reaction to reach completion in a reasonable time. Even this small amount of light ($\emptyset = 1 \times 10^{-6}$) is apparently not associated with equation (3) since there is evidence from several sources that a radical chain reaction is involved.

Hence in agreement with the known propensity of peroxides for intramolecular reaction, two conclusions seem inescapable. The cyclic route is enormously favoured (4 x 10^4 times*) over the linear route. It is also the only effective source of chemiluminescence. The effect of the enzyme on the pathway for reaction cannot be assessed from these studies. However, three points are noteworthy. The mechanisms employed by enzymes to catalyse carbonyl addition reactions are equally applicable to (2) or (3). Linear routes require bond making and breaking at many sites remote from the site of excitation, and heavy solvation would seem to provide an adequate means of energy dissipation. Lastly, in cases of energetic blue emission, it is highly likely that the loss of the 25-30 K.cal. energy associated with the four membered ring would prove fatal to the population of so energetic an excited state.

Reactions in Micelles as a Model for the Enzymic Reaction

It is well known that dipolar aprotic solvent are especially useful in chemiluminescent reactions of the type discussed above. Aqueous solutions very often do not allow chemiluminescence or are at best inferior. A significant feature of most luciferases is

---

* The rate increase observed[7] for the nucleophilic participation of a peroxide **ion** at saturated carbon, forming a dioxetane, is remarkably similar at 2.35 x 10^4.

that they appear to lack an oxidation co-factor. It is reasonable to assume that the anion is oxidised directly by oxygen to the peroxide, particularly since this is normally a very fast reaction. Such a suggestion has been made in an earlier report of the effect of micelles on Cypridina luciferin and related chemiluminescent compounds.[8] The acridans are much more easily studied, and we have a respectable knowledge of their chemistry. It would therefore be appropriate to study the effects of aggregation in this series in the hope of elucidating some of the features of the enzymic reaction. Our observations using cetyltrimethylammonium bromide (CTAB) are summarised by figures 1 and 2.

The most significant finding is that the efficiency of the reaction can be increased 130 times, bringing it to a level ($\phi_E = 6.5\%$) comparable to that of the reaction in aprotic solvents ($\phi_E = 10\%$). The source of this enhancement appears to be an increase in both the rate of removal of the C-9 proton and an increase in the rate of oxidation. The latter effect is very interesting and we are studying it further. The isotope effect is reduced compared to that in DMF, and the major competing reaction is hydrolysis before oxidation. Deuteration at C-9 lowers the light yield by a factor of 3.4 in agreement with the assumption that ionisation (followed by oxidation and chemiluminescence) is in competition with hydrolysis. The fluorescence yield of N-methylacridone is unaffected by (CTAB) so that the 10 fold increase in the yield of N-methylacridone must have another explanation.

Simplification of the system by use of the peroxide 9 in presence of CTAB does not provide the answer. Although the rate is increased about 10 times, the quantum yield is only very slightly increased. This might suggest that there are no competing reactions with rate differentials to be exploited, the relative inefficiency being inherent in the dioxetanone decomposition. At present we have no satisfactory explanation for the observation in the acridan case concerning the efficiency increase, and more work is required.

Cypridina and Aequorea Luciferin Synthesis

Although it is not proper to call the chromophore in the Aequorea photoprotein a luciferin, we believe that model studies on the suspected biosynthesis of the active molecule in both these organisms have interesting implications. Cypridinaluciferin is quite clearly biosynthesised from tryptophan, isoleucine and arginine. If it can be shown that a tripeptide of these acids can be easily cyclised to the required pyrazine derivative, then this could be reasonably considered a suitable synthetic and biosynthetic route. The sequence shown below has recently been achieved[10] in our laboratories, the very significant last step

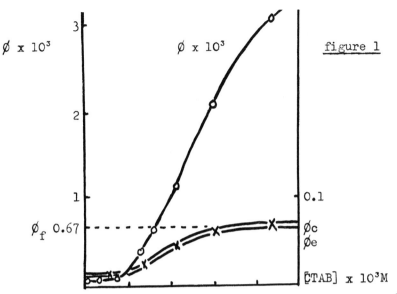

Reaction in 20% aqueous acetonitrile solution, acridan 1 x 10^{-5} M, with injection of NaOH solution such that final concentration of base is 0.05 M. ϕ is the chemiluminescence quantum yield, ϕ_c the yield of N-methylacridone, ϕ_f its fluorescence quantum yield, and ϕ_e is calculated from $\phi = \phi_c \times \phi_e \times \phi_f$.

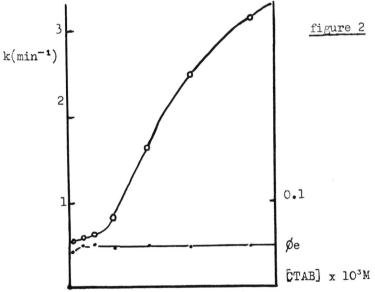

Peroxide 9 in 20% aqueous acetonitrile at pH 9 using borate buffer. Rate measured by decay of light emission and ϕ_e obtained as above.

occurring in quantitative yield. We have found that use of dehydrovaline, necessary in terms of the oxidation level, is absolutely essential for the final cyclisation.

The ultraviolet spectrum of the Aequorea photoprotein, as determined by the extensive work of Shimomura and Johnson,[8] does not display the characteristic long wavelength absorption of such imidazolopyrazines. Yet the molecule isolated is closely related to the oxidation and hydrolysis product of the Cypridina luciferin analogue. The scheme shown below accounts for all data available, and has the additional merit of placing both organisms on the same line of chemical evolution. We are also attempting a synthesis of the luciferin-like precursor to the presumed active molecule by our previous route although a synthesis of the peroxide will be much more difficult. We assume that the 'trigger-like' effect of calcium ion is associated with the catalysis of the peroxide decomposition which then forms the amide in an excited state. It is likely that this product is enzyme bound and that its removal from the enzyme results in hydrolysis with formation of the isolated product. Although it is possible that this chromophore can itself be excited by an as yet unsuspected mechanism, the suggestion made here is the most economical.

References

1. Frank McCapra and R. Wrigglesworth, Chem. Commun., 1969, 91.

2. M. DeLuca and M.E. Dempsey, Biochim. bisphys. Res. Commun., 40, 117 (1970).

3. W. Adam and J-C. Liu, J. Amer. Chem. Soc., 94, 2894, (1972).

4. Frank McCapra, Pure Appl. Chem., 24, 611 (1970).

5. Tables of Chemical Kinetics, Homogeneous Reaction Kinetics, National Bureau of Standards, Washington 1951.

6. W.P. Jencks and J. Carriuolo, J. Amer. Chem. Soc., 82, 1778, (1960).

7. W.H. Richardson and V.F. Hodge, J. Amer. Chem. Soc., 93, 3996 (1971).

8. T. Goto and H. Fukatsu, Tetrahedron Letters, 1969, 4299.

9. O. Shimomura and F.H. Johnson, Biochemistry, 10, 4149 (1971).

10. Frank McCapra and M. Roth, Chem. Commun., 1972, 894.

McELROY: I don't think there has ever been any argument that dioxetanes have been made and isolated and identified and that they can give rise to very low quantum yield luminescence. I think the real question is the nature of the environment in which you carry out these reactions. When you start using rate data and arguing against the linear concept I would argue that that is exactly what an enzyme does. An enzyme by definition is a chemical compound which increases the rate of a reaction and does it in an environment that can change a linear peroxide into a cyclic form depending on the nature of the groups on the enzyme.

McCAPRA: Yes, I am aware of that. This is only a comparison between two of the chemical functions a molecule has and I think it is pretty difficult to pick one rather than the other. Now there is the question of what the enzyme may be expected to do. The nucleophilicity of internal peroxides as found by us is 10^4 greater than that of hydroxyl ions. The way in which an enzyme catalyses nucleophilic attack on carbonyl is virtually all we are concerned with. In other words there must be some form of protonation at the carbonyl. It is very hard to imagine the enzyme passing up the opportunity to use the more nucleophilic sequence after having set up the catalysis. In other words it keeps the carbonyl in preparation for the nucleophilic attack. Then you have an intramolecular reaction which will also be assisted by exactly the same mechanism as for the hydroxyl in the picture. In other words you remove the proton from water or you remove it from peroxide. But peroxide has a pK_a of about 11 and water has one of 15, so peroxide ion will be in greater concentration. I am not saying that the electrophilic enzyme does nothing. But it's starting from the same base point in this molecule.

McELROY: I agree. That certainly looks like, as I said earlier, an attractive mechanism. But unfortunately there are data that have to be explained which cannot be accounted for by that. Isn't it conceivable that groups on the protein are essential to this. They are essential to high quantum yield to begin with which is very unusual compared to chemiluminescence. There may be groups on the enzyme which could interact with a linear peroxide to give a transitory intermediate of a cyclic nature. This could satisfy your symmetry rules and the isotopic data at the same time. I admit however that this does not explain the chemiluminescence data.

McCAPRA: You want to know what would convert the linear route into a cyclic one? Well you can't possibly extract what the enzymes does or doesn't do. I would guess in fact that the enzyme could catalyze both these reactions equally.

ADAM: The notion that the quantum yields are low is not really quite correct. It depends what one looks for. If one looks for a triplet quantum yield they're quite high.

McCAPRA: With bioluminescence in fact?

ADAM: Our conclusion regarding the yield, although we don't yet have quantitative data on the dioxetanone, or peroxylactone, is that it is very high. I wouldn't say unity but certainly better than 10%, maybe as good as 50%. So the fact that you see very little light doesn't mean that your quantum yields are low. It means that you are generating triplets which don't luminesce. They get destroyed in other ways. That's one point I'd like to make. The other point, comes back to the diradical which is concerted. Now if you were trying to say in your new scheme that you are 5 or 10% ahead in the O-O bond cleavage rather than the C-C bond cleavage, I call that still a concerted mechanism.

McCAPRA: So do I. I was trying to say they were one and the same.

SELIGER: Dr. McElroy brought up the question of whether the enzyme could conceivably accelerate the rate of this relatively energy forbidden linear hypothesis. I think for the sake of the record it ought to be indicated for the people who are going to be reading this in print that they should look at the data that Dempsey and DeLuca have obtained for the chemiluminescence of luciferyl adenylate in which there is no enzyme present and, then apparently, many hypotheses appear to explain the data.

McCAPRA: No, because we start with the peroxide in our case and the reduction is much slower than it is starting with the acridan. The large amount of oxidizable anion around in the latter case reduces the peroxide.

WHITE: On one of your slides you showed a tetrahedral analog. I guess it was the transition state of the hydrolysis going into the excited state?

McCAPRA: That was basically picking up the observation of Dempsey and DeLuca. Dr. McElroy and Dr. Seliger will have to account for that. I haven't seen the data.

ASPECTS OF THE MECHANISM OF BIOLUMINESCENCE

T. Goto, I. Kubota, N. Suzuki, and Y. Kishi

Department of Agricultural Chemistry, Nagoya University

S. Inoue

Faculty of Pharmacy, Meijo Univeristy

Nagoya, Japan

FIREFLY OXYLUCIFERIN

Firefly oxyluciferin (I) has been suggested as the emitter in the firefly bioluminescence and also in the chemiluminescence of firefly luciferin in DMSO (1). However attempted isolation or synthesis of the expected product (I) had been unsuccessful (2). Attempted condensation of ethyl thioglycolate and the appropriate nitrile under the usual condition, i.e. in ca 50% MeOH at pH 8 for a few hrs at r.t., did not give the desired compound (I), but a different compound, dioxyluciferin (II), containing one more oxygen atom in the molecule than the expected molecular formula. Its structure was assigned as II.

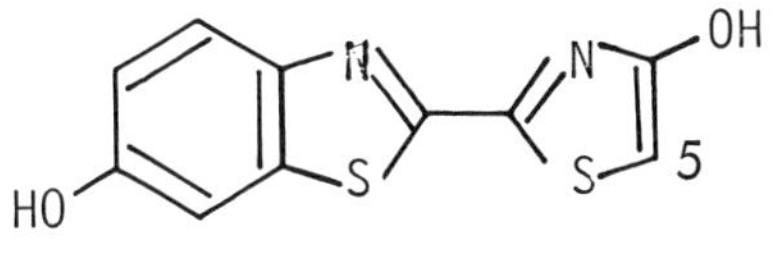

(I) 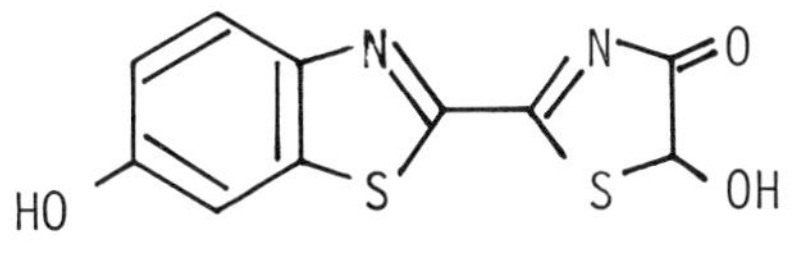   (II)

The observation that the reaction times necessary for production of 5-methyloxyluciferin is ca 10 times shorter than that for 5,5-dimethyloxyluciferin suggested that condensation of unsubstituted thioglycolate with the nitrile would proceed much faster than the case of the 5-methyl derivative. Indeed, oxyluciferin

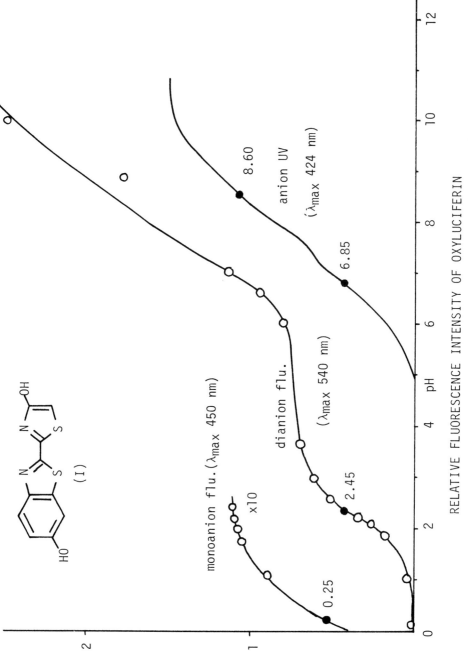

(I) was obtained when condensation was carried out in MeOH aq. for
2 min. under ice-cooling in nitrogen atmosphere. The orange-yellow
crystalline powder thus obtained was sufficiently pure for elemental
analysis and, when recrystallized, it became rather impure (3).
A DMSO solution of I was mixed with a DMSO solution of guanidine
carbonate under vacuum to give a red solution whose fluorescence
spectrum showed its maximum at 564 ± 3 nm. This fluorescence
spectrum is almost superimposable with the spectrum of the yellow-
green chemiluminescence of luciferin reported by White _et al._ (4)
supporting the assignment of the dianion of I as the emitter of
the yellow-green chemiluminescence. Relative quantum yields of
fluorescence of I in various pH were plotted at near λ(max) against
pH.

The formation of oxyluciferin (I) during _in-vivo_ bioluminescence
of firefly has been proved by the following experiments (5).
Fireflies stored at -20° were allowed to stand at r.t., when they
produced _in-vivo_ bioluminescence. After the emission ceased
their lanterns were extracted with MeOH. Tlc of the MeOH extracts
on Avicel gave the spot identical in Rf values with synthetic I.
For comparison, the same extraction was also made without allowing
bioluminescence. In this case, no I was detected. Attempted
isolation of I directly from firefly lanterns was unsuccessful
even if a large number of fireflies were used, since repeated tlc
of the oxyluciferin fractions led to the decomposition of I and
since large quantities of impurities disturbed chromatograms.
Hence, the isolation was done after acetylation of I. Repeated
chromatographic separations of the acetylated products afforded
I diacetate, which shows UV spectrum and Rf values identical with
those of authentic sample. Isolation of I from a spent solution
of chemiluminescence of luciferin was also carried out similarly.

Oxyluciferin (I), when heated with cysteine in basic media,
gave luciferin which was identified by comparison of Rf values of
tlc and UV spectrum with those of authentic specimen. As a by-
product was produced another fluorescent compound which was also
obtained from luciferin on hydrolysis with water. The structure
of this compound was tentatively assigned as III. This amide
could not be cyclized to luciferin by heating in aq. solutions at
any pH. Hence, the amide is not an intermediate of the production
of luciferin from I. This experiment may suggest that in firefly
lanterns the final product of luminescence, I, might be used for
re-synthesis of luciferin.

Inhibitory action of firefly dehydroluciferin on the firefly
luciferase has been extensively studied in connection with the firefly flashing (6). However there is a possibility that oxyluciferin
might be a true substance that acts as the inhibitor in the firefly
lanterns since it is produced stoichiometrically during the luminescence. Inhibition constant of I was measured and compared with

those of luciferin derivatives. Although inhibiting action of I is stronger than that of dehydroluciferin by a factor of 4, it is not activated with ATP and Mg ions in the presence of luciferase as in the case of dehydroluciferin which shows very strong inhibition after being activated. However, concentration of luciferase in firefly lanterns would be much higher than that in in-vitro experiments and hence the inhibitory action of oxyluciferin might be more important in the lanterns.

INHIBITION CONSTANTS

	pH	Ki [M]
Oxyluciferin	7.9	0.23×10^{-6}
Dehydroluciferin	7.7	1.0×10^{-6}*
Decarboxyluciferin	7.7	0.29×10^{-6}*

* J.L. Denburg, R.T. Lee & W.D. McElroy, A.B.B., 134, 38 (1969)

REACTION OF CYPRIDINA LUCIFERIN WITH MOLECULAR OXYGEN

Cypridina bioluminescence is produced in aqueous solution by oxidation of Cypridina luciferin with molecular oxygen in the presence of Cypridina luciferase which may be classified as one of the dioxygenases. Luciferin also chemiluminesces strongly in aprotic solvents such as diglyme without enzyme, and the mechanism of the bioluminescence is assumed to be identical with that of chemiluminescence with the exception that luciferin does not produce light in aqueous solution without enzyme (7). The mechanism has been assumed that luciferin anion is first oxidized by molecular oxygen to give a hydroperoxide anion which then decomposes through a dioxetane intermediate to form the acylaminopyrazine (Cypridina oxyluciferin(IV)) anion in a singlet excited state. In neutral

solution protonation occurs on the excited anion to give excited
neutral oxyluciferin molecule which subsequently gives light.
Decomposition of the hydroperoxide to the acylaminopyrazine (IV)
through the dioxetane is spontaneous reaction process and the
enzyme would not be necessary in this step. The steps in which
the enzyme may act important role are (i) reaction of luciferin
anion with molecular oxygen and (ii) emission of light from the
excited IV.

Although oxyluciferin gives strong fluorescence in aprotic
solvents, almost no fluorescence is observed in aqueous solution.
In the case of bioluminescence, the fluorescence quantum yield of
IV must be high, since the bioluminescence quantum yield is high.
It is one of the reasons that luciferin does not give light in
aqueous solution without enzyme. We have explained (7) this as
assuming the presence of an enzyme-substrate complex, in which the
emitter, IV in excited state, is in environment similar to that
in aprotic solvents (hydrophobic environment). Indeed, Shimomura
et al. (8) reported that addition of purified enzyme to an aqueous
solution of IV enhances the fluorescence intensity of oxyluciferin;
a 1:1 complex between IV and luciferase being formed. An oxyluciferin analog, 2-acetamido-5-phenylpyrazine (V), was found to
be strongly fluorescent in aqueous solution as well as in aprotic
solvents, and hence the corresponding luciferin analog, 2-methyl-
5-phenylimidazo[1,2-a]pyrazin-3-one (VI), should give light in
aqueous solution if its hydroperoxide is formed. Indeed, it gives
light in aqueous solutions in the presence of hydrogen peroxide and

(V)

(VI)

ferric or ferricyanide ions but not with molecular oxygen. Accordingly the major role of the enzyme is not only the enhancement
of fluorescence intensity of IV but also the enhancement of reaction
rate of luciferin with molecular oxygen.

When a catalytic amount of bis(salicylaldehyde)ethylenediimine-
cobalt(II) (salcomine) is added to an aqueous solution of VI luminescence is observed and V is produced in almost quantitative yield.
In the presence of 2.5×10^{-7}M of salcomine VI gives the apparent
Michaelis constant Km = 1.0×10^{-6}M in methanol. The reaction

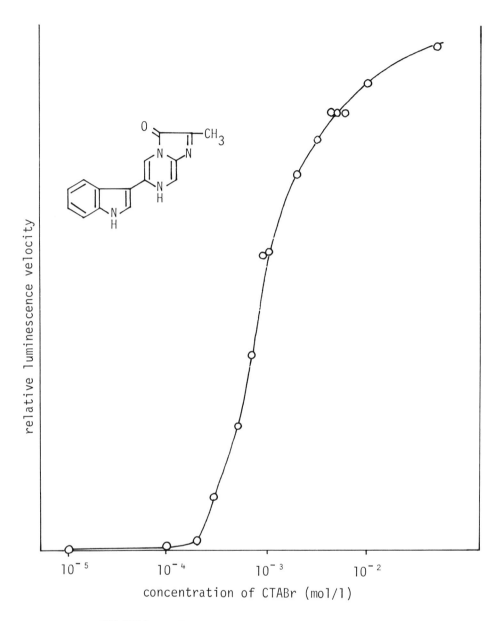

RELATION BETWEEN LUMINESCENCE VELOCITY
AND MICELLE CONCENTRATION

Solvent: 0.1M Tris buffer (pH 9.0)
Temp.: 24°

rate is proportional to oxygen pressure. Salcomine is known to absorb molecular oxygen reversibly to form 2:1 complex (9), and hence it is reasonable to assume that salcomine activates molecular oxygen rather than substrate. Di(3-salicylideneaminopropyl)amine-cobalt(II), a similar oxygen carrier which forms a complex with one molecule of oxygen (9), however, has little effect on the luminescence. Many other cobalt complexes show no effect. The fact that the apparent Michaelis constant is observed in spite of the formation of catalyst-oxygen complex rather than catalyst-substrate complex may be interpreted as

$$\text{salcomine} + O_2 \xrightarrow{\text{slow}} \text{salcomine} \cdot O_2$$

$$\text{salcomine} \cdot O_2 + L \longrightarrow \text{salcomine} + LO_2$$

Since luciferase has been reported (10) to contain no metal ions other than sodium and since it forms a complex with luciferin rather than oxygen, there scarcely remains the possibility that luciferase activates oxygen molecule as many oxygenases do, but it might be necessary to re-investigate this point. Recently Lynch et al. (11) reported that the luciferase may be a metalloenzyme containing calcium.

Consideration of these factors led us to the expectation that if micelles of a suitable surfactant are present, luciferin might be absorbed into micelle interior formed by the hydrophobic portion of the surfactant, and produce light in aqueous solution, as it does in the presence of the enzyme. Indeed, chemiluminescence of luciferin is observed in aqueous micelle solution of cetyltrimethyl-ammonium bromide (cationic), but not in the micelle solution of lauryl sulfate (anionic), although IV shows strong fluorescence in both of micelle solutions. The reason would be that the anionic surfactant forms micelles with highly charged negative surfaces which prevent luciferin anion to enter micelle interior, or in which anion-radical type oxidation reaction does not proceed. (15).

It has long been observed the presence of reversibly oxidized luciferin (luciferin-R). This is formed from luciferin by standing it in aqueous solution or by oxidation with ferricyanide or PbO_2. We have suggested that it is a luciferin dimer produced by combination of luciferyl radical formed by one electron oxidation of luciferin (12). It gives very dim luminescence with luciferase. Kinetic measurements gave the following values:

	Km [M]	$k_{rel.}$	total light emitted (rel.)
luciferin	6.6×10^{-7}	1	1
luciferin-R	8.5×10^{-8}	0.05	ca 0.5

Condition: 1/15 M phosphate buffer (pH 7.32) containing 0.1M NaCl + luciferase at 21°

Since the Km of luciferin-R is extremely small, only very dilute solutions of luciferin-R gives first-order reaction. We reported earlier (12) that the reaction is one-half order, but only under very restricted conditions it seems to fit one-half order. The relative reaction rate was measured in sufficiently dilute solution in which first order reaction was observed.

Luciferin-R is very unstable and cannot be isolated in a pure form and hence the structure determination is difficult. We recently have elucidated (13) the structures of three dimers of 2,4,5-triphenylimidazolyl (lophine dimer) which were obtained in pure forms. In this case phenyl groups stabilize the intermediate radical and also the dimer from further oxidation. Similarly, we have prepared a luciferin analog, 2,6,8-triphenylimidazo[1,2-a]-pyrazin-3-one (VII), and subjected it to oxidation. Reaction in methylene chloride with lead dioxide in a short time (20 sec) at 10° followed by evaporation of the solvent gave a mixture from which a reddish violet solid (VIII) was obtained by washing out of impurities with ether. Its mass spectrum shows the molecular peak at m/e 724 corresponding to the molecular weight of the dimer. Unfortunately it is very insoluble in most organic solvents which can be used for NMR measurements, its structure could not be elucidated. This dimer is easily isomerized on tlc plates to the second dimer (IX). Further oxidation of VII gave an orange product whose structure was elucidated as X. When the oxidation was carried out in methanol with ceric nitrate the yellow compound (XI) was obtained, which on heating was converted to the orange compound (XII). The reverse reaction was observed when the orange compound (XII) was dissolved in methanol containing hydrochloric acid. Treatment of the dimer (VIII) with ethanol containing HCl afforded monomer (VII) and the mixture of the yellow and the orange compounds. The ratio of monomer (VII) and oxidized compounds (XI & XII, Et instead of Me) are nearly 1:1, indicating the dimer structure for VIII.

Shimomura et al. (14) reported that low quantum yield in chemiluminescence of luciferin compared to bioluminescence is attributable to a side reaction which leads to the production of aminopyrazine and keto acid. Our experiment, however, shows that in a very low concentration of the substrate the product is exclusively acylaminopyrazine (V), whereas formation of other products are increased as the concentration is increased. Analysis of kinetics of luminescence also shows the increased production of the dimer as the substrate concentration is increased. This is interpretable by the radical non-chain mechanism. Thus, in low concentration of the substrate, luciferin anion gives its electron to molecular oxygen to form, in a cage, luciferyl radical and superoxide ion which combine rapidly to form hydroperoxide ion. When concentration of luciferin becomes high, the dimer formation becomes predominant.

conc. [M]	products		
0.33×10^{-4}	90%		
1.0×10^{-4}	60%	+	
3.0×10^{-4}	19%	22%	9%

solvent: diglyme containing 0.66% 0.1M acetate buffer (pH 5.6)

It is not a radical-chain mechanism since the first-order reaction const. is almost independent to the substrate concentration and since usual radical inhibitors have no effect on the rate of luminescence. Direct combination mechanism (non-radical) should give no dimer.

We thank prof. T. Matsuura, KyotoUniversity, for generous gift of cobalt complexes and valuable discussions.

REFERENCES

1) see E.H.White, E.Rapaport, H.H.Seliger and T.A.Hopkins, Bioorg. Chem., 1, 92 (1971).
2) P.J.Plant, E.H.White and W.D.McElroy, B.B.Res.Commun.31, 98 (1968).
3) N.Suzuki, M. Sato, K.Nishikawa and T.Goto, Tet. Lett. 4683 (1969).
4) E.H.White, E.Rapaport, J.A.Hopkins and H.H.Seliger, J. Amer.Chem. Soc., 91, 2178 (1969).
5) N.Suzuki and T.Goto, Tet. Lett., 2021 (1971).
6) J.L.Denburg, R.T.Lee and W.D.McElroy, Arch.Biochem.Biophys. 134, 381 (1969).
7) T.Goto, Pure Appl. Chem., 17, 421 (1968).
8) O.Shimomura, F.H.Johnson and T.Masugi, Science, 164, 1299 (1969).
9) R.H.Bailes and M. Calvin, J.Amer.Chem.Soc., 69, 1886 (1947).
10) O.Shimomura, F.H.Johnson and Y.Saiga, J.Cell.Comp.Physiol., 58, 113 (1961).
11) R.V.Lynch,III, F.I.Tsuji and D.H.Donald, B.B.Res.Commun., 46, 1544 (1972).
12) see ref. 7.
13) H.Tanino, T.Kondo, K.Okada and T.Goto, Bull.Chem.Soc.Japan, 45, 1474 (1972).
14) O.Shimomura, F.H.Johnson, Photochem.Photobiol., 12, 291 (1970).
15) T. Goto and H. Fukatsu, Tetr. Lett., 4299 (1969).

SELIGER: I'd just like to make one remark about the possible regeneration of luciferin in the firefly. I think Milt Cormier has run into this same observation with respect to Renilla. We have stayed out in the field very early in the summer and very carefully caught fireflies as soon as they have come up in their adult form before they flashed more than three or four times, and we have extracted from individual tails the enzyme and luciferin and knowing something about how long the firefly lives in the adult stage and knowing quantitatively the total quanta emitted per flash we can extrapolate as to how many flashes the firefly will emit during its lifetime. This comes to about 10^{15} quanta. We also measure in the male adult firefly as soon as it comes up out of the ground on the order of 10^{15} enzyme molecules and on the order of 10^{15} luciferin molecules as though it has everything it needs in order to flash until it can find a mate. Thus there doesn't seem to be a necessity for the regeneration of luciferin in the firefly. One more point in terms of light observed. We have been able to observe chemiluminescence of firefly luciferin of very low intensity in dimethyl sulfoxide and I have been talking with Drs. Stauff and Steele relative to the extremely sensitive means that they are using to detect weak chemiluminescence and I have an idea that if you spit up in the air you could probably measure it and so I sort of thought that the bucket full of product that Frank McCapra wants to put in front of a phototube that won't give light. This just happens to be a rather improbably dark reaction but I am sure there is some light associated with it.

GOTO: It must be a dark reaction in the case of the chemiluminescence because the quantum yield is very low but maybe the dark reaction leads also to our end product we are not sure about that. We could isolate oxyluciferin from the solution that chemiluminesced.

DeLUCA: I wondered whether your inhibition studies with the product appeared to be competitive with respect to luciferin. Is that correct?

GOTO: Oh, competitive with luciferin, yes.

DeLUCA: Did you observe any covalent or irreversible type of inhibition with this product in the enzyme, perhaps as a function of time?

GOTO: It's very similar to free deyhdroluciferin but not very different from dehydroluciferin attached to AMP.

DeLUCA: So your feeling is that it does not react covalently with the enzyme ever after long periods of time?

GOTO: No just between enzyme and oxyluciferin. If we preincubate oxyluciferin and enzyme there is no difference in the reaction rate.

MECHANISM OF THE LUMINESCENT OXIDATION OF CYPRIDINA LUCIFERIN

Osamu Shimomura and Frank H. Johnson

Biology Department, Princeton University
Princeton, New Jersey 08540

In an aqueous medium with Cypridina luciferase, or in an aprotic solvent without luciferase, the aerobic oxidation of Cypridina luciferin is accompanied by emission of light (1,2,3). The products of the luminescent oxidation have been reported as oxyluciferin and CO_2 (4,5,6), although the oxyluciferin can be further converted to etioluciferin by acid or slowly by luciferase (4,7).

On the basis of reaction of luciferin analogs, a mechanism of chemiluminescence of both Cypridina and firefly luciferins, involving the formation and decomposition of a four membered peroxide ring, has been suggested (8,9,10); the same mechanism has been thought to apply to luciferase-catalyzed luminescence of both luciferins (11, 12). In regard to the luciferase-catalyzed luminescence, however, the hypothesis has become questionable through the evidence reported by DeLuca and Dempsey (13) that molecular $^{18}O_2$ does not become incorporated into the CO_2 produced by oxidation of firefly luciferin. Moreover, in the chemiluminescence of Cypridina luciferin in organic solvents, the previously reported reaction scheme does not explain the direct formation of etioluciferin from luciferin (14), as well as the possible production of α-keto-β-methyl-n-valeric acid.

MATERIALS AND METHODS

All organic solvents were "Spectroquality", from Matheson, Coleman and Bell. Diglyme (Bis(2-methoxyethyl)ether) was used immediately after filtration through alumina. The Cypridina luciferin was the dihydrobromide, 3 times recrystallized, and almost colorless. Cypridina luciferin was of the same purity as used previously (7) and revealed a single band in polyacrylamide gel electrophoresis.

Quantum yields were measured, by a photomultiplier, Hamamatsu R-136, of which the spectral distribution of sensitivity had been calibrated.

The amounts of oxyluciferin and etioluciferin in reaction mixtures were calculated from UV absorbance measured at two wavelengths, 310 nm and 333 nm, after diluting the reaction mixture with 20 volumes of ethanolic 0.1 N HCl. The molecular extinction coefficients of oxyluciferin dihydrochloride (M.W. 478); 13,600 (310 nm) and 14,900 (333 nm), etioluciferin dihydrochloride (M.W. 382); 22,500 (310 nm) and 7,450 (333 nm) were used in the calculations. Some error would be expected due to possible presence of minor by-products; however, such error would be expected to be relatively small and should not affect conclusions drawn from the present study.

The amount of α-keto acid was estimated by the 2,4-dinitrophenyl hydrazine colorimetric method as modified by Rohrbough (16), and by further slight changes in the procedure, <u>viz</u>., by the replacement of 1 ml of 80% pyridine with the same volume of methanol, 25% KOH instead of 33% KOH, and measurement of color at 560 nm instead of 480 nm. Two ml of reaction mixture was used without pretreatment, except that (1) luciferase was first removed by ethanol, and (2) acetone was evaporated under vacuum, followed by redissolving the residue with methanol to the initial volume. A standard calibration curve was obtained using α-keto-β-methyl-<u>n</u>-valeric acid, sodium salt (95%, Sigma Chemical Company).

In the study regarding incorporation of ^{18}O into CO_2, 1.4 mg of luciferase was dissolved in 4.5 ml of 0.02 M glycylglycine, pH 7.8 (13), containing 0.04 M NaCl, prepared with regular water or $H_2^{18}O$, and placed in the bottom of a reaction vessel (Fig. 1). Two mg of luciferin dihydrobromide was dissolved in 0.12 ml of 60% methanol, and placed into the small side arm of the reaction vessel. The vessel was slowly evacuated <u>via</u> 2 traps without coolants (Fig. 1). When the solutions began bubbling, the vessel was placed in dry ice-acetone, and evacuation was continued until the pressure dropped to

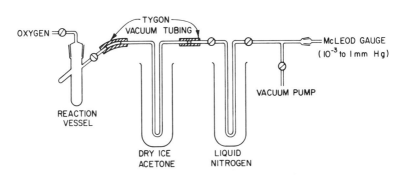

Fig. 1. Apparatus employed for collection of CO_2.

3 μHg or less. After introduction of $^{16}O_2$ or $^{18}O_2$, the vessel was warmed and contents equilibrated at 20°C, and then the two solutions were mixed. The resulting emission of bright light almost ceased in 30 seconds of reaction time, after which the vessel was frozen as before. It was evacuated through a trap in dry ice-acetone and a second trap in liquid nitrogen. The amount of CO_2 collected in the latter trap was estimated at room temperature by a McLeod gauge, then the CO_2 was again condensed in the same trap by liquid nitrogen for use in mass spectrometry to determine the ratio of m/e 46 and m/e 44 (by the Morgan-Schaffer Corporation, Montreal).

For the reaction in dimethylsulfoxide (DMSO), 2 mg of luciferin was mixed with 5 ml of DMSO containing 10 mg of potassium t-butoxide under $^{18}O_2$. By the same technique as described above for the luciferase reaction, this reaction was allowed to proceed for 5 min., and the CO_2 was then collected.

RESULTS AND DISCUSSION

Analysis for the luminescent oxidation products of <u>Cypridina</u> luciferin in several solvent systems are summarized in Table I. In the luciferase-catalyzed luminescence, oxyluciferin is the main product, whereas in organic solvents, or in aqueous solution without luciferase, the products were always a mixture of oxyluciferin, etioluciferin and α-keto acid, with the latter two compounds in approximately equimolar amounts. Moreover, in either diglyme or acetone, which gave relatively high quantum yields of chemiluminescence, etioluciferin rather than oxyluciferin was a predominant product. The possible hydrolysis of oxyluciferin to etioluciferin was negligible under the conditions of Table I.

Oxyluciferin and etioluciferin were confirmed by TLC (silicic acid - water saturated n-butanol). The presence of α-kety-β-methyl-n-valeric acid in each reaction mixture was proved as follows. First, the keto acid was extracted from acidified aqueous solutions with ether, and was then converted to 2,4-dinitrophenylhydrazone, and the hydrazone confirmed by mass spectrometry or by TLC (silicic acid - water saturated n-butanol) in comparison with an authentic sample. It is noteworthy that even the luciferase-catalyzed reaction gave some of the keto acid, although the amount was small.

The source of oxygen in the CO_2 product was studied with ^{18}O, with results shown in Table II. In the luciferase-catalyzed reaction, oxygen is incorporated into CO_2 largely from molecular O_2, and only small part from H_2O, whereas in DMSO, approximately 2/3 of oxygen incorporated into CO_2 is from molecular O_2. The yields of CO_2 in the luciferase-catalyzed reaction and in the reaction in DSMO were 90% and 60%, respectively, of the initial luciferin.

Table I. Products of the Luminescent Oxidation of Cypridina Luciferin.[a]

Solvents[b] (5 ml)	Reaction time (hours)	Oxy-luciferin (mole %)	Etio-luciferin (mole %)	α-Keto acid (mole %)
1.5 mg Luciferase in buffer[c]	0.05	86	10	10
Diglyme + 0.1 ml buffer, pH 5.6[d]	2	34	51	?[e]
Acetone + 0.1 ml buffer, pH 5.6[d]	2	31	57	?[e]
Pyridine + 50 μl buffer, pH 4.6[d]	4	23	65	67
DMSO + 150 μl buffer, pH 4.6[d]	4	25	69	67
DMSO + Potassium t-butoxide (10 mg)[f]	0.1	64	35	31
50% Methanol + 50 μl of 28% ammonia	24	55	29	29
H_2O + 5 mg $NaHCO_3$	24	27	49	51

a) Luciferin dihydrobromide (1 to 1.5 mg) was reacted at 22°C.
b) In sequence of greater quantum yield. The quantum yields at 22° for the first five solvents with optimum amounts of luciferin and other additives, typically with 30 μl of buffer and 10 μg of luciferin were: 0.26, 0.023, 0.021, 0.002, 0.0008.
c) 0.05 M Sodium phosphate containing 0.1 M NaCl, pH 6.8. Immediately after the reaction, luciferase was denatured by addition of ethanol and removed by centrifugation.
d) 0.3 M Sodium acetate buffer
e) Unreliable results because of residual carbonyl due to solvent.
f) Two drops of 12 N HCl were added at the end of the reaction.

The foregoing data suggest that at least 3 different pathways are involved in the luminescent oxidation of luciferin, as illustrated in Fig.2. Pathway A was first suggested by McCapra and Chang (8) for the chemiluminescence of a synthetic analog, pathway B is the same mechanism that DeLuca and Dempsey (13) proposed for the bioluminescence of the firefly to explain the incorporation of oxygen of H_2O into CO_2, and pathway C is proposed herewith to explain

Table II. Incorporation of Oxygen into CO_2 During the Luminescent Oxidation of Cypridina Luciferin.[a]

Conditions	^{18}O in CO_2 (Atom % excess)	Oxygens[b] Incorporated
$H_2^{16}O$ + $^{18}O_2$ (89 atom % excess)	35.8	0.80 from O_2
$H_2^{18}O$ (3.76 atom % excess) + $^{16}O_2$	0.29	0.15 from H_2O
$H_2^{18}O$ (3.76 atom % excess) + $C^{16}O_2$[c]	0.035	0.02 from H_2O
DMSO + Potassium t-butoxide + $^{18}O_2$ (93 atom % excess)	28.8	0.62 from O_2

a) Luciferase-catalyzed, except the last condition.
b) Half of the oxygens in the total CO_2 is taken as 1.0.
c) A control run without luciferin. CO_2 was produced by heat-decomposition of $NaHCO_3$ immediately before introduction of O_2.

the direct formation of etioluciferin and of the α-keto acid. In pathway C, an intermediary tertiary alcohol, formed probably by a base-catalyzed oxygen elimination or by a reaction of hydroperoxide with unreacted luciferin, is hydrolyzed.

From the data of Tables I and II, in the luciferase-catalyzed reaction, the pathway A is predominant and the ratio of luciferin oxidized by pathways A, B and C is approximately 7:1:1, whereas in the chemiluminescent reaction without luciferase, it appears that the 3 pathways are all significantly involved. Because the maximum quantum yield of luciferase-catalyzed luminescence is approxkmately 0.3 (14), pathway A must be a light-emitting reaction; however, no evidence is yet available whether pathways B and C are luminescent or dark.

In comparison with the luciferins of firefly and Renilla, which also liberate one molecule of CO_2 in luciferase-catalyzed bioluminescence reactions (12,17), the luminescent pathway A of Cypridina luciferin differs in that one oxygen in the liberated CO_2 comes from molecular O_2, in contrast to the firefly and Renilla systems, wherein one of the oxygens come from H_2O (13,17).

Bacterial bioluminescence, a flavin enzyme mono-oxygenase system which does not yield CO_2 as a product (18), bears only a remote resemblance to Cypridina luminescence. In the bioluminescence of Latia, which also appears to be a flavin enzyme system, the mechanism

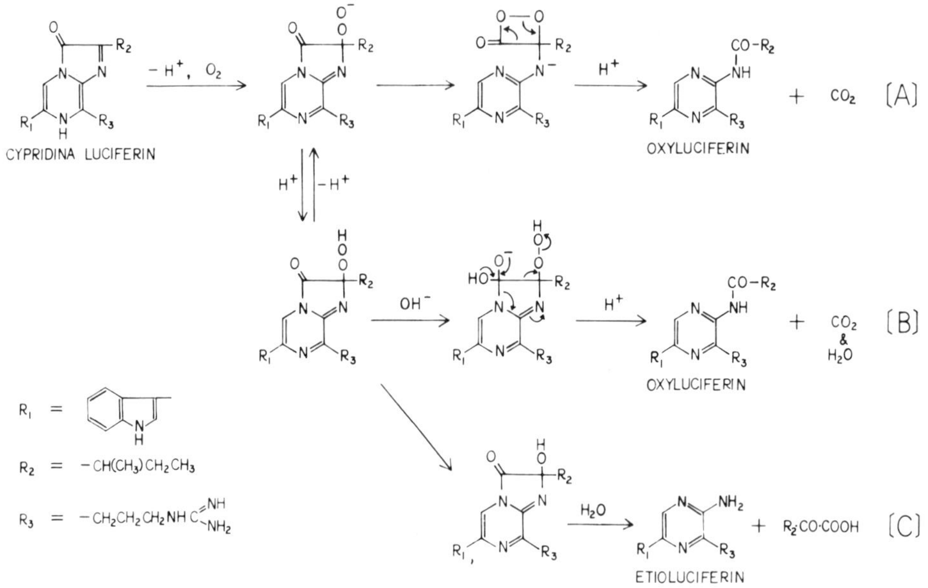

Fig. 2. Oxidative degradation of Cypridina luciferin.

of light emission is probably the same, or very similar to, that of bacterial bioluminescence, despite the production of CO_2 as a product in the Latia system (18).

The structure of the light emitting moiety separated from the photoprotein aequorin has been established (19,20), and the manner in which this moiety is bound to the protein residue has also been determined (21). Two kinds of principal reactions of aequorin, illustrated in Fig. 3, reveal a striking resemblance to the oxidation reaction of Cypridina luciferin shown in Fig. 2. Thus, BFP (blue fluorescent protein) formed in the luminescent reaction of aequorin

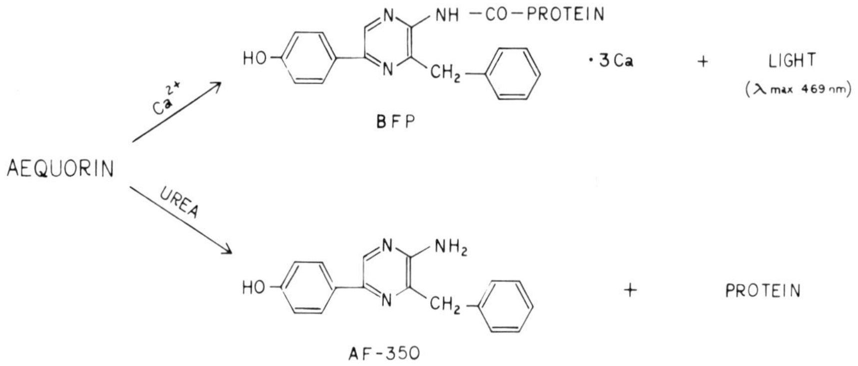

Fig. 3. Two principal reactions of aequorin.

corresponds to Cypridina oxyluciferin produced in the luminescent reaction of Cypridina luciferin, while AF-350 formed by urea-denaturation of aequorin corresponds to Cypridina etioluciferin produced in a side reaction (pathway C). Despite the resemblance of the schemes depicting the splitting of molecules, however, CO_2 is not formed in the luminescence reaction of aequorin with Ca^{2+} (21). Consequently, we would like to defer further comparisons between these two bioluminescence systems until the structure of the light-emitting group of unreacted aequorin is clarified.

Aided by NSF Grants GB 15092 and GB 30963X, and by ONR Contract N00014-67A-0151-0025.

REFERENCES

1. E.N. Harvey, Bioluminescence, Academic Press, New York, pp. 297-331 (1952).
2. F.H. Johnson, H.-D. Stachel, E.C. Taylor and O. Shimomura, Bioluminescence in Progress (Ed. F.H. Johnson and Y. Haneda), Princeton University Press, Princeton, N. J., p. 67 (1966).
3. T. Goto, S. Inoue and S. Sugiura, Tetrahedron Lett., 3875 (1968).
4. Y. Kishi, T. Goto, Y. Hirata, O. Shimomura and F.H. Johnson, Ref. 2, p. 89.
5. H. Stone, Biochem. Biophys. Res. Comm., 31, 386 (1968).
6. T. Goto, S. Inoue, S. Sugiura, K. Nishikawa, M. Isobe and Y. Abe, Tetrahedron Lett., 4035 (1968).
7. O. Shimomura, F.H. Johnson and T. Masugi, Science, 164, 1299 (1969).
8. F. McCapra and Y.C. Chang, Chem. Comm., 1011 (1967).
9. T.A. Hopkins, H.H. Seliger, E.H. White and M.W. Cass, J. Am. Soc., 89, 7148 (1967).
10. F. McCapra, Y.C. Chang and V.P. Francois, Chem. Comm., 22 (1968).
11. T. Goto, Pure and Applied Chemistry, 17, 421 (1968).
12. P.J. Plant, E.H. White and W.D. McElroy, Biochem. Biophys. Res. Comm., 31, 98 (1968).
13. M. DeLuca and M.E. Dempsey, Biochem. Biophys. Res. Comm., 44, 117 (1970).
14. O. Shimomura and F. H. Johnson, Photochemistry and Photobiology, 12, 291 (1970).
15. F.H. Johnson, O. Shimomura, Y. Saiga, L.C. Gershman, G.T. Reynolds and J.R. Waters, J. Cell. Comp. Physiol., 60, 85 (1962).
16. F.E. Critchfield, Organic Functional Group Analysis, Pergamon Press, p. 78 (1963).
17. M. DeLuca, M.E. Dempsey, K. Hori, J.E. Wampler and M.J. Cormier, Proc. Nat. Acad. Sci. USA, 68, 1658 (1971).
18. O. Shimomura, F.H. Johnson and Y. Kohama, Proc. Nat. Acad. Sci. USA, 69, 2086 (1972).
19. O. Shimomura and F.H. Johnson, Biochemistry, 11, 1602 (1972).
20. Y. Kishi, H. Tanio and T. Goto, Tetrahedron Lett., 2747 (1972).
21. O. Shimomura and F.H. Johnson, in preparation.

HASTINGS: Are you very sure that no CO_2 comes off during the calcium-triggered light reaction of aequorin? Are you going to try it again or do you feel that this is absolutely definitive.

SHIMOMURA: Oh, Aequorea, absolutely sure. About the exact amount of CO_2, yes none released. It stays in a bound state. No gaseous CO_2, and that's absolutely sure.

CORMIER: You indicated that AF-350 is linked to the protein by a peptide linkage. What evidence do you have for that?

SHIMOMURA: We don't. I have a comparative spectrum with acetate, butyrate, and acetyl glycine derivatives. The absorption spectrum showed very good agreement.

McELROY: Does urea break a covalent bond in this treatment of Aequorea?

SHIMOMURA: The chemical bond is not broken by urea. However it was separated.

WHITE: The hydrolyzed form of Latia is an aldehyde. Has it ever been tested in the bacterial system?

SHIMOMURA: Yes I have tested but found no activity. Almost none, I can't say zero but very low, certainly lower than C-6 aldehyde.

MECHANISM OF BIOLUMINESCENCE AND CHEMILUMINESCENCE ELUCIDATED BY USE OF OXYGEN-18*

Marlene DeLuca and Mary E. Dempsey

Department of Chemistry,
University of California, San Diego
and Department of Biochemistry,
University of Minnesota, Minneapolis, Minnesota

The purpose of this report is to summarize our recent studies designed to elucidate the mechanisms of several bio- and chemiluminescent systems. We are using oxygen-18 as tracer for events occurring on a molecular level during light emission. The accuracy and validity of the oxygen-18 methodology has been extensively examined. Some previously unpublished methodology is included in this report. We have studied the oxidation of firefly and sea pansy (<u>Renilla reniformis</u>) luciferin catalyzed by firefly and sea pansy luciferase (1,2) and also the oxidation of firefly luciferyl-adenylate in the presence of dimethyl sulfoxide and potassium t-butoxide, chemiluminescence. (3) Our data indicate that striking similarities must exist in the mechanisms of bio- and chemiluminescence. We found that CO_2 is produced during substrate oxidation and light emission in all these systems and that oxygen present in the CO_2 arises from water rather than molecular oxygen. These findings have permitted us to develop new mechanisms for the oxidative reactions occurring during luminescence in these systems.

Procedures: <u>Validation of the Oxygen-18 Methodology</u>.

Reaction Conditions and CO_2 Collection; The reaction vessel used for both the bio- and chemiluminescent studies is shown in Figure 1. The substrate,

*Supported by the National Science Foundation

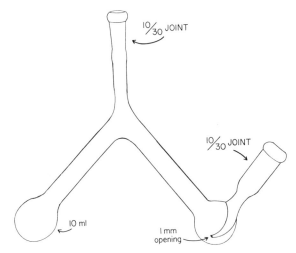

Fig. 1. Reaction vessel used for the nonenzymic experiments. The same type of vessel was used for the enzymic experiments, except without the device for CO_2 admission on the right-hand sidearm.

appropriate cofactors, and buffers (bioluminescence) or substrates and dimethyl sulfoxide (chemiluminescence) are placed in one sidearm by use of a syringe attached to a piece of small bore teflon tubing (12-14cm long). Enzyme and buffer (bioluminescence) or potassium t-butoxide in t-butanol (chemiluminescence) are placed in the other sidearm. The vessel is attached to a high vacuum line by way of the removable vacuum stopcock (for description of a typical high vacuum system see 4). The solutions are then frozen in dry ice and acetone and evacuated to less than 5 microns gas pressure, followed by thawing, refreezing and repeated evacuation. Alternatively, solutions are evacuated until all bubbling ceases. Control experiments have shown that once the vessel is evacuated to less than 5 microns pressure there is no air leakage into the closed vessel over a 24 hour period. In addition, related experiments demonstrated that our techniques of evacuating the reaction solution result in removal of all but a barely measurable level (less than 10 microns) of gas that behaves like CO_2. In other words, CO_2 released during luminescent reactions could not be diluted by CO_2 arising from leakage of air into the reaction vessel or present in the reaction solutions.

Following complete evacuation of the reaction solutions, oxygen (when $^{18}O_2$ is used its enrichment is greater than 95% atom percent excess) is admitted to the vessel at approximately 15 mm Hg pressure. To insure that no CO_2 is admitted to the reaction at this point, the vessel containing the oxygen is submerged in liquid N_2 for at least one hour prior to use. We found that all commercial preparations of $^{18}O_2$ contain low levels of $C^{18}O_2$ of high enrichment. Next, the reaction vessel (stopcock closed) is removed from the vacuum line and the solutions in the sidearms mixed. When light emission ceases, 10-30 seconds for firefly bioluminescence and chemiluminescence (1,3), both sidearms of the vessel are plunged into liquid N_2. During sea pansy luminescence (2) reaction times are much longer (40 minutes) therefore CO_2 was trapped as it was released by allowing the reaction to occur in one sidearm of the vessel and maintaining the other sidearm in liquid N_2. CO_2 released during the luminescent reactions is collected for analysis by first attaching the reaction vessel to the vacuum line. The reaction vessel (contents still frozen in liquid N_2) is opened and evacuated to less than 5 microns pressure. Then the contents of the vessel are warmed to the temperature of dry-ice acetone to release CO_2 gas, but not H_2O vapor. The CO_2 is trapped in another glass bulb by use of a liquid N_2 trap (5).

Mass Spectral Analysis of CO_2; Calculation of Oxygen-18 Enrichment.

Each CO_2 sample is admitted to a Hitachi-Perkin-Elmer RMU-6D mass spectrometer via an all glass inlet system. The sample is scanned at least four times from mass 43 to 49. $C^{18}O_2$ standard samples of known enrichment (4) are included with each batch of unknown samples. The heights of each mass peak are measured manually and extrapolated to zero time using a least squares computer program. In addition, each sample is scanned broadly to determine the occurrence of air, water, or volatile fragments. Results obtained on samples contaminated with gases other than CO_2 are discarded. Atom percent oxygen-18 enrichment of the CO_2 sample is calculated from equilibrium and probability considerations by the following formula (4,5):

$\frac{R}{2+R} \times 100$, where R = $\frac{\text{mass 46 peak height at zero time}}{\text{mass 44 peak height at zero time}}$

Atom percent excess enrichment of the sample is obtained by subtracting the normal abundance of $C^{18}O_2$ (0.2%) from the calculated enrichment of the sample. The level of enrichment of the $^{18}O_2$ gas or $H_2^{18}O$ used for the luminescent reaction is determined similarly (5). To estimate the expected enrichment of CO_2 produced during a reaction if one oxygen were incorporated from oxygen or water, the atom percent excess of the $^{18}O_2$ or medium $H_2^{18}O$ is divided by two. Thus, if the medium is prepared in $H_2^{16}O$ and exposed to $^{18}O_2$ (96 atom %), incorporation of one oxygen from O_2 into CO_2 would result in an enrichment of 48 atom % in the CO_2. This value is far in excess of the limits of the mass spectral techniques, e.g. we are able to detect < 1% or lower enrichment of CO_2 with $\pm$ 2% accuracy.

Table I

LACK OF APPRECIABLE EXCHANGE OF WATER OXYGENS WITH THOSE OF CO_2

Conditions	Oxygens Incorporated
Rapid Mixing of $H_2^{18}O$ and $C^{16}O_2$	
16 seconds	0
30 seconds	<0.1
80 seconds	<0.1
Release of Fine Bubbles of $C^{16}O_2$ into $H_2^{18}O$*	
15 minutes	0.1
20 minutes	0.2
40 minutes	0.2

* CO_2 was trapped in liquid N_2 after passing through the $H_2^{18}O$.

Lack of Appreciable Exchange of CO_2 Oxygens with Medium.

The results presented in Table I demonstrate the lack of appreciable nonenzymic exchange of CO_2 oxygens with those of water under conditions simulating the enzymic reactions. These results support the validity of our findings with the enzyme systems. In addition, the partial pressure of CO_2 during the enzymic reactions was low (micron range) and this would not favor solution in the medium of appreciable quantities of CO_2. Thus, rapid hydration and dehydration of CO_2 known to occur at neutral pH was minimized (6-9). Furthermore, the firefly experiments are conducted at pH 7.8 and in control experiments conducted at that pH we were unable to detect dehydration of $NaHCO_3$. Thus CO_2 hydrated during the enzymic reaction would not be reversibly dehydrated under these conditions. Also exchange of the oxygens of CO_2 with those of the medium could not be detected in the highly basic conditions of chemiluminescence (Table II). In this regard, exchange of the oxygens of $C^{18}O_2$ with those of the medium could not be detected under the highly alkaline conditions of the chemiluminescent reaction (Table II.).

Results

The data presented for bioluminescence in Table III show that one oxygen from water is incorporated into CO_2 released during firefly bioluminescence. No incorporation of 18-oxygen into CO_2 is observed in the presence of $^{18}O_2$. Similar results are obtained during bioluminescence of the sea pansy system (Table III) except that two oxygens from water are incorporated into the CO_2 released. The additional oxygen is probably contributed by the non-enzymic exchange of water oxygen with the ketone oxygen of luciferin (2).

In contrast, bioluminescence catalyzed by the Cypridina luciferase and luciferin (gifts of Drs. Shimomura and Johnson) measurable incorporation of 18-oxygen occurs from $^{18}O_2$. These results are in accord with those of Shimomura and Johnson (10) and indicate differences in the mechanism of Cypridina luminescence from that of the firefly and sea pansy. We did not have sufficient quantities of Cypridina luciferin and enzyme to carry out an experiment with $H_2^{18}O$.

Table II

LACK OF APPRECIABLE EXCHANGE OF

CO_2 OXYGENS DURING CHEMILUMINESCENCE

Conditions	Oxygens Exchanged
$(CH_3)_2SO$; K-t-butoxide* $C^{18}O_2$	< 0.1

* CO_2 was trapped in liquid N_2 after passing through the solvents.

Finally in accord with the firefly and sea pansy bioluminescence experiments, recent studies of the chemiluminescence of firefly luciferyl-adenylate in dimethyl sulfoxide and potassium t-butoxide offer strong evidence that the mechanism of bio and chemiluminescence are similar. Table IV gives the data from the chemiluminescent experiments and demonstrates that incorporation of labelled oxygen into CO_2 is from water not molecular oxygen.

Discussion

Since there are many examples in the literature which demonstrate that four membered peroxide rings, dioxetanes, decompose with the emission of light, it seemed reasonable to suggest the participation of dioxetanes as intermediates in various bioluminescent reactions (11, 12). However, the $^{18}O_2$ studies do not appear to be consistent with the dioxetane mechanism.

The sea pansy system is perhaps the most susceptible to non-enzymic exchange due to the long reaction time. However, if we look at the proposed intermediate (Fig. II - Structure III) which is an α keto peroxide, it has been shown previously by other workers that α keto peroxides exist in solution as the linear peroxide. Infrared spectroscopy, Ultraviolet spectroscopy and nuclear magnetic resonance studies gave no indication of any cyclic peroxide structure (13,14). Therefore these data are consistent with the mechanism proposed in Fig. II. If larger amounts of sea pansy luciferase and luciferin become available it would be nice to repeat these experiments at higher concentrations.

Table III

SOURCE OF OXYGEN IN CO_2 RELEASED

DURING BIOLUMINESCENCE

Preparations and Conditions	Oxygens Incorporated
Firefly Luciferase + Luciferin	
$H_2{}^{18}O$; $^{16}O_2$	1.0
$H_2{}^{16}O$; $^{18}O_2$	0
Sea Pansy Luciferase + Luciferin	
$H_2{}^{18}O$; $^{16}O_2$	1.9
$H_2{}^{16}O$; $^{18}O_2$	<0.1
Cypridina Luciferase + Luciferin	
$H_2{}^{16}O$; $^{18}O_2$	0.3

In the case of the firefly bioluminescence two points are important in establishing the validity of the results. First the quantum yield with respect to luciferin is approximately 0.9 (15). This is important since we are assured that essentially all of the CO_2 formed arises from a light producing pathway. Secondly, the same isotope incorporation occurs at two different concentrations of enzyme and substrate which rules out a fortuitous dilution of CO_2 by any residual dissolved gas (3).

Table IV

SOURCE OF OXYGEN IN CO_2 RELEASED

DURING BIO- AND CHEMILUMINESCENCE

Preparation and Conditions	Oxygens Incorporated
Luciferyl-Adenylate + Luciferase	
$H_2{}^{18}O$; $^{16}O_2$	1.0
$H_2{}^{16}O$; $^{18}O_2$	0
Luciferyl-Adenylate + Solvents*	
$H_2{}^{18}O$; $^{16}O_2$	1.0
$^{18}O_2$	<0.1

* Dimethyl sulfoxide + K-t-butoxide

The experiments in which the chemiluminescence of firefly luciferyl-adenylate were studied lend further support to the linear peroxide mechanism. The reaction is carried out in DMSO which has been dried over anhydrous $CaCl_2$. However, this solvent still contains of the order of 1×10^{-2} M water (16) which can supply the necessary OH^- for reaction.

Upon addition of $H_2{}^{18}O$ to a final concentration of 1.6M the incorporation of 1 oxygen into the CO_2 was observed. This is the expected result since 1×10^{-2} M residual water would not significantly alter the $H_2{}^{18}O$ concentration.

If the linear peroxide mechanism is correct we might expect that increasing the H_2O content of the DMSO would effect the rate of the chemiluminescent reaction. This was shown to be true since the time required to reach maximal light intensity was less when the DMSO was made approximately 2M in H_2O (3). It should be noted that another system N-methylacridone described by Rauhut et al. (17) also requires water for chemiluminescence. This effect of H_2O is not consistent with the α-peroxy lactone intermediate. However, a detailed interpretation of the effect of water certainly requires further experiments.

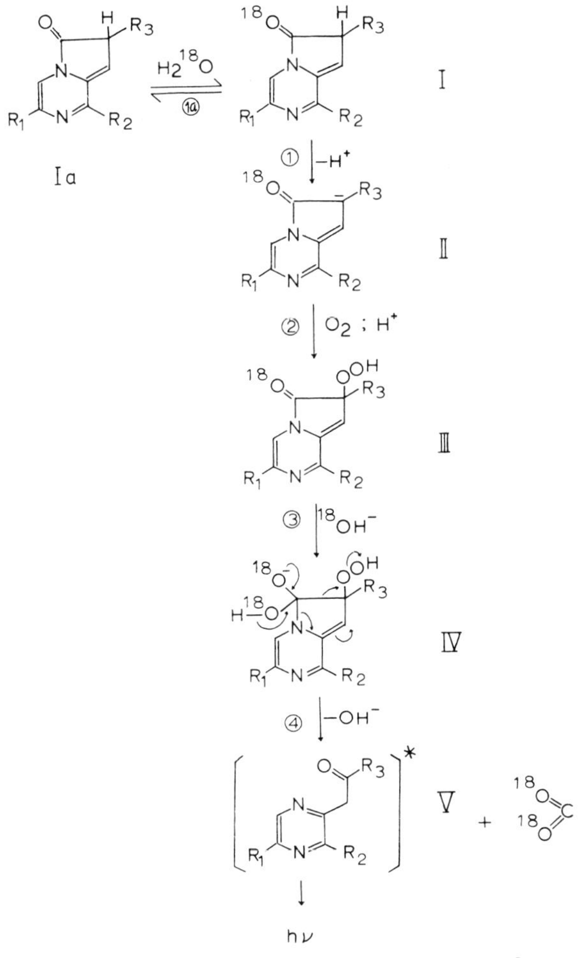

Fig. II. Proposed mechanism for oxidative CO_2 production during Renilla reniformis bioluminescence.

In Fig. III we present an alternate possible mechanism which includes an α-peroxylactone intermediate and is consistent with the $^{18}O_2$ studies. Here the α-peroxylactone is formed followed by attack of either OH^- or H_2O. Such an attack would be favored by a good leaving group, as well as a strained ring system. It is not known whether such a decomposition of the α-peroxylactone would produce light.

In conclusion the isotope experiments with the firefly and sea pansy systems do not support the formation of a cyclic oxygen intermediate. However, further experiments must be done in order to establish whether the reaction sequence shown in Figure III is operative.

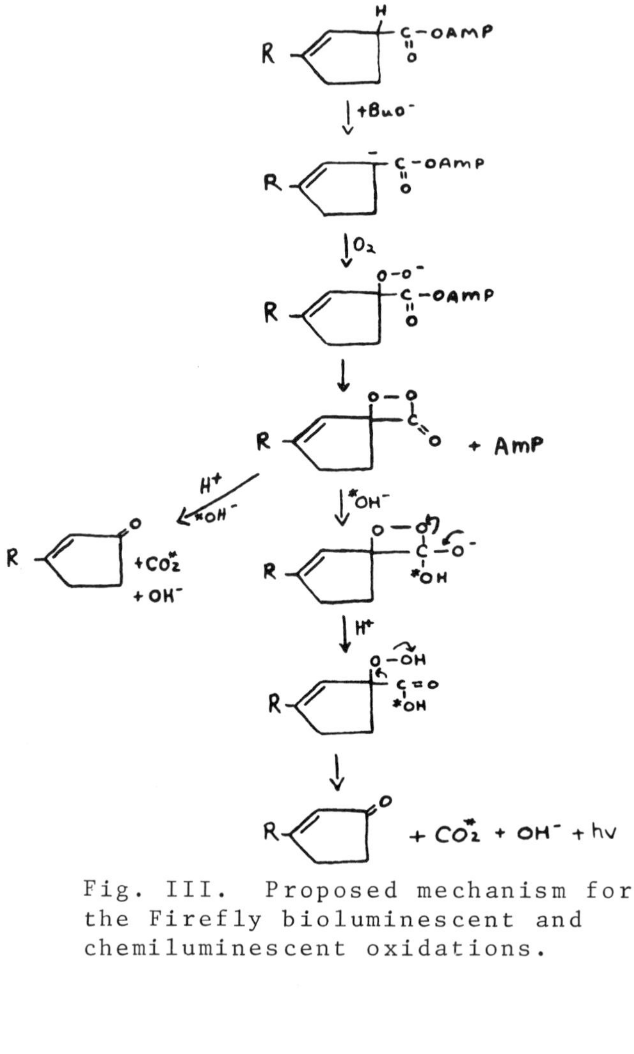

Fig. III. Proposed mechanism for the Firefly bioluminescent and chemiluminescent oxidations.

References

1. DeLuca, M., and Dempsey, M.E., Biochem. Biophys. Res. Commun., $\underline{40}$, 117 (1970)
2. DeLuca, M., Dempsey, M.E., Hori, K., Wampler, J.E. and Cormier, M.J., Proc. Nat. Acad. Sci. U.S.A., $\underline{68}$ 1658 (1971).
3. DeLuca, M.,and Dempsey, M.E., manuscript in preparation.
4. Boyer, P.D. and Bryan, D.M., in "Methods in Enzymology", Vol. 10 (Estabrook, R.W. and Pullman, M.E., Eds.), New York, Academic Press, Inc., 1967, p.60.
5. Boyer, P.D., Graves, D.J., Suelter, C.H., and Dempsey, M.E., Anal. Chem., $\underline{33}$, 1906 (1961).
6. Krebs, H.A., and Roughton, F.J.W., Biochem. J., $\underline{43}$ 550 (1948).
7. Ho, C., and Sturtevant, J.M., J. Biol. Chem., $\underline{238}$, 3499 (1963).
8. Gibbons, B.H., and Edsall, J.T., J. Biol. Chem. $\underline{238}$, 3502 (1963).
9. Cooper, T.G., Tchen, T.T., Wood, H.G., and Benedict, C.R., J. Biol. Chem., $\underline{243}$, 3857 (1968).
10. Shimomura, O., and Johnson, F.H., Biochem. Biophys. Res. Comm. $\underline{44}$, 340 (1971).
11. McCapra, F., Chem. Comm. 155 (1968).
12. Hopkins, T.A., Seliger, H.H., White, E.H. and Cass, M.W., J. Amer. Chem. Soc. $\underline{89}$, 7148 (1967).
13. Richardson, W.H., and Steed, R.F., J. Amer. Chem. Soc. $\underline{32}$, 771 (1967).
14. Fuson, R.C., and Jackson, H.L., J. Amer. Chem. Soc. $\underline{72}$, 1637 (1950).
15. Seliger, H.H., and McElroy, W.D., Arch. Biochem. Biophys. $\underline{88}$, 136 (1960).
16. Martin, D., Weise, A., and Niclas, H., Angew, J. Chem. Intl. Ed. $\underline{6}$, 318 (1967).
17. Rauhut, M.M., Accts. of Chem. Res. $\underline{2}$, 80 (1969).

ADAM: We have observed that you must vigorously exclude water or even alcohols from the reactions because water or alcohol very quickly results in hydrolysis leading to decomposition. So it is likely in either of your alternative mechanisms that you have formed the α-peroxylactone but you have competing hydrolysis of the intermediate.

DeLUCA: Is your hydrolysis a dark reaction?

ADAM: Well we haven't investigated this in detail yet but in other words the peroxylactone decomposition by itself is barely visible to the eye even in a very dark room. We haven't yet looked at that in the presence of fluorescers so I don't really know at this moment. I know that you rigorously must exclude water or alcohol for the peroxylactone to survive.

DeLUCA: There will be a competing hydrolytic reaction but it looks as though we are seeing an additional effect on the light reaction with water here.

J. LEE: Matheson and I have been studying the generation of superoxide in air-saturated DMSO and in the presence of t-butoxide both by absorption and EPR. The data so far suggest that under these conditions there is almost no free oxygen in the solvent. It is all in the form of superoxide. The oxygen that is there spontaneously forms superoxides and the equilibrium lies far to this side. I wonder if the water effect therefore is an effect on the equilibrium between molecular oxygen and superoxide merely driving it back and forming free oxygen and thereby increasing the reaction rate that way?

DeLUCA: That is certainly a possibility.

RAUHUT: Is it possible to recover luciferin from an incomplete reaction to determine whether the recovered luciferin has picked O^{18} from the labeled water extract?

DeLUCA: With the enzyme that would be almost impossible, the reaction occurs so rapidly. Now with the chemiluminescence system you could probably adjust conditions such that we could let part of the light reaction proceed and then freeze it very rapidly and try to reextract the luciferin and look for O^{18}. We haven't done that.

RAUHUT: That might be worth doing if it is possible experimentally because there are mechanisms in this ester hydrolysis area of course with that exact same thing happening.

DeLUCA: Right. The thing that makes me think that it is most unlikely is that we are going from 55 molar water in the bioluminescence system, to 10^{-2} molar in the chemiluminescent system. I would certainly have expected that, if you are looking at the tetrahedral intermediate of the hydrolytic reactions you would see some difference in the isotope incorporations.

HOPKINS: The results with the enzymes are always subject to the possibility that some kind of selective exchange might occur on the enzyme. The results of the chemiluminescence is the subject of the questions that I have and the first of these is related to the conditions under which you did the exchange reactions. Did you measure the chemiluminescence yield?

DeLUCA: No we did not. It's under conditions where you are getting red light, not the very basic conditions under which you get the yellow-green so I assume that the quantum yield would be that which is reported in the literature of approximately 0.2. We have no evidence of any different reaction paths. Again it appears as though all the CO_2 produced is coming via the same organic mechanism. Now why one of these pathways is dark versus one being light at this point I don't think that we can say.

HOPKINS: Okay the next question. These are sort of a bunch of questions together which have to do with the use of t-butoxide which is a stronger base than is necessary for the red chemiluminescence and which is known to produce artifacts in this high basicity with organic molecules of all kinds. The question isn't so much as why you use that. It has to do with the addition of water. With the addition of water there are all kinds of possibilities with t-butoxide. The one outstanding result that I seem to get from your last result was that the yield in fact decreases with the addition of water.

DeLUCA: Not all that much. Actually if I look at the total area under the entire curve until it comes back down to the baseline there is only maybe 10% difference in total light.

HOPKINS: Can you tell me the difference between those two curves in terms of the concentration of water? Was one exactly two molar versus the 55 M?

DeLUCA: Right, exactly. One was whatever was endogenous in DMSO which I am assuming from the literature value 10^{-2} M. When I added water it was to make a final concentration of approximately 2 M. The further addition of any more water did not make any difference.

HOPKINS: OK, the last thing has to do with the quantum yield of the chemiluminescence. One of the problems that I found when I was doing the chemiluminescence using t-butoxide was that the use of t-butoxide in DMSO is tricky because of the rate at which that solution picks up water. In other words I am worried about the amount of water in DMSO and it seems that somehow we ought to be able to do some kind of experiment to show whether the water is in fact necessary.

DeLUCA: I agree. I think perhaps the best experiment to try to do is to get a solvent which is really anhydrous say down to 10^{-8} or 10^{-9} molar. It turns out that that's not all that experimentally easy to achieve. I think ideally that would be a nice experiment.

McCAPRA: The mechanism of addition of water to the dioxetane ring is in fact interesting. We also thought that was reasonable before we did the work. Actually this was one of the alternatives and could account for some of these reactions especially the rate of decomposition of whatever it is, dioxetane or peroxyacid. It is very much faster when you have an unstable carbonyl conjugated with other groups and rings. Another point is are you really saying that the addition of water to the reaction whose rate determining step is known to be the removal of the proton, says that much about the subsequent hydrolysis reaction, when you are two reactions down the line? That you simply have a medium effect is a much simpler explanation.

DeLUCA: I am not sure under these conditions that proton removal is the rate determining step. We had about 5,000 fold excess t-butoxide versus products. The addition of more t-butoxide did not increase my rate of reaction. We certainly need to do a lot more to clarify this water effect.

WHITE: Let me play the devil's advocate with respect to our dioxetanone mechanism for firefly bioluminescence; I don't feel that the evidence presented is convincing enough to discard it yet. Can you really rule out a reaction similar to the one that Frank McCapra mentioned, that the prime reaction is hydrolysis of your active ester followed, at least in the case of chemiluminescence, by decarboxylation in a dark reaction?

DeLUCA: Could I rule that out entirely?

WHITE: Yes, and that's connected with the fact that do you really know what the products of this reaction are? You say the quantum yield is 0.2 for the chemiluminescence?

DeLUCA: I think Dr. Seliger and Dr. McElroy mentioned that in their discussions earlier. All I am saying is that again the oxygen incorporation is stoichiometric. It is one coming from water and zero coming from oxygen. Now even if 80% of the CO_2 arises by a dark pathway at least 20% of it we would still have detected if for example we were incorporating only 0.8 oxygen rather than one. It is certainly within the range of what we should see if the CO_2 produced by the light reaction is not going by the same organic path.

WHITE: Yes that's provided the quantum yield is 20% under your reaction conditions. Then the majority of your CO_2 (80%) could come from dark reactions.

DeLUCA: Yes.

WHITE: But I find from experience with repeating literature procedures that that's a dangerous gamble. It seems to me you really have to get the quantum yield from the same experiment from which you measured the CO_2.

DeLUCA: This certainly would be a good thing, I agree. However, the enzyme quantum yield is 0.9.

WHITE: You know that its 0.9 under optimum conditions; you know that some people have gotten 0.9. But my point is that if I went to the lab I bet that it would take me a long time to really get conditions right so that I would get 0.9.

DeLUCA: We used exactly the same conditions that Dr. Plant used where he had C^{14} luciferin and he measured production of $C^{14}O_2$ and Dr. Seliger measured the quantum yield. That's why we chose these conditions.

WHITE: Dr. Plant didn't measure the quantum yield either.

DeLUCA: Yes he did at the same time he did the isotope experiments.

WHITE: It wasn't in the paper that he published.

DeLUCA: Oh, it may not have been in the paper, but it was done.

WHITE: The last question is one of technique. You mentioned something which was ominous – that under your conditions bicarbonate ion didn't release any CO_2. If that's true then one wonders about a few things, because here you have roughly 0.8 of a microliter of CO_2 being generated in 3.9 ml of water and pH 7.8. Now at that pH the ratio of carbonate to CO_2 is about 40 and the rate constant is sufficiently high that the half life for hydration of CO_2 is seconds. I can't really believe that you could get 0.8 microliters our of 3.9 ml of water in sufficient time to compete with the carbonate formation.

DeLUCA: The half life for isotopic equilibration of CO_2 under these conditions as reported by Dr. Urey is 16 minutes. You've got to remember that every time you hydrate a molecule of CO_2 and then reverse it you've got one out of three chances there of losing an O^{18} label since the isotopic equilibration half life is 16 minutes. In answer to your other question on low amounts of CO_2 that we are collecting we have now done the experiments over approximately an 8 fold range of concentration. The 0.8 microliter was the very lowest one and the first experiment we did. We have since gond up to 8 times that concentration and are generating 8 times as much CO_2 and again we find no difference in the isotope incorporation.

WHITE: Have you measured the absolute amount of CO_2 formed?

DeLUCA: No we don't have an absolute measure but we do have a relative measure. We get this from the manometer on the high vacuum line and we also get a number back from the mass spectral analysis. I think these numbers could be off by 10% but probably no more than that. So it looks as though we are getting a quantitative recovery of the CO_2.

STRUCTURE AND SYNTHESIS OF A LUCIFERIN ACTIVE IN THE BIOLUMINES-
CENT SYSTEMS OF THE SEA PANSY (*RENILLA*) AND CERTAIN OTHER
BIOLUMINESCENT COELENTERATES

Kazuo Hori and Milton J. Cormier

Department of Biochemistry, University of Georgia

Athens, Georgia 30601

Due to modern technology the elucidation of the chemical structure of luciferins is not a difficult problem *per se*. However the problem is made difficult by the tremendous effort required to obtain small amounts of purified material. For example from about one ton of sea pansies (*Renilla reniformis*), obtained by dredging the ocean bottom at depths of 10-20 meters, one can obtain about one mg of pure *Renilla* luciferin.

From about a half a mg of pure luciferin we obtained chemical and physical data which led us to propose a recent tentative structure for *Renilla* luciferin (I in Fig. 1). Because of the sample size and the fact that the compound is easily autooxidizable it

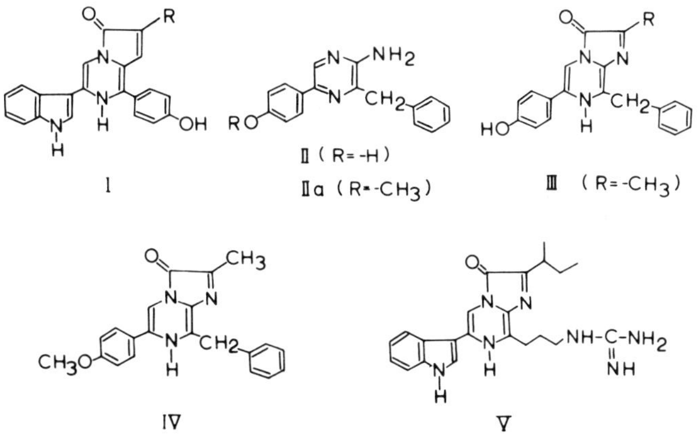

FIGURE 1

was impossible to obtain good NMR data. Although this fact made it difficult to decide between alternative structures a number of observations have been made recently by us and Drs. Shimomura and Johnson (Princeton University) which led us to predict the correct structure of a biologically active form of *Renilla* luciferin. The key observations which led to the proper choice of structure are outlined below.

When the luminous tissues of bioluminescent jellyfish such as *Aequorea* are extracted with EDTA-containing buffers a protein can be isolated which exhibits a bluish luminescence upon the addition of calcium ions (1,2,3). Such proteins were termed photoproteins by Drs. Shimomura and Johnson. Protein preparations which produce light upon the addition of calcium ions have also been demonstrated recently in extracts of a wide variety of coelenterates such as *Obelia, Aequorea, Pelagia, Renilla, Mnemiopsis, Campanularia, Clytia, Phialidium, Lavenella, Ptilosarcus*, and *Diphyes* (4). All of these calcium-triggered light emissions have been attributed to the existence of "photoproteins" in these extracts by Hastings and co-workers (4). These and other observations led us to look for possible biochemical similarities in the bioluminescence of coelenterates. Such similarities were indeed found. For example components required for luminescence in *Renilla* were also found in a number of bioluminescent coelenterates examined such as *Aequorea, Obelia, Cavernularia, Ptilosarcus, Stylatula, Acanthoptilum, Parazoanthus* and *Mnemiopsis* (5). Depending on the organism these included one or more of the following: luciferyl sulfate, luciferase, "photoprotein" and luciferin sulfokinase. These findings suggested that the luciferins and luciferases from these coelenterates were either identical or very similar.

When *Aequorea* photoprotein is treated with urea and mercaptoethanol a compound is released which is designated AF-350 (6). When native *Renilla* luciferin is allowed to autooxidize at room temperature for 24 hours at pH 10 a characteristic product is formed that can be isolated on LH-20 columns as previously described (10). We have noted that the absorption characteristics of AF-350 were identical to that observed for *Renilla* autooxidized luciferin (Fig. 2). The fluorescence properties of *Renilla* autooxidized luciferin (λ_F = 426 nm) was also found to be identical with that of AF-350. The reported molecular weight of AF-350 is 277 which is also the strongest peak observed when *Renilla* luciferin is analyzed in the mass spectrometer. Further, the mass spectroscopy patterns of AF-350 and *Renilla* luciferin are essentially the same up to a mass of 277.

The structure of AF-350 (II in Fig. 1) was reported recently (7) and this structure has been confirmed by synthesis (8).

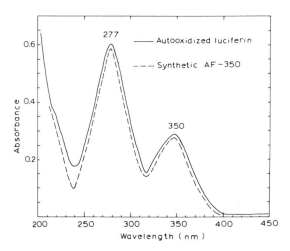

FIGURE 2. Comparison of absorption spectra of AF-350 and autooxidized *Renilla* luciferin in methanol.

We noted that with minor modifications of our tentative structure of *Renilla* luciferin (I in Fig. 1), AF-350 becomes an integral part of the structure of *Renilla* luciferin. For example a C_8H_7N fragment found in the mass spectrum could be explained either by an indole nucleus as shown in structure I or by a benzyl moiety as shown in structure III. Further, there are restrictions placed on structures involved in luminescent reactions. These restrictions are based in part on considerations of energy requirements for creation of an electronically excited state. For these reasons and others outlined above we predicted that a likely ring structure for *Renilla* luciferin would involve fusing an imidazole ring to structure II thus converting it to structure III. On the assumption that the R group of III would not significantly affect its biological activity we synthesized this compound where R is methyl. Structure III was found to be biologically active. This paper describes the synthesis and some of the biological properties of this newly synthesized luciferin.

RESULTS AND DISCUSSION

Synthesis of Luciferin

The methyl ether of AF-350 (IIa in Fig. 1) was synthesized as recently described (8). Luciferin (III in Fig. 1) was synthesized from IIa by modification of the final steps involved in the synthesis of *Cypridina* luciferin (9). The method used here was as follows: IIa was reacted with methyl glyoxal in ethanol

-12NHCL solution (1:1) under an argon atmosphere in a sealed tube at 110°C for 1.5 hours. The hydrochloride of IV was obtained in good yield (85%). It was converted to III by treatment with pyridine-HCl at 210°C for 2 hours under an argon atmosphere in a sealed tube. The hydrochloride of III was isolated from the reaction mixture in good yield (90%) by chromatography of the reaction mixture through LH-20 as previously described (10). The elution solvent was 80% methanol made 0.1 N with HCl. The hydrochloride of III was converted to free luciferin (III) by neutralizing with phosphate buffer. Yellow crystals of III were obtained by heating an aqueous suspension of III at pH 2 under a hydrogen atmosphere and allowing the solution to slowly cool. It should be noted that by using derivatives of glyoxal other than methyl any desired substitution can be made at the 2-position of luciferin (III).

We have referred to our synthetic compound III as luciferin. It should be emphasized that the R group in synthetic luciferin (III) is methyl whereas in native luciferin it is not. The molecular weight of synthetic luciferin (III) is 331 while that of the native compound is 513. As outlined below the difference between the two compounds lies in the R group. In native luciferin it is obviously bulky and at the present time unknown. However as indicated below the synthetic compound (III) is biologically active.

Properties of Native and Synthetic Luciferin

The absorption properties of native *Renilla* luciferin at different pH values are illustrated in Fig. 3. At neutral pH in

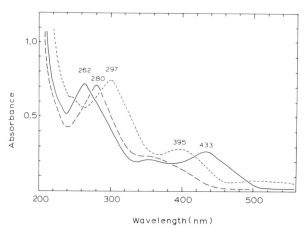

FIGURE 3. Absorption spectrum of methanol solutions of *Renilla* luciferin as a function of pH. pH 7.0 (—); pH 1.0 (--). pH 11.0 (---).

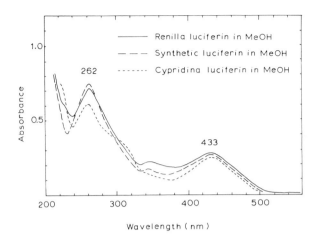

FIGURE 4. Comparison of absorption spectra of native *Renilla* luciferin, synthetic luciferin and *Cypridina* luciferin.

methanol, the millimolar extinction coefficients at 262 and 433 nm were found to be 22.8 and 9.0 respectively. As shown in Fig. 4 the absorption properties of synthetic luciferin (III) are similar to those of the native compound. Identical pH shifts as noted in Fig. 3 are also observed with synthetic luciferin. In addition the two compounds have similar millimolar extinction coefficients for the two transitions listed above. Furthermore both compounds exhibit identical yellow-green fluorescence emissions (λ_F = 538 nm).

Both synthetic luciferin (III) and *Cypridina* luciferin (V in Fig. 1) contain a pyrazine nucleus with a fused imidazole ring. The side chains, however, are considerably different. Side chains derived from tryptophan and arginine in *Cypridina* luciferin (9,11, 12) are replaced by those derived from tyrosine and phenylalanine in III. Since the heterocyclic ring structure of both compounds are the same, it is not surprising that we noted striking similarities in their absorption properties (Fig. 4). The fluorescence characteristics of both compounds, as well as that of native luciferin, are also similar.

The mass spectral pattern of III was found to be identical to that of native luciferin up to a mass of 331 which represents the molecular ion of III. Although the R group in native luciferin is bulky (197 mass units) it does not influence the spectral characteristics of the compound as judged by the absorption, fluorescence and bioluminescence emission of native luciferin when compared with the synthetic compound.

Biological Activity of Synthetic Luciferin

Synthetic luciferin (III) was found to be biologically active while II was totally inactive. When III was added to *Renilla* luciferase light production occurred. When equal amounts of both native and synthetic luciferins were used the kinetics of the corresponding bioluminescent reactions were similar. Furthermore the color of bioluminescence was identical in both cases as shown in Fig. 5.

By using the flash peak as a measure of initial intensity III was found to be 10% as active as native luciferin in producing light with *Renilla* luciferase. This difference in activities must

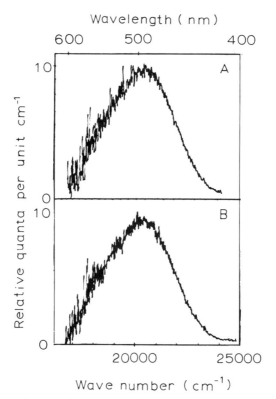

FIGURE 5. Comparison of the color of bioluminescence upon initiation of the reaction with synthetic luciferin (A) and native *Renilla* luciferin (B).

reside in the R group of native luciferin. It was interesting to note that the methylated form of III (IV in Fig. 1) was totally inactive in producing light with luciferase. Upon demethylation biological activity was restored.

Cormier, et al. (5) recently demonstrated that *Renilla*-like luciferase could be extracted from the tissues of eight different species of bioluminescent coelenterates. These included three species of *Renilla*, *Cavernularia*, *Ptilosarcus*, *Stylatula*, *Acanthoptilum* & *Parazoanthus*. Thus identical light emissions occurred when native *Renilla* luciferin was added to each of these luciferases. Synthetic luciferin (III) will replace native luciferin in these reactions as well. Whereas luciferase activity could not be demonstrated in the coelenterates *Aequorea* and *Obelia* we did extract *Renilla*-like luciferyl sulfates from these bioluminescent organisms. These experiments suggested that the luciferases and luciferins of the various bioluminescent coelenterates were similar or identical and that the chemistry leading to light emission among the coelenterates was also similar. Our finding that AF-350 (II), which was isolated from *Aequorea* photoprotein (6), is an integral part of the structure of *Renilla* luciferin reinforces this concept. These findings further suggest that the native substrate in the *Aequorea* photoprotein consist of AF-350 (II) with a fused imidazole ring like that found in *Renilla* luciferin (III). The energy requirements alone would argue for such a structure. Whether the R group at the 2-position is the same or different will have to await further studies.

Another logical and interesting conclusion one can draw here is that a *Renilla*-like luciferin somehow becomes an integral part of the photoprotein complex in certain bioluminescent coelenterates.

Mechanism of the Reaction

Based on structure III and on the mechanism of the luminescent reaction in *Renilla* employing ^{18}O (13), a mechanism consistent with the results and one which predicts the structure of the emitter (oxyluciferin) is shown in Fig. 6. Note that hydrolysis of the peptide bond in oxyluciferin would lead to the formation of AF-350 (II).

Preliminary results on the chemiluminescence of III in organic solvents show that oxyluciferin is indeed the product of the chemiluminescence reaction and that emission occurs from the monoanion of oxyluciferin as shown in Fig. 6 (14).

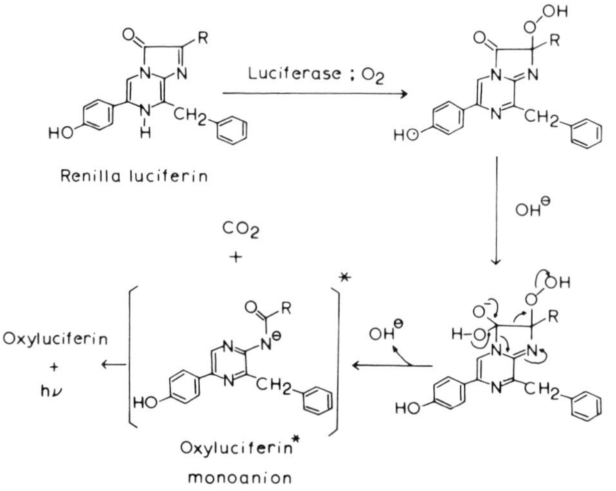

FIGURE 6

Mechanism of the *Renilla* bioluminescent reaction.

ACKNOWLEDGEMENTS

This work was supported in part by the National Science Foundation and the U.S. Atomic Energy Commission.

REFERENCES
1. Shimomura, O., Johnson, F. H. and Saiga, Y. (1962) *J. Cell. and Comp. Physiol. 59*, 223-240.
2. Shimomura, O., Johnson, F. H. and Saiga, Y. (1963) *J. Cell. and Comp. Physiol. 62*, 1-8.
3. Shimomura, O., Johnson, F. H. and Saiga, Y. (1963) *J. Cell. and Comp. Physiol. 62*, 9-16.
4. Morin, J. G. and Hastings, J. W. (1971) *J. Cell. Physiol. 77*, 305-311.
5. Cormier, M. J., Hori, K., Karkhanis, Y. D., Anderson, J. M., Wampler, J. E., Morin, J. G. and Hastings, J. W. (1972) *J. Cell. Physiol.*, in press.
6. Shimomura, O. and Johnson, F. H. (1969) *Biochemistry 8*, 3991-3997.
7. Shimomura, O. and Johnson, F. H. (1972) *Biochemistry 11*, 1602-1608.
8. Kishi, Y., Tanino, H. and Goto, T. (1972) *Tetrahedron Letters 27*, 2747-2749.
9. Inoue, S., Sugiura, S., Kakoi, H. and Goto, T. (1969) *Tetrahedron Letters 20*, 1609-1610.
10. Hori, K., Nakano, Y. and Cormier, M. J. (1972) *Biochim. Biophys. Acta 256*, 638-644.
11. Kishi, Y., Goto, T., Hirata, Y., Shimomura, O. and Johnson, F. H. (1966) *Tetrahedron Letters 29*, 3427-3436.
12. Kishi, Y., Goto, T., Inoue, S., Sugiura, S. and Kishimoto, H. (1966) *Tetrahedron Letters 29*, 3445-3450.
13. DeLuca, M., Dempsey, M. E., Hori, K., Wampler, J. E. and Cormier, M. J. (1971) *Proc. Nat'l. Acad. Sci. USA 68*, 1658-1660.
14. Hori, K., Wampler, J. E. and Cormier, M. J. Unpublished results.

BACTERIAL BIOLUMINESCENCE. MECHANISTIC IMPLICATIONS OF ACTIVE

CENTER CHEMISTRY OF LUCIFERASE

J. W. Hastings, A. Eberhard, T. O. Baldwin, M. Z. Nicoli,
T. W. Cline, and K. H. Nealson

Harvard University, Ithaca College, and the University
of Massachusetts

Viewed as a chemiluminescent reaction, the enzyme luciferase plays a truly important role with regard to the reaction pathway. The non-enzymatic oxidation of reduced flavin mononucleotide ($FMNH_2$) has a very low quantum yield: less than 10^{-10} of the photons which can be evoked enzymatically are observed in the non-enzymatic oxidation of $FMNH_2$ in the presence of a long chain aldehyde.

The non-enzymatic oxidation of $FMNH_2$ has been shown to be autocatalytic (1); the products of the reaction are FMN and H_2O_2. A complex between oxygen and reduced flavin occurs as an intermediate in this reaction, which is essentially complete within 300 msec (1,2).

The enzymatic light emitting reaction, as studied in vitro, is known to involve the oxidation of $FMNH_2$ by molecular oxygen; a long chain saturated aldehyde is required for a high quantum yield. The standard assay procedure involves the rapid mixing of a solution containing luciferase, aldehyde, and oxygen with reduced flavin. Molecular oxygen and reduced flavin are required substrates for the reaction. Aldehyde has been postulated to be oxidized to the corresponding acid in a mixed-function oxidase type reaction (3,4,5).

$$FMNH_2 + O_2 + RCHO \xrightarrow{\text{luciferase}} FMN^* + H_2O + RCOOH$$

There are two peculiar features of the reaction which should be underlined (6). First, in the in vitro system the turnover time for a luciferase molecule is long, several seconds or more, while the lifetime of the free substrate, $FMNH_2$, is short — less than one second. The reaction may thus be viewed as a situation in which there is a "pulse" of substrate, which gives rise to enzymatic intermediate(s) with relatively long lifetimes. Secondly, aldehyde is <u>not</u> required for the enzymatic oxidation of the $FMNH_2$, but only for a high photon yield. Moreover, the lifetime of the enzymatic intermediate may be different with aldehydes of different chain lengths. A hypothetical scheme which accomodates these and other observations is shown in Figure 1, where the long lived intermediate is II, either with or without aldehyde.

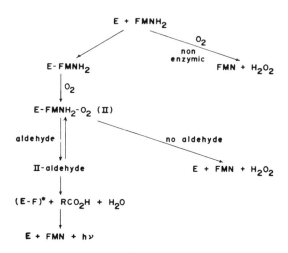

Figure 1

Bacterial luciferase is composed of two non-identical subunits (7,8) which may be readily and quantitatively separated by DEAE-Sephadex column chromatography in 5 M urea (Figure 2). Dimeric structure ($\alpha\beta$) and full activity is recovered upon dilution of the two subunits into buffer without urea. A single flavin binding site per dimer has been demonstrated by kinetic techniques (9). By both chemical modification studies (10) and by luciferase mutant analysis (11) the flavin binding site has been shown to be on the α subunit. A single flavin binding site per luciferase molecule strongly supports the mixed function oxidase type mechanism suggested above.

Dim mutants in which the luciferase itself has suffered a lesion affecting the kinetics of the reaction were selected. Upon analysis, all of these (40 to date, 13 described here) were shown to possess lesions in the α subunit. These lesions resulted in an alteration in the observed decay rate of the reaction (Table 1, column 1) which was interpreted as an effect upon the lifetime of the long lived intermediate (II). Furthermore, at least one of

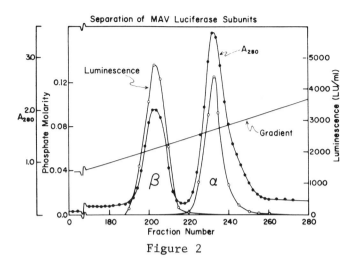

Figure 2

these mutants exhibits a substantial red shift in the in vivo
bioluminescence emission spectrum (12). Complementation of mutant
enzymes with wild type α subunit resulted in the recovery of wild
type activity, kinetics and other properties (columns 2 and 3).
The lesions which result in altered kinetics may cause, at the same
time, very large differences in the enzymes' binding properties for
one or both of the substrates (columns 4 and 5 and Figure 3).
Lesions in the luciferase not selected on the basis of kinetic
alterations, possessing instead lesions causing a loss in thermal
stability, occur with about equal frequency in both subunits.

Table 1

LUCIFERASE	$k_1(sec^{-1})$ CONTROL	$k_1(sec^{-1})$ RENATURED WITH β	$k_1(sec^{-1})$ RENATURED WITH α	$K_D(Mx10^4)$ $FMNH_2$	$K_D(Mx10^4)$ DECANAL
MAV (WT)	.21	.20	.20	0.0069	0.068
AK-6	?	?	.20	3.2	0.069
AK-20	.036	.043	.20	0.0013	0.51
AK-16	.14	.15	.20	0.0086	0.066
AK-15	.87	.85	.19	0.46	0.040
AK-7	.090	.095	.20	0.083	2.2
AK-24A	.30	.29	.20	0.12	0.086
AK-9	.11	.11	.21	0.0063	0.65
AK-17	.16	?	.21	0.0072	0.16
AK-11	.35	?	.20	0.046	0.23
AK-3	.28	.27	.21	0.057	0.41
AK-18	.16	.17	.21	0.080	0.10
AK-2H	.55	.55	.20	0.025	3.8
AK-1H	.71	.69	.21	0.36	0.57

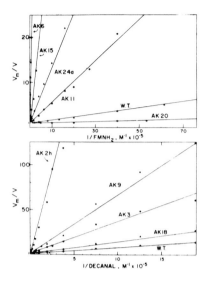

Figure 3

Similar conclusions were obtained from studies involving chemical modification of the luciferase (10). Succinylation of the α subunit, while not disrupting the three-dimensional integrity of the protein, results in an enzyme with very low activity, affecting both the stability of the intermediates (i.e., the apparent first-order rate of decay of light emission) and the binding of $FMNH_2$. Hybrid luciferase with the β subunit succinylated retains over 50% activity, and is like native enzyme in its catalytic parameters.

Several lines of evidence suggest the presence of an essential cysteinyl residue in the luciferase active center (13). Modification of a single thiol group with either N-ethylmaleimide (NEM) or iodoacetamide inactivates the enzyme; no $FMNH_2$ binding site can be measured for those enzyme molecules which have been modified. The rate of inactivation of luciferase by N-alkyl maleimides shows a strong dependence on the chainlength of the alkyl group, longer chainlengths reacting much more rapidly than shorter chainlengths. This chainlength effect is indicative of a highly hydrophobic active center. Additional support for this suggestion comes from the observation that longer chain aldehydes offer better protection from NEM inactivation. N-(4-dimethylamino-3,5-dinitrophenyl)-maleimide (DDPM), a bright yellow alkylating reagent, inactivates luciferase extremely rapidly; an enzyme preparation reacted with one molar equivalent of DDPM is less than 10% active, and more than 90% of the yellow color is associated with the α subunit, as shown in Figure 4. Studies with ^{14}C-N-ethylmaleimide have confirmed that the reactive sulfhydryl is on the α subunit; the tryptic peptide containing this cysteinyl residue has the amino acid sequence Phe-Gly-Ilu-Cys-Arg (14).

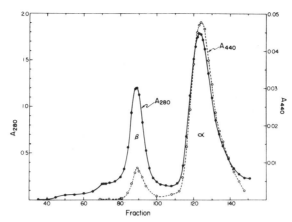

Figure 4: DEAE-Sephadex chromatography of DDPM-inactivated luciferase subunits in 5 M urea (see Figure 2). A_{280}, ●——●; A_{440}, ○----○.

The emission spectrum of the in vivo bioluminescence is a broad band peaking at about 490 nm, varying somewhat, depending upon species. Although the most obvious candidate for the emitter is oxidized flavin in an electronically excited state, there is a large discrepancy between the fluorescence emission spectrum of FMN (λ_{max}, 530 nm) and the emission spectrum of bioluminescence in vitro (λ_{max}, 490 nm). This discrepancy does not rule out FMN as the emitter, as the active center of luciferase has been shown to be hydrophobic (13), and flavin in nonpolar solvents exhibits a substantial blue shift relative to the fluorescence emission in water (15).

A number of different reduced flavins, and isomers and analogues thereof, were found to exhibit weak but authentic bioluminescence activity with luciferase (15). Significant differences were observed between the bioluminescence emission spectra with these flavins which could not be correlated with differences in their fluorescence emission spectra (Table 2). For example, the bioluminescence emission spectrum with iso-FMN is blue shifted relative to the bioluminescence of FMN, while the fluorescence emission of iso-FMN is red shifted relative to FMN. Bioluminescence using 2-thio-FMN is centered at 534 nm, although the compound itself is non-fluorescent in aqueous solution.

Recent studies of fluorescence emission of flavin protonated at the N-1 position demonstrated an emission spectrum similar to bacterial bioluminescence, suggesting $FMNH^+$ as the emitter (16). The best matching of spectra occurred for FMN in 18 N H_2SO_4 at 77°K (λ_{max}, 493 nm) while the emission in a less polar medium,

Table 2

	λ max, nm	
	Bioluminescence	Fluorescence
FMN	492	524
FAD	484	524
Riboflavin	492	524
iso-FMN	472	543
2-thio-FMN	534	non fluorescent
2 morpholino, 2 desoxy FMN	484	536
3 acetyl FMN	488	524

ethanol-HCl (1:1) showed a substantial blue shift (λ_{max}, 480 nm). The correspondence between flavin cation fluorescence and *in vitro* bioluminescence was less convincing for the other flavin analogues studied (FAD, iso-FMN, and 2-thio-FMN) than for FMN, the discrepancy being nearly 60 nm for iso-FMN. However, the luciferase-bound N-1 protonated flavin cation remains an attractive candidate for the emitter in bacterial bioluminescence.

A chemical mechanism accomodating the flavin cation hypothesis, which also accounts for several known features of the bacterial bioluminescence reaction was recently proposed (5). It should be noted that an essentially identical mechanism can be written with the loss of the N-1 proton, which would result in FMN itself as the emitter. The first step of the postulated mechanism (Figure 5) is the binding of $FMNH_2$ to the enzyme to give Intermediate I. The second step involves the addition of oxygen at carbon atom 4a of the isoalloxazine ring to give the half-oxidized flavin peroxy anion, Intermediate II. Intermediate II is common to both the enzymatic pathways (aldehyde and non-aldehyde) and to the non-enzymatic pathway. Luciferase binds fully reduced flavin; half-oxidation must occur with the flavin enzyme-bound. This suggestion is supported by the observation that the free substrate ($FMNH_2$) is lost by autoxidation via radical intermediates, the overall process having a half-time of about 100 msec (1,2) and also by the observation that scavengers of free radicals do not have an effect on the bioluminescence reaction. Therefore, the enzyme binds $FMNH_2$ before the substrate interacts with oxygen.

Intermediate II appears to be able to react via two alternative pathways, the "aldehyde" pathway leading to a high quantum yield and the "non-aldehyde" dark pathway. Intermediate II of luciferase prepared from the MAV strain (8) decays with a half-time of about 14 sec at 22°C. Intermediate IIA, the Intermediate II-aldehyde complex, has a life-time dependent on the aldehyde chainlength; the half-times at 22°C with octanal, decanal, and dodecanal are about 28 sec, 3.5 sec, and 26 sec, respectively (11,17).

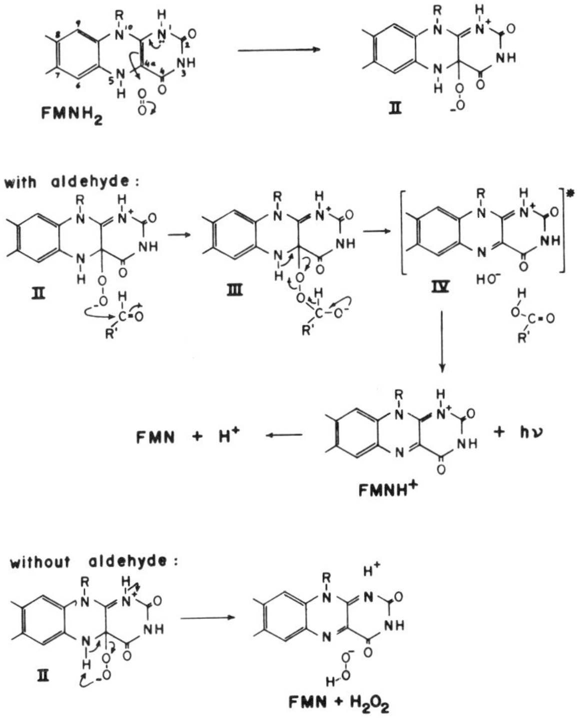

Figure 5

The binding of aldehyde to Intermediate II is reversible. If an assay is initiated in the presence of octanal and during the exponential decay of Intermediate IIA a secondary injection of decanal is made, the decay rate will shift to a value intermediate between the decay rates of the individual aldehyde complexes (18).

The reversible binding of aldehyde to Intermediate II plays an important role in the stability of Intermediate II. This must be true because the decay of Intermediate II of the MAV strain is faster than the decay in the presence of either octanal or dodecanal; the decay of IIA with decanal is much faster, but photons are still produced in high yield.

Intermediate IIA is then assumed to form a covalent complex, Intermediate III, by the attack of the peroxy anion (II) on the aldehyde, giving the anion of a peroxy hemi-acetal. The aldehyde is then oxidized to acid in a step similar to a Baeyer-Villiger shift, hydroxide is lost, and an excited state of protonated flavin, IV, remains. Intermediate IV can then return to ground state via the emission of a photon. Since the emission at 490 nm corresponds to about 60 kcal per einstein, the excitation energy of 95 kcal imparted to IV in the transition from III is sufficient to account for the bioluminescence. In the absence of aldehyde a total of only 27 kcal is available, not enough to result in an excited state of flavin.

The proposed mechanism accounts for several known characteristics of the reaction. However, experimental evidence for the details of the proposed mechanism is meager. The stoichiometry of the reaction remains uncertain; the product from the aldehyde has not been rigorously demonstrated; and an unambiguous identification of flavin cation as the emitter has not been achieved. However, recent studies by Shimomura et al. (19) indicate that the aldehyde is indeed oxidized to the corresponding acid. Studies by Nakamura and Matsuda (20) are also in agreement with the mixed function oxidase type reaction, but those of Lee (21) which indicate a two flavin stoichiometry, are not.

REFERENCES

1. Gibson, Q. H., and Hastings, J. W., Biochem. J., 83, 368 (1962).
2. Massey, V., Palmer, G., and Ballou, D., in "Flavins and Flavoproteins," (H. Kamin, ed.), p. 349, University Park Press, Baltimore, Md. (1971).
3. Mc Elroy, W. D., et al., Arch. Biochem. Biophys. 56, 240 (1955).
4. Nealson, K. H., et al., J. Biol. Chem., 247, 888 (1972).
5. Eberhard, A., and Hastings, J. W., Biochem. Biophys. Res. Comm., 47, 348 (1972).
6. Hastings, J. W., et al., J. Biol. Chem., 238, 2537 (1963).
7. Friedland, J. M., and Hastings, J. W., PNAS, 58, 2336 (1967).
8. Gunsalus-Miguel, A., et al., J. Biol. Chem., 247, 398 (1972).
9. Meighen, E. A., et. al., J. Biol. Chem., 246, 7666 (1971).
10. Meighen, E. A., Nicoli, M. Z., and Hastings, J. W., Biochemistry, 10, 4062, 4069 (1971).
11. Cline, T. W., and Hastings, J. W., Biochemistry 11, 3359 (1972).
12. Cline, T. W., PhD thesis, Harvard University, 1973.
13. Nicoli, M. Z., PhD thesis, Harvard University, 1973.
14. Baldwin, T. O., and Nicoli, M. Z., manuscript in preparation.
15. Mitchell, G. W., and Hastings, J. W., J. Biol. Chem., 244, 2572 (1969).
16. Eley, M., Lee, J., Lhoste. J.-M., Cormier, M. J., and Hemmerich, P., Biochemistry, 9, 2902 (1970).
17. Hastings, J. W., et al., Biochemistry, 8, 4681 (1969).
18. Hastings, J. W., Gibson, Q. H., Friedland, J., and Spudich, J., in "Bioluminescence in Progress," (F. H. Johnson and Y. Haneda, eds.), p. 151, Princeton University Press (1966).
19. Shimomura, O., Johnson, F. H., and Kohama, Y., PNAS, 69, 2086 (1972).
20. Nakamura, T. and Matsuda, K., J. Biochem., 70, 35 (1971).
21. Lee, J., Biochemistry, 11, 3350 (1972).

SELIGER: I think your studies on the association and dissocition of subunits are really beautiful in terms of these prolific luciferases. This seems to be the only system where subunits are available that are able to be reassociated. Now I'd like to ask one specific question related to mechanism in terms of energetics. What concrete evidence do you have that you can achieve light emission in the complete absence of aldehyde?

HASTINGS: This question relates very specifically to the detailed mechanism in the pathway where no aldehyde is added. Although we indicated this as a dark pathway, we have and still do think of it as a low quantum yield pathway. The gain of the photomultiplier can always be turned up, and experimentally one always gets light in the absence of aldehyde. Several people have tried to purify this system by various techniques to remove aldehyde to demonstrate that the system doesn't have the capacity to emit light in the absence of aldehyde and to my knowledge, all these experiments indicate that we can get some light. Our values, I think, go down to a value of about 10^{-3} of that with aldehyde. Dialysis of luciferase for a week against repeated changes of an aldehyde trapping agent fails to reduce it more. If one varies the concentration of a trapping reagent and measures the light intensity this "endogenous" value exhibits an apparent limiting value as a function of concentration. This result is concrete but not definitive evidence that the "dark" pathway actually has a very low quantum yield emission. That's all I can say unless you have some other new points that I haven't heard about.

SELIGER: No, I was just trying now to use some type of argument about energetics in terms that this might be unreasonable. You mentioned the only available energy for the exergonic reaction would be the formation of hydrogen peroxide.

HASTINGS: And that's only 27 Kcal which is pretty far down the street.

SELIGER: I think we can get to this conclusion too equally well by Boltzman distribution probabilities.

ELEY: I would like to comment on the flavin cation as the bacterial emitter. Dr. Hastings and associates have in recent papers and in this lecture stated that the best match for bacterial bioluminescence was the FMN cation in 18 N H_2SO_4 and that the data presented in the original paper on the flavin cation (Eley, et al., 1970) were less convincing for the cation fluorescence of other flavin analogs relative to the wavelengths of maximum bioluminescence intensity (λ_B) reported by Mitchell and Hastings (1969). Particular emphasis was placed on a 60 nm difference in the wavelength of maximum fluorescence intensity (λ_F) of the iso-flavin cation as compared to the λ_B elicited by reduced iso-FMN. I would like to show that no such discrepancy in the λ_F and λ_B for iso-flavin or for any of the other flavin analogs by using two slides which were made from the data presented in the original paper. In the table with the exception of the riboflavin data, is a condensed form of the data presented in 1970. As Dr. Hastings has already pointed out, the FMN cation λ_F in 18N H_2SO_4 is indeed a good match for the λ_B of <u>Photobacterium fischeri</u>, but it should be pointed out also that by simply changing the solvent polarity, the λ_F of the FMN cation can be shifted over the complete range of λ_B for all species of luminous bacteria (478-505nm). It should be obvious that similar results can be obtained with the FAD cation by using a solvent with polarity between the two given in the Table that is, the λ_F of the FAD cation would match the λ_B in a solvent of proper polarity. The data for the riboflavin cation is very similar, if not identical, to the data for the FMN cation. Although lumiflavin is probably not active in bacterial bioluminescence, it does have absorption and fluorescence properties very similar to FMN, and since FMN is not soluble in non-polar solvents, lumiflavin should provide data similar to that expected for FMN in non-polar solvents, possibly similar to the polarity of active center of many proteins including bacterial luciferase. The same argument can be made for using iso-lumiflavin to represent iso-FMN in non-polar solvents. In the non-polar solvents, the λ_F of the iso-flavin cation in EPA-HCl is 490nm and in ethyl ether-HCl is 450nm. From this data it should be obvious that a solvent polarity between those in the Table would provide the proper environment for the λ_F of the iso-flavin cation to be 472nm. Although the data are not as impressive for the 2-thio-FMN cation due to its very weak fluorescence, it is reasonable to expect that in a solvent of the proper polarity the λ_F of the cation would match the λ_B as the data in the Table would indicate. One other piece of data from the 1970 paper is shown in the Figure. These two spectra are the fluorescence emission spectra of the oxidized neutral FMN in propylene glycol: water (1:1) in liquid and glassy solutions. In this case there is indeed a blue shift in the λ_F of FMN in the rigid glass, but there is a definite structural feature in the emission spectra in both cases. This structured emission of FMN, along with other data, should eliminate a perturbed FMN as the bacterial emitter. On the other hand, bacterial bioluminescence emission spectra and the fluorescence emission spectra of flavin cations, particularly $FMNH^+$, show no structural features, and the λ_B of all species of luminous bacteria can be matched to the λ_F of the FMN cation by simply changing the polarity of the solvent.

HASTINGS: I'd like to ask a question to this interesting comment. We have observed structure from bacterial bioluminescence and I wondered if you had looked at this recently?

ELEY: No, not recently.

CORMIER: Which species of bacteria did you see structure in?

Comparison of Intensity Maxima of Flavin Cation Fluorescence (λ_F) and Bioluminescence Emission (λ_B).

Flavin	Flavin Cation Fluorescence, 77°K				Bioluminescence, 9°C*
	EtOH^a-HCl(1:1) λ_F (nm)	18 N H$_2$SO$_4$ λ_F (nm)	EPA^b-HCl λ_F (nm)	Ether-HCl λ_F (nm)	λ_B (nm)
FMN	480	493	–	–	492
FAD	480	493	–	–	484
Riboflavin	480	490	–	–	492
Lumiflavin	475	–	443	–	–
Iso-FMN	530	530	–	–	472
Iso-lumiflavin	530	–	490	450	–
2-Thio-FMN	510–530	510–530	–	–	534

* Bioluminescence data are uncorrected values (Mitchell and Hastings, 1969).
^{a}95% ethanol.
^{b}ethyl ether-isopentane-ethanol (5:5:2).

HASTINGS: MAV, PF, and an unidentified species.

CORMIER: That's a very interesting observation. We'll have to look at that on our machine.

HASTINGS: Yes I'm sure you will know it by tomorrow morning.

McELROY: A point of clarification on the riboflavin and other substrates. Several years ago when we cleaned up the enzyme and really got rid of all the FMN, reduced riboflavin, reduced FAD and a number of other compounds failed to work. Now I hear that riboflavin does work. How active is it and are you sure that all the FMN is removed?

HASTINGS: The species of bacterium from which the enzyme is purified makes a big difference. All luciferases appear to give a small but authentic response with various flavin analogues and isomers as well as with riboflavin and FAD. In addition Meighen and McCormick (Biochemistry, in press, 1973) have recently shown that the activity of reduced riboflavin may be greatly enhanced by added inorganic phosphate.

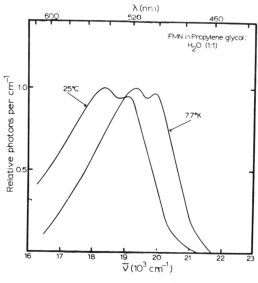

Figure (Eley)

Effects of Aldehyde Carbon Chain Length and Type of Luciferase on the Quantum Yields of Bacterial Bioluminescence

John Lee and Charles L. Murphy

Department of Biochemistry, University of Georgia

Athens, Georgia 30601

Bacterial bioluminescence results from the oxidation of reduced flavin mononucleotide (FMNH$_2$) and an aliphatic aldehyde (RCHO) by molecular oxygen in the presence of the enzyme bacterial luciferase. The final product of the reaction is FMN but as shown in Figure 1, perturbation of its fluorescence, for instance by an environment of low polarity, cannot blue shift its fluorescence to overlap the spectral emission distribution of the bioluminescence. In addition the low polarity enhances a vibrational structure in the FMN fluorescence whereas the properly corrected bioluminescence is unstructured. Lumiflavin is used here since FMN is insoluble in solvents of low polarity.

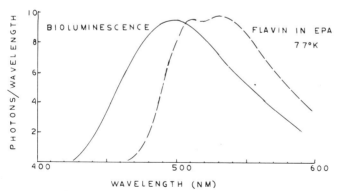

Figure 1. Bioluminescence emission spectrum from the reaction of FMNH$_2$, O$_2$ and dodecanal with luciferase from <u>Photobacterium fischeri</u> (23°C) and the fluorescence of lumiflavin in EPA glass (77°K), excited at 390 nm.

We have proposed[1] that the emitter is $FMNH^+$ within the active site of the enzyme. Under suitable perturbation conditions the fluorescence emission from $FMNH^+$ or its iso analog $i\text{-}FMNH^+$, precisely overlap the bioluminescence induced by using $FMNH_2$ or $i\text{-}FMNH_2$ in the reaction.

The stoichiometry of the luciferase reaction is[3]

$$2 FMNH_2 + 2 O_2 + RCHO \longrightarrow 2 FMN + H_2O + H_2O_2 + RCOOH$$

This was inferred from quantum yield (Q_B) measurements of each component except the $RCHOOH$ which has been shown by others to be formed as a major product[4]. From this stoichiometry it can be seen that $Q_B(RCHO) = 2 Q_B(FMNH_2) = 2 Q_B(O_2) = Q_B(H_2O_2)$ and these integral ratios are exactly what are observed.

The question arises as to why twice as much $FMNH_2$ is needed to give light as $RCHO$. The possibility to be examined here is that there may be a point in the reaction sequence where, independent of the light reaction, a dark path of $FMNH_2$ oxidation competes equally with the light reaction.

The reaction studied for the stoichiometry measurement used dodecanal and the luciferase from <u>Photobacterium fischeri</u> (PF) (23°C). This paper extends these observations to another luciferase ("MAV") and to other aldehydes at 10°C.

In spite of differences in Q_B's and large changes in overall reaction rates, the ratio $Q_B(RCHO)/Q_B(FMNH_2)$ remains at 2.0. This indicates that the above stoichiometry is also that of the light reaction and that any mechanism that is invoked must provide a two flavin reaction sequence.

Results

The first step in the sequence is the reaction of $FMNH_2$ with luciferase and O_2, which must be competitive with homogeneous oxidation of $FMNH_2$ by molecular O_2. This competition can be overcome by using a sufficient excess of luciferase and results in a saturation behavior of $Q_B(FMNH_2)$ with increasing concentrations of luciferase. In Figure 2 the curve for PF taken from previous work is compared with the results using MAV luciferase.

Two things are apparent from these results. Firstly, the saturation $Q_B(FMNH_2)$ with MAV luciferase is about 65% that with PF, yet there is an order of magnitude difference in their specific activities. Secondly, the points of half-saturation, where the rates of reaction through the luciferase and of homogeneous reaction with molecular oxygen are the same, occurs at exactly the same concentration for each luciferase. Therefore even though they differ so much in specific activity, the rate of reaction of $FMNH_2$ through the first steps in the sequence must be the same, and the overall reaction rate is controlled by a step far removed from the $FMNH_2$ interaction.

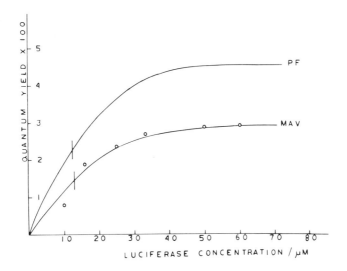

Figure 2. Bioluminescence quantum yield of $FMNH_2$.

The overall reaction rate is measured by the inital light intensity I_0 which, as has been noted before[5], is sensitive to the chain length of the aldehyde used. Figure 3 shows the marked effect of aldehyde carbon chain length (n) and I_0 for the MAV luciferase; the PF behaves similarly. All aldehydes were purified by vacuum fractionation and the concentration used was in an optimum range, since excess causes marked inhibition[3].

The results differ in some detail with those reported by others[5]. Longer aldehydes show an increase, not a decrease as previously reported[5]. However this kinetic data is difficult to interpret clearly. Figure 4 shows that, in spite of the marked variability in I_0, the Q_B ratio is invariant at 2.0 for n > 8.

The Q_B(RCHO) shown by the top line, was determined using the appropriate chain length aldehyde in limiting quantities and reacting to completion by adding $FMNH_2$ and luciferase in excess, with optimal concentration of luciferase[3]. For $Q_B(FMNH_2)$ the lower line, the $FMNH_2$ was added in limiting quantities and the other reagents in excess using the aldehyde of indicated chain length.

A slower reaction with aldehydes of shorter chain length probably favors a dark oxidation of aldehyde with hydrogen peroxide and oxygen and the Q_B(RCHO) is consequently lowered. This does not affect Q_B(FMNH$_2$) so much since the aldehyde is present in sufficient excess to overcome this problem. Eventually as n decrease the Q_B(FMNH$_2$) drops too.

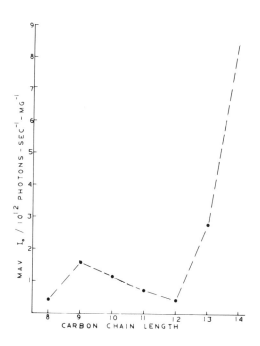

Figure 3. Effect of aldehyde carbon chain length on the maximum light intensity of the bioluminescence reaction between FMNH$_2$ and luciferase (MAV, 10°C).

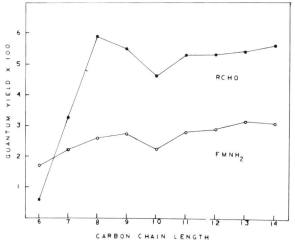

Figure 4. Bioluminescence quantum yield of aldehyde (•) and FMNH$_2$(-o-o-) as the chain length of aldehyde is changed from hexanal (6) to tetradecanal (14) (MAV luciferase, 10°C).

Discussion

Since it would be difficult to conceive of a branching point in the sequence that would provide a dark reaction independent of the light reaction but competing with it in a manner unaffected by aldehyde chain length, the type of enzyme and temperature, it must be concluded as the simplest explanation, that the light reaction itself is a two-flavin process. Since no decarboxylation occurs a dioxetanone excitation mechanism, which invokes concerted decomposition of a cyclic intermediate, is eliminated. Efficient excitation by one electron transfer however, has now become well documented[6] and a simple radical process in bacterial bioluminescence can be supported by several analogous mechanisms.

The homogeneous path involves free radical intermediates and the most important pathway appears to require kinetic intermediates of the type [FH_2O_2] and a two flavin complex [$FMNH_2FMN$]. In model studies of hydroxylation mechanisms, Mager and Behrends[8] have demonstrated that under acid conditions, hydroxylation occurs via the hydroxyl radical, which is generated by the two flavin reaction:

$$FH_2 + FH_2O_2 + H^+ \longrightarrow FH^+ + FH + HO\cdot + H_2O$$

In a similar sequence, with or without the intermediacy of free $\cdot OH$, the RCHO on bacterial luciferase could generate peracyl and then annihilate $FH\cdot$ in a single step:

$$RCHO + O_2 + HO\cdot \longrightarrow RC(O)O\cdot + H_2O_2$$

$$RC(O)O\cdot + FH\cdot \longrightarrow RC(O)O^- + FH^{+*}$$

An analogous process has recently been proposed for the mechanism of acridine aldehyde chemiluminescence

$$RCHO + O_2 \longrightarrow RC(O)O-O(O)R \longrightarrow RC(O)O\cdot$$

$$RC(O)O\cdot + D^- \longrightarrow RC(O)O^{-*} + D$$

where D is an added electron donor.

A complete report of this work will be published elsewhere.

Acknowledgement. This work was supported in part by a Summer Student Research Grant to C.M. from Mr. William N. Creasy, via Fight for Sight, Inc. We thank John Turrentine for his assistance.

---

1. M. Eley, J. Lee, J.-M. Lhoste, C.Y. Lee, M.J. Cormier and P. Hemmerich, Biochemistry 9 2902 (1970).

2. John Lee and C.L. Murphy, unpublished observations.

3. John Lee, Biochemistry 11 3350 (1972).

4. O. Shimomura, F.H. Johnson and Y. Kohama, Proc. Natl. Acad. Sci. (U.S.) 69 2086 (1972).

5. J.W. Hastings, K. Weber, J. Friedland, A. Eberhard, G.W. Mitchell and A. Gunsalus, Biochemistry 8 4681 (1969).

6. These Proceedings.

7. Q.H. Gibson and J.W. Hastings, Biochem. J. 83 368 (1962).

8. H.I.X. Mager and W. Berends, Rec. Trav. Chim. 91 611, 630 (1972).

9. E. Rapaport, M.W. Cass and E.H. White, J. Amer. Chem. Soc. 94 3160 (1972).

HASTINGS: Did you measure the quantum yield with respect to the enzyme?

J.LEE: Dr. Eley and I have studied this and found that $Q_B(E)$ is at least 50%, that is five times greater than aldehyde. This indicates that the enzyme can turn over but we would like it to be greater than 100% to be sure of this.

HASTINGS: What is the quantum yield under non-turnover conditions?

J.LEE: We have not measured this. I remember you estimated it to be the same as for $FMNH_2$.

HASTINGS: You suggest that the light reaction is a two flavin reaction. Where does the second flavin come in?

J.LEE: I think the Mager and Behrends mechanism is a good analogy. Now in regard to the observed first-order dependence of light intensity on $FMNH_2$ concentration, a simple sequential addition of two $FMNH_2$ can be written where one of these is rate-limiting and results in the observed first-order behavior.

HASTINGS: Your mechanism then strongly predicts that when you add more and more enzyme you will come to a point where the enzyme cannot get two flavin molecules and you will get less light.

J.LEE: I don't agree the mechanism strongly predicts this. The $FMNH_2$ is not strongly bound for one thing (J. Lee and C.L. Murphy, Biophys. J. 13, 274a, 1973). We do observe effects at high enzyme concentration but I don't know what the interpretation is.

LUMISOMES: A BIOLUMINESCENT PARTICLE ISOLATED FROM THE SEA PANSY *RENILLA RENIFORMIS*

James M. Anderson and Milton J. Cormier

Department of Biochemistry, University of Georgia

Athens, Georgia 30601

INTRODUCTION

The biochemical pathway leading to light emission in the sea pansy, *Renilla reniformis*, is now fairly well understood. Luciferyl sulfate, an inactive storage form of luciferin, is converted to luciferin by the enzyme luciferin sulfokinase (1). Luciferin is oxidized in the presence of oxygen and *Renilla* luciferase to CO_2 and oxyluciferin in an electronically excited state (2,3). Return to the ground state of oxyluciferin results in the production of blue light (4). Furthermore, the structure determination and chemical synthesis of a biologically active form of *Renilla* luciferin has recently been accomplished (3).

A major problem in the study of *Renilla* bioluminescence has been in relating the *in vitro* reaction described above to the *in vivo* bioluminescence. The *in vitro* reaction produces a blue emission (λ_{max} = 490 nm, half band width = 78 nm) which is in contrast to the *in vivo* green emission (λ_{max} = 509 nm, half bandwidth = 20 nm) (4,5). The green *in vivo* emission has been postulated to be due to energy transfer from the electronic excited state of oxyluciferin to a second chromophore (4,5,6). A protein bound chromophore has been isolated from *Renilla* which exhibits fluorescence characteristics identical to the *in vivo* bioluminescence (4) and thus it must be the *in vivo* emitter.

In the Hydrozoan, *Aequorea*, bioluminescence arises from the interaction of calcium ions and a protein, termed photoprotein (7). Luciferase, such as is found in the Anthozoan, *Renilla*, is not found in *Aequorea* (8). *Aequorea* does, however, contain luciferyl sulfate and a green fluorescent protein (8). A calcium activated "photoprotein" has also been extracted from *Renilla* (8,9). The data, then, suggest that a common biochemical mechanism underlies the bioluminescence in coelenterates. The recent finding by Hori and Cormier (3) that a compound involved in the luminescence of *Aequorea* photoprotein is an integral part of the structure of *Renilla* luciferin reinforces this concept.

It is obvious from the above discussion that the individual components of the *Renilla* bioluminescent system leading to light emission *in vitro* have been isolated and separately studied. The organization of these components into a biochemical system which explains the production and control of *Renilla in vivo* bioluminescence remains to be determined. The transfer of energy from the excited state of oxyluciferin to the green emitting chromophore must require protein-protein interaction between the two proteins involved (10). Further, the existence of a calcium activated protein in a cell requires protection of that protein from cellular supplies of calcium ion. An obvious mechanism for organizing the biochemical reactions would be to associate them with a membrane system in a way similar to the energy transfering systems of mitochondria and chloroplasts. Enclosing the bioluminescent system inside a membrane bound vesicle would sequester the calcium activated protein away from cellular sources of calcium ions. Morin and Hastings (5) presented evidence that particulate "photoprotein" activity could be sedimented from crude extracts of the Hydrozoan, *Obelia*. The particles apparently required lysis, as well as the presence of calcium ions, prior to light emission. Although no spectral data was presented, the authors indicated that the light emitted was green.

In light of the above discussion we felt it prudent to look for bioluminescent particles in *Renilla*. We indeed, found such particles in *Renilla* and present evidence that these highly purified particles contain all the necessary components required for producing the typical calcium-activated, green bioluminescence observed *in vivo*. We propose to call these light producing particles, lumisomes.

RESULTS AND DISCUSSION

A fraction could be obtained from *Renilla* which, in similarity to the photoproteins from Hydrozoans, yielded light upon specific addition of calcium ions. *Renilla* "photoprotein," however, required oxygen as well as calcium for activity. The "photoprotein" activity was found to be of two types: a soluble fraction which produced a blue emission (λ_{max} = 490 nm) identical to the luciferase catalyzed oxidation of luciferin; and a particulate (lumisome) fraction which could be sedimented from *Renilla* extracts at 1,000 to 20,000 g. The lumisome fraction produced green light (λ_{max} = 509 nm) with spectral characteristics identical to the *in vivo Renilla* emission (Figure 1).

The lumisome fraction required a drop in ionic strength as well as calcium ions to produce light. The particles apparently were lysed, osmotically, by the change in ionic strength. Lumisomes could also be disrupted by sonication with light being emitted on addition of calcium ions. The lumisomes also required oxygen for activity in similarity to the soluble "photoprotein." Light was emitted from lumisomes with rapid kinetics, in contrast to the kinetics of the soluble "photoprotein" emission (Figure 2). The kinetics of the soluble "photoprotein" emission resembled those of the luciferin-luciferase reaction while the kinetics of the lumisome emission resembled those of the *Renilla in vivo* emission. Thus, the data strongly suggest that the lumisome is involved in producing the *in vivo Renilla* bioluminescence.

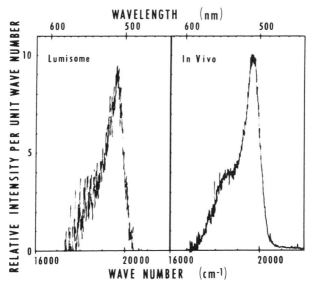

FIGURE 1. The corrected emission spectra of the *in vitro* "photoprotein" reaction of lumisomes and the *in vivo* emission of *Renilla mülleri*.

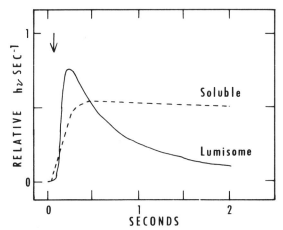

FIGURE 2. Kinetics of *Renilla* "photoprotein" bioluminescence initiated with addition of calcium ions at the arrow.

The lumisome fraction could be purified by use of sucrose density gradients. Lumisomes banded in one major (1.7 g cm^{-3}) and two minor bands (1.13 and 1.10 g cm^{-3}) on the sucrose gradients (Figure 3C,B,A). The minor bands varied in amount from preparation to preparation. Purification of lumisomes was accomplished by taking lumisome material from 1.10 g cm^{-3} density region of a

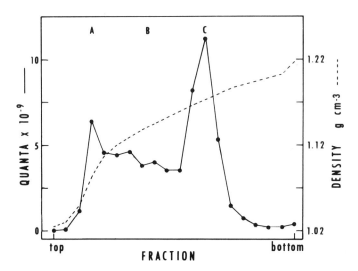

FIGURE 3. Distribution of lumisome "photoprotein" activity from *Renilla reniformis* on sucrose density gradients.

gradient (Figure 3A) and rebanding it on a second sucrose gradient. The lumisome activity rebanded at density 1.17 g cm^{-3} while contaminating particles rebanded at their original density regions. Lumisome activity from the 1.13 g cm^{-3} region also rebanded at density 1.17 g cm^{-3} on a second sucrose gradient. Lumisome preparations prepared by this method were shown by electron microscopy to be fairly uniform. The micrographs showed membrane enclosed vesicles of approximately 0.2 μm in diameter.

Lumisomes were found to contain the following proteins: luciferase (11), "photoprotein," and the green fluorescent protein (4,5,8,9). The proteins appeared to be tightly associated with the lumisome. They did not solubilize upon lysis of the particles and could only be slowly removed with 2.5 M NaCl treatment. In fact, lumisomes which were lysed, followed either by reisolation by centrifugation or not, still produced green light (λ_{max} = 509 nm) upon addition of calcium ions.

Lumisomes were found to occur in all the species of bioluminescent Anthozoan (*Renilla mülleri*, *R. reniformis*, *R. kollikeri*, *Ptilosarcus guernyi*, *Stylatula elongata*, *Acanthoptilum gracile*, and *Parazoanthus lucificum*), and Hydrozoan (*Obelia longissima*, *O. geniculata*, and *Clytia edwardsia*) coelenterates studied. Although material from each species was not always sufficient to test for all the lumisome characteristics, where material was available each species was found to have lumisomes with characteristics similar to those found earlier for *R. reniformis*. For example, they all required a drop in ionic strength as well as calcium for activity, they all banded on sucrose gradients with a major peak at a density close to 1.17 g cm^{-3}, and they all produced green light (λ_{max} = 509 nm) which was typical of the *in vivo* emissions of these animals. The widespread occurrence of lumisomes suggest that they represent the cellular site of bioluminescence in coelenterates.

The relationship between *Renilla* luciferase and "photoprotein" remains unclear. The similarities between the light reactions of both proteins suggests that "photoprotein" is a luciferin-luciferase complex. The difference in kinetics between the light reactions of soluble and lumisome "photoproteins" make it clear, however, that this explanation of "photoprotein" is not sufficient. A clue may come from the role the lumisome plays in stabilizing the "photoprotein"-green fluorescent protein complex allowing for the production of green light by energy transfer. It is also apparent that the lumisome must play an important role in controlling the light reaction by controlling calcium access to the "photoprotein." A possible mechanism for such control involves active transport of calcium ions across the lumisome membrane to a membrane stabilized, light emitting protein complex.

REFERENCES
1. Cormier, M. J., Hori, K., and Karkhanis, Y. D. (1970) *Biochemistry 9*, 1184.
2. DeLuca, M., Dempsey, M. E., Hori, K., Wampler, J. E., and Cormier, M. J. (1971) *Proc. Nat'l. Acad. Sci. USA 61*, 1658.
3. Hori, K., and Cormier, M. J., *Proc. Nat'l. Acad. Sci. USA*, in press.
4. Wampler, J. E., Hori, K., Lee, J. W., and Cormier, M. J. (1971) *Biochemistry 10*, 2903.
5. Morin, J. G., and Hastings, J. W. (1971) *J. Cell. Physiol. 77*, 313.
6. Hastings, J. W., and Morin, J. G. (1969) *Biol. Bull. 137*, 402.
7. Shimomura, O., and Johnson, F. H. (1969) *Biochemistry 8*, 3991.
8. Cormier, M. J., Hori, K., Karkhanis, Y. D., Anderson, J. M., Wampler, J. E., Morin, J. G., and Hastings, J. W. (1972) *J. Cell. Physiol.*, in press.
9. Morin, J. G., and Hastings, J. W. (1971) *J. Cell. Physiol. 77*, 305.
10. Wampler, J. E., Karkhanis, Y. D., Hori, K., and Cormier, M. J. (1972) *Fed. Proc. 31*, 1133.
11. Karkhanis, Y. D., and Cormier, M. J. (1971) *Biochemistry 10*, 317.

ACKNOWLEDGEMENTS

This work was supported in part by the National Science Foundation and the U.S. Atomic Energy Commission.

A REVIEW OF EXPERIMENTAL MEASUREMENT METHODS BASED ON GAS-PHASE
CHEMILUMINESCENCE

>Arthur Fontijn[†]
>AeroChem Research Laboratories, Inc., P.O. Box 12,
>Princeton, N.J. 08540
>
>Dan Golomb
>Air Force Cambridge Research Laboratories, L.G. Hanscom
>Field, Bedford, Mass. 01730
>
>Jimmie A. Hodgeson
>Environmental Protection Agency, Technical Center,
>Research Triangle Park, N.C. 27711

I.	INTRODUCTION	394
II.	LABORATORY METHODS	395
	A. Elementary Processes at Moderate Temperatures	395
	B. Flame and Shock Tube Studies	400
	C. Flow Visualization and Supersonic Flow Studies	404
III.	MONITORING OF AIR POLLUTANTS	405
	A. Ozone	406
	B. Nitrogen Oxides, Ammonia and Amines	407
	C. Sulfur Compounds	409
	D. Carbon Monoxide and Phosphorus, Boron, Chlorine Compounds	411
IV.	UPPER ATMOSPHERE APPLICATIONS	412
	A. Artificial Chemiluminescence	412
	B. Natural Chemiluminescence (Airglow)	416
V.	EPILOGUE	418
	REFERENCES	419

[†] This work was sponsored by Project SQUID, which is supported by the U.S. Office of Naval Research, Department of the Navy, under Contract N00014-67-A-0226-0005, NR-098-038.

I. INTRODUCTION

The first application of gas-phase chemiluminescence could have been the contribution to the illumination of a cave by the blue portion of a wood flame. Qualitative flame analysis has been in use since at least the 16th Century.[1] For many present-day applications, observation of chemiluminescence has the advantages that (i) measurement of such radiation in no way interferes with the reacting environment, (ii) the measuring tool does not have to be exposed directly to the active reaction environment, i.e. remote sensing is possible, (iii) transient species can be readily identified and (iv) in its most elementary form the observer's eye is the only tool required. For these reasons chemiluminescence measurements represent an obvious and often immediately available method for investigating a given medium. When coupled with modern reliable sensitive detection methods, which cover the optical spectrum from the vacuum ultraviolet through the near infrared, chemiluminescence provides a major tool for species concentration measurements. Chemiluminescence measurements are also used in studies of homogeneous reaction kinetics, thermochemistry, gas-surface interactions, temperatures of reaction environments, atmospheric mass transport, and gas dynamics.

It is the purpose of the present paper to summarize these various methods and to discuss them in the context of the environments to which they have been applied, in the expectation that such a juxtaposition of knowledge from different disciplines and research areas can lead to further useful developments based on chemiluminescence. While many examples pertaining to specific species and reactions will be given, these are not intended to include all such observations but rather to illustrate the methods being discussed. To further delineate the scope of this review, two definitions are in order (i) chemiluminescence is the emission of radiation from a chemi-excited species and (ii) chemi-excitation is a process by which excited species are formed as a direct result of the formation of new chemical bonds. Thus radiation following collisional excitation (e.g. $A + B$ + kinetic energy $\rightarrow A + B^*$)--a principal radiation source in high temperature reaction environments, or dissociative reactions (e.g. $AB^+ + e^- \rightarrow A^* + B$), or radiation resulting from charged species impact or the absorption of radiant energy (e.g. fluorescence) will not be considered. Emphasis is placed on optical emission measurement methods which are specifically based on chemiluminescence.

From the above definition a formalized description of chemiluminescence may be written as:

$$A + B \overset{(M)}{\rightarrow} C^* + D \qquad (1)$$

$$C^* \rightarrow C + h\nu \qquad (2)$$

By virtue of the implied occurrence of a chemical reaction chemiluminescence measurements thus often have the important characteristic that the species being observed is a reaction product and not a reactant. However, since processes such as

$$A + B + C \xrightarrow{(M)} AB + C^* \quad (3)$$

followed by light emission are also examples of chemiluminescence this does not always apply. In reactions such as (3) C^* may be formed directly in an excited state or be excited by collisions with AB^*. We will refer to the latter case as indirect chemiluminescence.

Chemiluminescence measurement methods have found their widest variety of uses in the laboratory. Most methods for applied usage have been derived from laboratory methods and more such methods await application. Laboratory methods and their limitations are discussed in Section II. The newest, most rapidly growing field of gas-phase chemiluminescence application is pollutant monitoring. Air or exhaust gases are continuously sampled and the concentration of a selected pollutant is determined from the intensity of the chemiluminescence produced by its reaction inside a monitor; these methods are discussed in Section III. Releases of chemical tracers leading to artificial upper atmospheric chemiluminescence are a very practical, relatively inexpensive means for studying that environment; the status of this area and that of natural airglows is reviewed in Section IV.

II. LABORATORY METHODS

A. Elementary Processes at Moderate Temperatures

Chemiluminescence is a widely used indicator in titration reactions of many atomic and free radical species in tubular fast-flow reactors. Such reactors have been used mainly in the 200 to 750 K regime. (For a description of their use see e.g. Refs. 2-4.) The technique may be illustrated by the air afterglow reaction:[2,5]

Indicator Reaction $O + NO + M \rightarrow NO_2 + M + h\nu$ (4)

O atoms at a fixed point in the reactor are titrated by addition of NO_2 via the fast reaction

Titration Reaction $O + NO_2 \rightarrow NO + O_2$ (5)

When the (volume) flow of NO_2 is less than that of the O atoms, NO is produced in the presence of O and the whitish-green glow from Reaction (4) is observed. However, when the flow of NO_2 equals or exceeds that of O no O atoms are present downstream from the NO_2 inlet and the glow contracts to the small mixing region surrounding the nozzle. The important requirement for a titration reaction is that it be fast and complete near the titrant inlet. The indicator reaction should be slow so that it can be observed downstream from the inlet. Other methods besides chemiluminescence can be used to follow the titration reaction, e.g. mass spectrometric determination of the NO_2-consumption plateau[6] and E.P.R.[7] or optical absorption measurements[8] of the O-atom disappearance. These methods have confirmed the validity of the chemiluminescence measurements but generally require a more elaborate experimental set-up. It is useful to keep in mind that in the sequence of Reactions (5) and (4) chemiluminescence is used as the indicator reaction not as the titration reaction, contrary to an unfortunately rather common misuse of terms.

Indicator reactions are also used to measure the change in reactant concentration with time obtained from the distance along the reactor and the average gas velocity. If a trace of NO is present in or added to an atomic-O flow, the relative intensity of the air afterglow is a measure of the relative [O], since the combination of Reactions (4) and (5) results in recycling of NO and hence a constant [NO]. Examples of quantitative kinetic studies in which the O-atom consumption rate has been determined in this manner include

$O + NO + M \rightarrow NO_2 + M$	(6)	see e.g. Refs. 2,5
$O + O_2 + M \rightarrow O_3 + M$	(7)	see e.g. Refs. 2,5,9
$O + SO + M \rightarrow SO_2 + M$	(8)	Refs. 10, 11
$O + C_xH_y \rightarrow$ Products	(9)	see e.g. Refs. 12-14
$O + XCN \rightarrow NCO + X$	(10)	Ref. 15

in which X = H, Cl, Br. While consumption of NO by free radicals does not seem to have interfered with measurements in O/hydrocarbon systems (Reaction 9), NO is known[16] to be a very efficient free radical scavenger and some caution in the use of the air afterglow in such a reaction system is necessary. Reactions (8) and (10) serve to illustrate uses of chemiluminescence in the measurement of free radicals and reaction products. Reaction (8) is accompanied by emission of the SO_2 afterglow

$$O + SO + M \rightarrow SO_2 + M + h\nu \qquad (11)$$

By simultaneously measuring the relative intensities of both the O/NO and O/SO afterglows a measure of the consumption of both O and SO is obtained.[10,11] Reaction (10) is followed by:

$$O + NCO \rightarrow NO + CO , \qquad (12)$$

i.e. by NO production. By adding known small flows of NO to the O/XCN mixture and plotting the air afterglow intensity versus NO flow, a negative intercept on the abcissa (NO flow axis) is obtained, the magnitude of which directly yields the NO-production rate.[15]

Absolute concentrations can also be obtained from the relative air afterglow intensity. This requires that the detector (photomultiplier plus light filter or monochromator) output be calibrated in terms of [O][NO] for a given reactor detector geometry and that the absolute concentration of one of these reactants be known. This approach has been found useful in static reactors in which the NO_2 titration because of its flow nature cannot be carried out,[17] in systems in which ground state O atoms are produced (such as by quenching of singlet excited O atoms)[18] and where flow metering of NO_2 is found to be inconvenient.

Fontijn, Meyer and Schiff[19] (FMS) have accurately measured the absolute rate coefficient for light emission of the air afterglow reaction, k_4, together with the spectral distribution of the glow, which is a continuum extending from 388 nm well into the infrared. This glow is commonly used as a secondary standard for absolute quantum yield determinations of light emitting processes in the gas phase. The relative intensity emitted in a given wavelength region by the investigated process need only be compared to that of the O/NO reaction using the same conditions of geometry and optics. Fontijn and Lee[20] have extended such comparisons to the liquid phase and found good agreement with the common liquid-phase (luminol oxidation) chemiluminescence light intensity standard. A check on k_4 by Vanpee et al.[21] also gave good agreement, but subsequent work on the spectral distribution[21-23] has shown deviation from the FMS data in the (infra)red, though excellent agreement exists for $\lambda \leq 650$ nm. For quantum yield measurements above 650 nm it would be best to use a combination of the FMS data below 650 nm with the more recent ir spectral distribution measurements. Stair and Kennealy[23] have shown the emission to extend to at least 3700 nm. Since the O/NO continuum does not extend below 388 nm, it cannot be used directly as a standard for that wavelength region. However, the quantum yields of other glows, which partially overlap the O/NO wavelength region, have been determined by comparison to the O/NO glow and can be used as secondary standards in the ultraviolet. In particular, the SO_2 afterglow continuum[11,24] (Reaction 11) appears useful down to about 260 nm. Recently Mandelman, Carrington and Young[26] have accurately determined the absolute rate coefficient for N-atom/O-atom pre-association leading to NO δ(0,0) emission (Reaction 16, below) near 191 nm. No standard reactions at shorter wavelengths have yet been provided.

The NO/O standard can be used directly in the 1 to 10 Torr regime, at 300 K, in an Ar or O_2 carrier. It must be remembered that the detailed reaction mechanism[25] involves participation of a third body, M, and that, as a result, the intensity depends on the nature of M and becomes [M]-dependent at lower pressures where quenching no longer competes successfully with emission. For the M-dependences of k_4 see Clyne and Thrush[27] and Kaufman and Kelso;[28] for the temperature dependence see Hartunian, Thompson and Hewitt.[29]

The reaction

$$H + NO + M \rightarrow HNO + M + h\nu \qquad (13)$$

gives rise to a banded HNO spectrum in the 600-800 nm region. The intensity of this emission is[30] proportional to $[H][NO]$ and the reaction has been used in H-atom studies. The titration reaction for which (13) serves as the indicator is

$$H + NO_2 \rightarrow OH + NO \qquad (14)$$

Subsequent rapid reactions of OH regenerate H and as a result the NO_2 end point flow is 1.5 times the initial H-atom flow.[8,31]

Further examples of laboratory afterglow methods may be illustrated by the nitrogen, nitric oxide and oxygen afterglows. Absolute N-atom concentrations can be determined via titration[7,8,32,33] with NO:

$$N + NO \rightarrow N_2 + O \qquad (15)$$

When an insufficient flow of NO is added to remove all N atoms a mixture of N and O atoms exists downstream from the NO inlet and the blue NO afterglow can be observed[34-37]:

$$N + O \xrightarrow{(M)} NO + h\nu \qquad (16)$$

When the NO flow is in excess of the N-atom flow, the downstream gases contain O and NO and the air afterglow appears again. At the end point the tube visually appears dark. However, in the ultraviolet O_2 Herzberg bands and in the infrared O_2 Atmospheric bands, both due to[36-39]:

$$O + O + M \rightarrow O_2 + M + h\nu, \qquad (17)$$

can still be observed. In the nitrogen afterglow itself several band systems, due to[36,37,40-44]

$$N + N + M \rightarrow N_2 + M + h\nu \qquad (18)$$

are observed, of which the yellow First Positive Bands are most commonly used for measurements. These bands and the NO β, γ and δ bands can be used to follow atom concentrations. The light intensities are proportional to $[N]^2$ and $[N][O]$, respectively. For rate coefficient data and restrictions of pressure see the references quoted. The temperature dependences of k_{16} and k_{18} have been measured over a wide range by Gross and Cohen,[45,46] (see also Campbell and Thrush[34]). The reader is referred to review papers by Carrington and Garvin[47] and by Thrush[48] for the detailed mechanisms of Reactions (16) to (18), as well as of several further reactions which may have potential for practical applications e.g., the O/CO glow and the halogen atom recombination afterglows.

The oxygen afterglow intensity from Reaction (17) is proportional to $[O]^2$ and in principle can also be used for concentration measurements in O-atom reaction studies. However, both band systems are strongly forbidden (radiative lifetimes > 1 sec)[37,38] and consequently have a low intensity. Moreover, even though quenching will reduce the effective lifetime of the emitters (O_2 $A^3\Sigma_u^+$ and $b^1\Sigma_g^+$ respectively), the emission in flow tubes may occur well downstream from the point of emitter formation and the intensity then will no longer be a true indication of local $[O]$. Radiative lifetimes of the other emitters discussed above are on the order of 10^{-6} sec or less; since typical linear velocities in laboratory flow tubes are on the order of 10^2 to 10^4 cm sec^{-1} this transport problem does not arise in their use. However, care must also be exercised with short-lived emitters, for example in the use of the $I \propto [N]^2$ and $\propto [N][O]$ relationships in mixtures containing a second reactant. In a few instances it has been shown that such reactants or a reaction intermediate can enhance afterglow intensities by catalytically increasing the rate of atom recombination into the emitting state. Thus, I_2 addition can enhance the Lewis-Rayleigh afterglow intensity,[49] and C_2F_4 and C_2H_4 have been observed to increase the NO γ and $\beta(v' \leq 2)$ intensities.[50] The NO β emission intensity also has been shown recently[51] to be not strictly proportional to $[N][O]$, due to N, O and O_2 action on a precursor of the emitting B $^2\Pi$ state. For these reasons it appears safest to use the NO $\delta(v' = 0)$ bands, whose emitter is produced directly by a two-body pre-association reaction[35,36] unlike the other emissions discussed above which are produced by three-body reactions.

The nitrogen afterglow also offers several examples of the use of indirect chemiluminescence. $N_2(A\ ^3\Sigma_u^+)$ molecules are formed via N-atom recombination but, due to their long radiative and short chemical lifetime, these excited molecules cannot normally be observed in emission. However, when a trace of Hg is added to the flow, rapid energy transfer occurs and the resulting 253.7 nm emission can be used to monitor $N_2(A\ ^3\Sigma_u^+)$ and estimate its concentration.[52] Other N_2^* species formed in the afterglow apparently do not excite the emitter, $Hg(6\ ^3P_1)$.[52] $CO(a^3\Pi)$ has been similarly monitored in various reaction mixtures i.e. by energy transfer to Hg and to $NO(X\ ^2\Pi_r)$, leading to Hg 253.7 nm and NO β and γ band emission, respectively.[53,54] Vibrationally excited ground-state N_2 molecules can also be formed in nitrogen afterglows. Addition of CO_2 leads to energy transfer and emission of the 4300 nm vibrational fundamental band of CO_2. Electronically excited species present do not produce this band. Similar observations have been made using N_2O.[55]

Chemiluminescence from atom reactions has also been used to determine bond dissociation energies. The general spectroscopic methods for determining bond energies have been discussed by Gaydon.[56a] Chemiluminescence is sometimes the preferred technique

for producing clean interpretable spectra, for example in the
determination of the lower limit to the BaO bond energy from the
Ba + NO_2 chemiluminescence in a crossed beam experiment.[57] Chemiluminescence measurements of changes in atom concentrations near
catalytic surfaces have been employed to measure atom-surface
recombination coefficients,[58-60] γ, the efficiency of catalytic
probes in destroying free atoms[61] and atomic diffusion coefficients.[58,61-63] Atom-surface interactions can also give rise to
luminescent surface phenomena, i.e. to candoluminescence of lumophoric substances such as CaO and MgO. The intensity of the candoluminescence associated with the impingement of N and O atoms has
enabled determinations of γ.[64-66]

B. Flame and Shock Tube Studies

Several self-sustaining flame types are known and are
discussed in several texts, e.g. Gaydon and Wolfhard.[67] Of these
premixed laminar flames are probably most widely used in present-day laboratory practice. In such flames reaction time and distance
are (as in flow tubes) directly proportional to each other, which
is an important advantage in (i) kinetic studies and (ii) all
studies in which it is desirable to distinguish between flame zones.
Two distinct major reaction zones are present in premixed flames;
(i) the primary combustion zone characterized by wide departures
from equilibrium and in many flames intense chemiluminescence and
(ii) the burned gas zone in which equilibrium conditions are more
closely approached and in which reactions of free radicals, etc.,
formed in the primary zone can readily be studied. Several of the
atomic and radical species present in flames are identical to those
discussed in the previous section and some of the same chemiluminescent reactions can be used in a similar manner. However, since
flame environments usually contain a larger number of reactants,
errors due to side reactions or overlapping emissions are more
likely and considerable caution in the use of chemiluminescent
emissions in measuring e.g. reactant-time profiles is needed. Moreover, as a consequence of the high temperature reaction environment
represented by flames, emissions from other sources, particularly
collisionally excited species, often dominate and background emissions may underlie some of the emissions being measured.

Distinguishing between chemi-excitation and collisional
excitation is sometimes difficult and may require extensive research.
Both processes can contribute to the establishment of thermal equilibrium levels of excitation. However, in systems in which collisional excitation dominates the excitation cannot exceed equilibrium
levels. Hence, non- or extra-equilibrium radiation (also referred
to as "suprathermal" radiation) is usually chemiluminescent in origin. For general reviews of flame excitation processes, see, e.g.
Alkemade and Zeegers[68a] and Sugden.[69] For general discussions of
flame emission methods the reader is referred to Refs. 68-73.

Extensive studies have been performed in the burned gases of C_xH_y/O_2 and H_2/O_2 flames with or without added diluents. Little "natural" emission from these gases occurs in the uv and visible part of the spectrum; as a result emission due to additives can readily be studied. Chemiluminescence is the dominant emission mechanism at "low" temperatures, achieved e.g. by heavy admixture of diluent; increasing temperature favors collisional excitation. The definition of low temperature in this connection varies with excitation energy. Sugden et al.[74] estimate that for excitation of 600 nm radiation, collisional effects predominate above 1900 K, while emissions at 300 nm become primarily collisional only at temperatures above 2300 K.[†] In these burned gas mixtures H, OH and O are the principal naturally-occurring radicals. H and OH are in equilibrium with each other via

$$H + H_2O \rightleftarrows H_2 + OH \qquad (19)$$

Since the equilibrium constant for this reaction and the $[H_2]$ and $[H_2O]$ at equilibrium are known, determination of the concentration of one of these reactants suffices for the determination of both. Padley and Sugden[76] added small quantities of Na to fuel-rich $H_2/O_2/N_2$ flames and monitored the Na D-line emission resulting from

$$H + H + Na \rightarrow H_2 + Na^* \qquad (20)$$

The emission intensity is proportional to $[H]^2$; $[H]$-profiles thus obtained agree well with those observed by the CuH method.[77] The latter can also be used at high temperatures and is based on the equilibrium

$$Cu + H + M \rightleftarrows CuH + M \qquad (21)$$

and the accompanying CuH (0,0) 428.8 nm emission. Rosenfeld[78] has used the Pb equivalent of Reaction (20) at temperatures down to 1350 K. The $[H]$-profiles obtained in such studies have made it possible to measure the heat of dissociation of a large number of metal oxides and hydroxides in flames.[56b,69] It should be emphasized that Reaction (18) is a practical method for $[H]$-determination in O-atom lean flames. In O-rich flames the faster reaction

$$O + O + Na \rightarrow O_2 + Na^* \qquad (22)$$

dominates.[79]

[†] This statement should not be construed to indicate that no chemiluminescence can be observed in flames at high temperatures. For example, Bulewicz[75] has found strong indications that most visible and ultraviolet emission from the primary reaction zone of some subatmospheric C_2N_2/O_2 flames, in which temperatures in excess of 4000 K are obtained, is attributable to chemiluminescence.

In addition to this type of kinetic/thermochemical study, the secondary reaction zone of premixed hydrocarbon or H_2 flames is also widely used in emission flame spectrophotometry applied to analytical chemistry. This type of analysis generally yields no information on the kind of chemical compound being analyzed but rather on the elements present in these compounds. It is primarily a quantitative (and qualitative) analysis method for metallic compounds present in liquid samples which may be delivered to the flame by a variety of means (see Alkemade[70a] for details). At the temperatures used the atomic line emissions observed are often the result of collisional excitation. However, chemiluminescence can be made to be the dominant metal atom excitation source using the other major burner type for this sort of work, the turbulent diffusion type burner.[70a] Unlike premixed flames at atmospheric pressure where the primary reaction zone is very thin, in turbulent diffusion flames primary combustion reactions can occur over several cm. By using a C_2H_2 flame, or introducing the sample with an organic solvent into an H_2 flame, intensity increases several orders of magnitude over the thermal equilibrium intensity have been obtained for many metallic lines.[80] This chemi-excitation is apparently related to the formation of the energetic (11.1 eV) CO bond,[68,72a,80] which is also thought to be responsible for most of the prominent natural chemiluminescence features of organic flames,[81] i.e. CH, CHO, CO and OH.

While not yet much exploited for quantitative analysis many emission bands of compounds formed in flames from metallic[69,82] and non-metallic[70b] elements have been observed, many of which are probably chemiluminescent in origin (for a compendium of such emissions see Pearse and Gaydon[83]). Examples of the utilization of such chemiluminescent emissions include measurements of gaseous samples of sulfur and phosphorus compounds with the Brody-Chaney H_2/O_2 flame photometer. In the class of photometers to which this device belongs, the secondary zone is chilled to reduce the background emission.[70b] This is achieved by the use of a chimney (like a Smithsells separator). The Brody-Chaney instrument was originally developed as a detector for gas chromatographic analyses and has now also been applied to direct sampling of air pollutants (see further Section III). In flame kinetic studies the NO/O continuum (Reaction 4) is widely used to measure [O] and/or [NO].[69,74] The uv-visible ($\lambda \geq 210$ nm) CO/O continuum[68a,73] is similarly used for [O] and/or [CO].

Chemiluminescence may also be used in a very different manner to determine concentrations of reaction intermediates or other unstable species in high temperature environments, i.e. by the use of chemical scavenger probes. In such continuous flow probes, the sampled gases are expanded adiabatically (or cooled otherwise) to freeze the composition. A scavenger gas is introduced immediately downstream from the sample inlet to form stable products which can then be analyzed. Fristrom and Westenberg[84] used the O/NO_2 titration

(Reaction 5) in this manner to determine [O] from flame gases; although they used mass spectrometric detection, chemiluminescence could probably have been used equally well. This was in fact done for N atoms via the N/NO titration (Reaction 15) by Fontijn, Rosner and Kurzius[85] who sampled from a supersonic plasma jet.

In shock tubes—the other major device for high temperature kinetic studies—a number of chemiluminescent reactions have been observed and used. The intensity of the uv O_2 Schumann-Runge system (B $^3\Sigma_u^-$ -X$^3\Sigma_g^-$) due to the two-body radiative pre-association reaction

$$O + O \rightarrow O_2 + h\nu \quad (23)$$

is proportional[86,87] to $[O]^2$ and has been used to study O-atom reactions.[88] Such pre-association reactions are of an endothermic nature.[89] Reaction (23) has been used at temperatures above 2500 K where it leads to strong emission. By contrast, Reaction (17) which near room temperature gives rise to the weak Herzberg bands in the same wavelength region has (as a typical three-body reaction) a negative temperature dependence and contributes negligibly[86] in high temperature environments. Wray and Fried[90] using an atmospheric pressure arc have demonstrated that the emission from (23) is a discrete spectrum and that the spectral distribution is in accord with equilibrium theory. The emission can therefore also be used to determine reaction temperatures. The B $^3\Sigma_u^-$ -X$^3\Sigma_g^-$ emission from the S-atom equivalent of Reaction (23)

$$S + S \rightarrow S_2 + h\nu \quad (24)$$

can be used in a similar manner for kinetic studies (and presumably for temperature determinations).[91] The radiative association of ground-state chlorine atoms gives rise to emission from the $Cl_2\,^1\Pi_u$ and $^3\Pi_{0^+u}$ states,[92] which has been used to study the rate of Cl_2 dissociation behind shock waves.[93] Some other chemiluminescent reactions used in shock tube studies are the CO_2 continuum, due to the O + CO + M reaction, for which[94] $I \propto [O][CO]$, the OH 306.4 nm system, due to the O + H + [M] reaction, for which[95] $I = [O][H]$ and other emissions typical of organic combustion reactions such as CH, CHO, C_2, CO chemiluminescence. The latter emissions though not simply related to reactant concentrations have nonetheless proven quite helpful in unraveling combustion mechanisms in shock tubes and flames and atom reactions in flow tubes.[81]

In closing this section, mention should be made of the glow discharge shock tube. In this device a flow tube, such as discussed in Section II.A., is used in such a fashion that a shock wave can be fired through the glowing gas thus heating the gas mixture so that the temperature dependence of chemiluminescent reactions can be studied. The temperature range covered thus far with this apparatus is from about 300 to 3300 K. The temperature dependence of the rate coefficients of Reactions (4), (11), (16), and (18) and the CO/O glow have been determined in this manner.[29,45,46,96]

C. Flow Visualization and Supersonic Flow Studies

One of the implicit advantages of using transparent vessels in which to study reactions leading to chemiluminescence is that flow patterns can easily be observed, thereby verifying that good mixing and smooth laminar flow, required for many kinetic studies, is obtained. Flow visualization is also used in studying supersonic flows. In wind tunnel tests at pressures below a few Torr, Schlieren photography is no longer practical for obtaining density variations around fixed objects.[97a] Here, chemiluminescence, particularly the nitrogen afterglow, has been found to be a very useful technique.[97b] Nitrogen is subjected to an electrical discharge before expansion through a nozzle and the shock waves around objects can readily be photographed.

The fluid dynamics of wakes and trails behind planetary atmospheric entry bodies has been studied in ground-based laboratories by shooting ≈ 1 cm diam spheres at ≈ 6 km sec^{-1} through air at reduced pressures. Temperatures in the shock layers around these spheres are typically on the order of 6000-8000 K. As the gas expands around the body into the wake, its temperature decreases but its chemical composition relaxes relatively slowly, thus creating concentrations of atoms in excess of equilibrium and setting up conditions for chemiluminescence to occur. While the shock layer radiation is essentially thermal, the radiation from the wake is chemiluminescent in origin and consists of the O_2 Schumann-Runge Bands (Reaction 23), the air afterglow NO_2 continuum (Reaction 4), the NO β, γ and δ Bands, the O_2 Herzberg and Atmospheric Bands and the N_2 First Positive Bands (Reactions 16-18). The emissions and their intensities depend on the flux of chemically active species into the (turbulent) wakes (i.e. the mixing rates with ambient air), local temperatures and the rates of those reactions which deplete the chemically active species. Models describing the wake chemistry and fluid mechanics have been tested by comparing calculated emission properties with measured properties.[98-100] Emission from the relatively hot near wake is dominated by the strongly temperature-dependent O_2 Schumann-Runge Bands which last for several hundred body diameters.[99] At larger distances (the far wake) the O/NO air afterglow reaction emission dominates and has been used for verifying models.[100]

High speed entry bodies are subject to considerable heating and as a result, evaporation (ablation) from their surfaces occurs. The rate of such evaporation depends on the chemical composition of the body's exposed surface. Experimental studies have been performed to verify flow models taking into account large rates of evaporation ("massive blowing").[101,102] Objects of various configurations have been placed in arc tunnels, with a stream of O atoms impinging upon them. The objects have a porous surface through which NO is injected. The streamlines separating the blown gas from the oncoming stream are readily visualized by emission from O/NO (Reaction 4).

Upper-atmospheric NO releases[103] have been simulated in a low density wind tunnel. A free jet of NO is injected into a supersonic flow containing O atoms and the so-called headglow is generated. The O-atom flux is determined by titrating O with NO_2 (Reaction 5) until the headglow is extinguished. The measured headglow intensity at known reactant fluxes is used to analyze atmospheric releases (see Section IV.A.1). The known O-fluxes can also be used to calibrate other upper atmospheric sampling devices such as mass spectrometers.[104]

III. MONITORING OF AIR POLLUTANTS

The first application of chemiluminescence to monitoring of an air pollutant--detection of atmospheric O_3 by (heterogeneous) chemiluminescence[105]--was reported more than a decade past. Active interest in chemiluminescence as a method for monitoring air pollution dates from about 1968 and at present chemiluminescence monitors for O_3, SO_2 and oxides of nitrogen (NO, NO_2 and NO + NO_2 = NO_x) are being employed at a rapidly expanding rate.

The majority of detectors in practical use are based on homogeneous gas-phase chemiluminescence. Two types of such detectors may be distinguished. The ambient temperature detector employs the chemiluminescent reaction between the small molecule of interest, X, in air or in the exhaust gas, and a second reactant species, R, which is added in excess to the flow reactor. For the reactions used, the intensity of the chemiluminescence is directly proportional to the product of reactant concentrations, $I = k[R][X]$. Since the second reactant gas is normally in large excess and its concentration is constant, I is directly proportional to sample concentration, $I = k'[X]$. In the other type of detector, chemiluminescence resulting from reactions between atomic or molecular fragments produced from primary molecules introduced into a flame is observed in the secondary combustion zone. Flame chemiluminescence methods are generally less specific than room temperature methods and are more applicable to the detection of classes of compounds, e.g. nitrogen or sulfur compounds.

Chemiluminescence detectors have some common advantages and disadvantages for air pollutant monitoring. They inherently possess a high degree of sensitivity, specificity and simplicity. With the use of high gain, low dark current photomultiplier tubes, extremely low levels of chemiluminescent emission can be detected. Trace concentrations are more readily detected in emission, in which a positive quantity is measured, than in absorption, in which concentration is proportional to a small differential quantity. In order for another molecule to be a positive interference in the chemiluminescence detection of X it must react with R, this reaction must produce chemiluminescence, and this chemiluminescence must overlap considerably with the spectral region in which X produces emission. Such interference would be a rare happenstance. A third body which quenches the excited state responsible for emission is a potential negative interference. In trace atmospheric detection, however, the predominant quenching agents are O_2 and N_2, which do not vary in concentration.

Chemiluminescence detectors can be simple, compact and constructed from commercially available components. The common components of the detectors include (i) the flow train--gas inlet system, flow meters, reaction chamber, and sample pump, and (ii) the electronics--a photomultiplier tube closely coupled to the reaction chamber through an optical filter (if needed), photomultiplier high voltage supply and amplifier and an analog read-out.

A limitation of chemiluminescence detection is that not all pollutants participate in useable reactions. In common with all optical techniques absolute measurements are too difficult for routine use and instrument calibration is required.

A. Ozone

Regener[105] developed the first practical chemiluminescence O_3 detector, which has been applied in both upper and lower atmospheric measurements. In Regener's procedure, the chemiluminescence obtained from the reaction between O_3 and Rhodamine-B adsorbed on activated silica gel is measured as function of $[O_3]$. The technique is specific and is the most sensitive method known for O_3, with a lower limit of detection well below one part per billion (1 ppb). The Environmental Protection Agency (EPA) became interested in Regener's approach for routine detection of O_3 in polluted atmospheres. Preliminary field studies were conducted in urban locations,[106] as a result of which preparation of the Rhodamine-B surface was modified to provide improved lifetime and stability.[107] The improved monitors have since shown excellent reliability in extended field studies[108] and form the basis for a commercially available monitor. The principal limitation of the Regener approach is the variable and slowly decaying sensitivity of the Rhodamine-B surface. An internal O_3 source is necessary to provide frequent, periodic calibrations of the surface and compensate for any sensitivity changes. Although the internal calibration source works satisfactorily, a Regener-type instrument is more complicated than the homogeneous gas-phase chemiluminescence analyzers discussed below which have shown better response characteristics in comparative studies.[109]

In 1965, Nederbragt et al[110] published a note on the application of a homogeneous gas-phase chemiluminescence technique based on the atmospheric pressure chemiluminescent reaction between O_3 and C_2H_4. No information was available then on the mechanism of the reaction. The emission is a broad continuum centered near 435 nm and has been tentatively assigned to an excited aldehyde linkage (e.g. formaldehyde, glyoxal).[109,111] Scant attention was paid this technique until interest was revived by a 1970 investigation by Warren and Babcock.[112] Prototype detectors based on Nederbragt's concept were constructed and evaluated by EPA and were shown to have more than adequate sensitivity and specificity for ambient O_3 measurements.[109] In addition, the simplicity of the detector implied that a low cost monitor could be made. The Nederbragt method has been so successful in laboratory and field applications that it has been designated as the reference method for the routine O_3 measurements required by recent Federal air quality standards.[113] Several instrument companies now offer commercial versions. Good agreement has been obtained between Regener and Nederbragt O_3 monitors in extensive field studies.[108,114-116] In such field tests atmospheric data are collected with a variety of prototype and commercial units to determine reliability of instrument performance (e.g. downtime, maintenance requirements) and stability of instrument operating parameters (e.g. zero drift, span drift) and to compare data from different types of instruments.

The gas-phase chemiluminescent reaction between O_3 and NO may be used for the detection of either component when the other is present in excess.[117] For O_3 measurement the characteristics are essentially equivalent to those of the Nederbragt detector.[109] The reaction has been used almost exclusively for the measurement of NO and NO_x as discussed below.

B. Nitrogen Oxides, Ammonia and Amines

Although oxides of nitrogen (NO_x) play an important role in atmospheric pollution, methods of measuring NO_x were until recently the least satisfactory among various atmospheric monitoring techniques. Chemiluminescent reactions of NO_x have been adapted to fill this gap in measurement technology. The chemiluminescence system most frequently used is:

$$NO_2 \xrightarrow{\text{Energy}} NO + 1/2\ O_2 \tag{25}$$

$$NO + O_3 \rightarrow NO_2 + O_2 + h\nu\ (\lambda \geq 600\ nm) \tag{26}$$

Reaction (26) is a red-shifted ($\lambda \geq 600$ nm) modification of the air afterglow reaction (4).

Applications of gas-phase chemiluminescent reactions for detection of oxides of nitrogen were first discussed and experimentally demonstrated by Fontijn et al.[117] at AeroChem under an EPA contract. They subsequently constructed a prototype chemiluminescence NO monitor based on the NO–O_3 reaction.[118] Sample air containing NO mixes with excess O_3 (0.5% in O_2 from an internal O_3 source) in a reactor cell, which is maintained at a total pressure of approximately 2 Torr with a small mechanical pump. A thermoelectrically-cooled, infrared sensitive photomultiplier (e.g. EMI-9558A) and a filter cutting off radiation below 600 nm are closely coupled to the reactor vessel. The limit of sensitivity is approximately 0.002 ppm and the linear range of response extends up to 1000 ppm. Fontijn's prototype instrument, which is typical of many of the later commercial models, has been evaluated and compared with other NO_x instruments in extended field studies.[114-116]

The application of the chemiluminescent NO/O_3 reaction has occupied several other investigators.[119-121] Prototype chemiluminescence NO monitors suitable for source and ambient measurements were developed independently at Ford Research Laboratories.[119,120] The chemiluminescence NO monitor has in fact proved to be an ideal method for measuring NO_x in automotive exhausts.[121,122] This technique is now specified by EPA as the test procedure for NO_x analytical measurements in determing whether vehicular emissions meet published standards.[123] Recently a chemiluminescence NO/O_3 detector which operates at atmospheric pressure was demonstrated.[124] Sensitivity for ambient concentrations (0.001-1 ppm) was achieved by changing reactor geometry and sample flow conditions. This development has led to more compact and less costly detectors.

The means for carrying out Reaction (25) have occupied several investigators.[121,124,125] The initial studies were by Sigsby et al.,[121] who used a stainless steel tube heated to temperatures greater than 900 K and observed quantitative conversion of NO_2 to NO. Thermal dissociation cannot account for this quantitative conversion and some contribution must be attributed to reduction of NO_2 at the hot metal surface. NH_3 can be oxidized to NO at high temperatures and is a potential interference since it is present in the atmosphere and may be present in certain sources. Acidic scrubbers which quantitatively remove NH_3 and pass NO_2 have been used.[121,124] Breitenbach and Shelef[125] have described a number of carbon impregnated metals which may be operated at two different temperatures for the conversion of NO_2 or NO_2 + NH_3 and recommended a carbon-molybdenum composite heated to 750 K. Application of such converters has not yet been reported.

When a converter is used NO_2 concentrations can be determined by difference from the NO_x and NO measurements. For the direct measurement of NO_2 a "photofragment" technique has been used. In this detector NO_2 is photolyzed and the resulting O atoms are measured via Reaction (4), by addition of excess NO.[126] For concentrations below 1 ppm (ambient air concentrations) the response is linear, with a lower limit of detection of 1 ppb.

NH_3 is a constituent of the atmosphere which is produced by natural processes and by man's activities. Measurements of atmosperic NH_3 are of interest to determine its chemical fate and to assess the potential role of NH_3 in photochemical air pollution. A pyrolytic converter and a phosphoric acid pre-scrubber have been used by Hodgeson et al.[124] to measure non-urban NH_3 concentrations from 0.001 to 0.01 ppm via $\{ [NH_3] + [NO_x] \} - [NO_x]$.

The use of the O/NO reaction (4) can be advantageous in applications such as most source emissions where only a total NO_x measurement is required. Since the O/NO emission falls partly in the visible region, less expensive photomultipliers may be used. The original work here was performed by Snyder and Wooten[127] of Monsanto under an EPA contract. Equivalent responses were obtained for NO and NO_2 and the detector sensitivity was found to be a few parts per billion. However, the electrical discharge O-atom sources used yielded a fluctuating background O/NO signal and O concentration. These problems have hindered application of this reaction. Work on a thermal O source, which could be more useful, has been reported.[128]

Reaction (13) also appears practical for NO_x measurements. Although no ambient temperature applications of this reaction have been reported, Krost et al.[129] have observed the characteristic H/NO chemiluminescence in the secondary combustion zone of an H_2-rich flame, into which nitrogen compounds (NO, NO_2, NH_3, organic amines) were introduced. By incorporating a near infrared interference filter (690 nm) between the flame chemiluminescent zone and a photomultiplier, a detector for gas-phase nitrogen compounds was developed. The chemiluminescence intensity has been found to be directly proportional to the total concentration of nitrogen compounds, and the lower limit of sensitivity was established as approximately 0.1 ppm.

C. Sulfur Compounds

Atmospheric sulfur-containing pollutants include SO_2, H_2S, organic sulfides and mercaptans, sulfuric acid and sulfates. Of these SO_2 has been the major pollutant of concern in atmospheric measurements. H_2S and organic sulfur comprise the class of malodorous sulfur compounds in the localized pollution associated with the Kraft paper industry. Sulfuric acid and sulfates occur predominantly in the particulate form.

The only chemiluminescence approach which has been applied to date is a flame method, which is applicable to the detection of gas-phase sulfur compounds as a class. When sulfur compounds are burned in an H_2 rich flame, a strong blue chemiluminescence is emitted. The emitting molecule is $S_2(B\ ^3\Sigma_u^-)$, formed via S-atom recombination either directly (Reaction 24) or indirectly (energy transfer).[130,131] Since two atoms are required, the intensity of the chemiluminescence is proportional to the square of the concentration of sulfur compound in the flame (for compounds containing only one sulfur atom). The original flame photometric detector (FPD) for sulfur compounds in air was revealed in a patent by Draeger.[132] A modified version by Brody and Chaney[133] has been incorporated in a commercial sulfur monitor which has found extensive application in atmospheric monitoring. The calibration and application of this monitor for detecting atmospheric concentrations of SO_2 has been discussed by Stevens et al.[134] The typical FPD uses as fuel a mixture of 200 cc/min of air and 200 cc/min H_2. An inert diluent gas (N_2, He) can be used to lower the flame temperature and the flame emission background (compare Section II.B). Since air already contains N_2, additional inert gas is not added to the atmospheric flame detector. The primary combustion zone is recessed in a barrel which functions as a light shield. The photomultiplier views the (cool) secondary combustion zone through an interference filter which transmits only the strong band at 394 nm. Although the emission is detected above the flame, the background signal obtained is predominantly due to flame background and is an order of magnitude greater than the photomultiplier dark current.[133] Thus the detector performance and sensitivity are determined by the magnitude of the drift and noise associated with the flame background emission. These are in turn strongly affected by flame temperature and gas flow rates. The flame noise for a typical flame photometric detector is equivalent to an SO_2 concentration of approximately 0.005 ppm. Efforts are underway to improve the performance of the FPD by correlation techniques.[135,136] The correlation technique used involves viewing and comparing one or more S_2 emission peaks and adjacent background wavelengths.

Considerable effort has been expended to develop a flame photometric based system which is specific for individual sulfur compounds. The most successful approach for atmospheric monitoring is the gas chromatographic (GC)-system developed by Stevens et al.[137] for the quantitative elution of the sub-ppm concentrations of sulfur compounds found in the atmosphere (SO_2, H_2S, simple mercaptans and organic sulfides). Commercial instruments based on the GC-FPD combination are now available.

Both FPD and GC-FPD instruments have been extensively evaluated in field studies.[108,114,115] One of the most important observations of these studies was the close agreement obtained between the total sulfur measurement obtained with the FPD and the specific measurement of SO_2 by the GC-FPD. These results confirm those of Stevens et al,[137] who concluded that for most applications SO_2 accounts for 90% or more of the total gaseous S.

Ambient temperature chemiluminescent reactions have been suggested for measurement of SO_2 and other sulfur compounds, but applications have not yet been reported. In addition to their work

on NO_x detection, Snyder and Wooten[127] investigated the use of
O-atom chemiluminescence for the detection of SO_2. Reaction of SO_2
and C atoms produces SO_2 chemiluminescence in the ultraviolet
via[138]

$$O + O + SO_2 \rightarrow O_2 + SO_2 + h\nu \quad (\lambda_{max} = 280 \text{ nm}) \quad (27)$$

The sensitivity obtained for SO_2 detection using this chemiluminescence was 0.001 ppm, which is quite adequate for atmospheric detection. Since potential interfering molecules are known to produce chemiluminescence with O atoms in the ultraviolet, an interference filter is required to select a region of the SO_2 chemiluminescence spectrum. As is the case with NO_x measurements, problems with O-atom sources have hindered development of a chemiluminescence SO_2 detector. The use of thermal O-atom sources may stimulate practical developments.

Finally, Kummer, Pitts and Steer[139] have observed chemiluminescence from the gas-phase reactions of O_3 with a variety of sulfur compounds, including H_2S and organic sulfides, and suggested use of this chemiluminescence for measuring atmospheric sulfur. The emission observed from both H_2S and organic sulfides was due to electronically excited SO_2. Therefore the use of O_3 as a second reactant gas should lead to a non-specific detector for gaseous sulfur compounds. Moreover, since the reaction rate of O_3 with different sulfur compounds varies considerably, the sentivity of such a detector would depend upon the compounds measured.

D. Carbon Monoxide and Phosphorus, Boron, Chlorine Compounds

The reaction between O and CO produces a weak chemiluminescence in the 300-500 nm range.[140] Comparison[117] to the O_3/NO reaction suggests a predicted limit of sensitivity of 1 ppb. A preliminary study[127] was not successful but further investigation appears warranted.

In addition to measuring sulfur and nitrogen compounds, flame chemiluminescence methods have been used for the detection of phosphorus,[132,133] boron[141] and halogen[142] compounds in air. HPO emission is observed in the 540 nm region when phosphorus compounds are burned; the sensitivity is in the sub-ppm range. Such detectors have been applied to the detection of phosphorus-containing pesticides. Boron compounds (used as catalysts in industrial applications and as high quality rocket fuels) are detected via BO_2 emission near 550 nm;[141] this FPD method is sensitive to concentrations of less than 0.1 ppm in air. Gilbert[142] has developed a sensitive flame emission method for chlorine compounds, in which indium metal is placed in the flame barrel. The presence of a small amount of indium vapor in the secondary combustion zone results in an intense chemiluminescent emission from InCl whenever chlorinated compounds are introduced into the flame. Application of this technique to atmosperic analysis has not been reported.

IV. UPPER ATMOSPHERE APPLICATIONS

Chemiluminescent reactions in the upper atmosphere have been chiefly observed at heights above 90 km. The pressure at 90 km is about 10^{-3} Torr and falls to about 5×10^{-7} Torr at 200 km. Here most of the solar ultraviolet radiation is absorbed, dissociating molecular oxygen into atomic oxygen. Since there are few three-body collisions, O does not appreciably recombine with O_2 to form O_3, as is the case at lower altitudes. At about 105 km, the number density of O equals that of O_2; at about 150 km, $[O] = [N_2]$. At 200 km, the atmosphere is composed of 70% O, 29% N_2, 0.2% O_2 and small quantities of He, Ar and H_2.[143] Atomic O plays the major role in producing chemiluminescence, especially when electronic excitation is involved.

A. Artificial Chemiluminescence

Luminescent clouds are produced by releasing from a rocket-borne container a reactive chemical that interacts with an atmospheric constituent to produce chemiluminescence. Such clouds are confined to the 90-200 km region of the upper atmosphere. Below 90 km not only the absence of O, but also the rapid condensation of the released chemicals prevents the observation of chemiluminescence, although it is possible to produce solar resonance excitation of atomic Na and Li as low as 80 km, and one can observe solar ray scattering from smoke trails at still lower altitudes. Above 200 km the released chemicals diffuse too rapidly to produce observable chemiluminescence. However, resonance fluorescence† of Na, released from a Soviet Sputnik [144] has been observed at 156,000 km altitude; and of Ba ions, released in a joint US-German rocket experiment, at 32,000 km.[145]

The artificial glow clouds or trails are photographed against a star background in the night or twilight sky. Time lapse photography allows measurement of drift velocity; thus, the horizontal and vertical components of the wind vectors can be established.[146] Up to about 110 km the clouds develop turbulent eddies and motion; above this altitude the clouds usually grow by molecular diffusion. Hence, eddy and molecular diffusion coefficients can be obtained by measuring the rate of growth of chemiluminescent clouds.[147,148] Atmospheric density can be derived from the molecular diffusion rates (see e.g. Ref. 149). Atomic O concentrations can be deduced from the measurement of light intensity of relatively simple chemiluminescent processes, such as the NO/O glow (Reaction 4).[150] Other applications of artificial chemiluminescence include the simulation of luminescent phenomena associated with missile exhaust gases and natural airglow

---

† This paper deals with chemiluminescence; therefore the numerous rocket experiments in which solar excitation of atomic resonance lines or molecular fluorescence is observed will be mentioned only in passing.

processes. In the following, we shall discuss the release techniques, reaction mechanisms and data interpretation of those chemiluminescent release agents found most useful in upper atmospheric studies. Doubtlessly, the ingenious atmospheric chemist will generate further ideas for suitable release chemicals and applications of chemiluminescent reactions for investigating this important region of the atmosphere.

1. <u>Nitric Oxide</u>. The air afterglow reaction was an obvious first choice for generating artificial chemiluminescence at night in the upper atmosphere. The first successful release of NO occurred in 1956.[151] This was a "point" release in which the NO was vented instantaneously by rupturing the container with a high explosive at 106 km altitude. An intense glow immediately appeared at the release point, growing in size and diminishing in intensity until 10 min after release when it was no longer observable. The second series of NO releases occurred in 1962.[152] The gas was released from a high pressure container through two sideward pointing orifices by opening a squib valve at 90 km altitude till the gas supply was exhausted at about 150 km. Here an unexpected effect was noticed. A bright annular headglow surrounded the rocket as it traversed the sky, followed by a much dimmer afterglow. The afterglow was observed for only about 100 sec after rocket passage in the 90 to 120 km region where it followed the ambient wind and turbulent motions. Both the headglow and afterglow showed a spectrum similar to the laboratory NO/O reaction.[19] Analysis of the headglow brightness revealed that with reasonable values for atomic O densities the volumetric emission rate exceeded by 3-4 orders of magnitude the expected emission rate from the normal three-body reaction (4).

Fontijn and Rosner have plausibly explained the anomalous intensity.[153] During the adiabatic expansion, the temperature of the NO drops rapidly and the saturation curve is passed. Therefore, it is expected that part of the gas forms molecular clusters and possibly, condensed particles. Before entering the mixing zone, the NO passes through a shock wave in which the temperature of the gas is nearly restored to the reservoir value. It is probable, however, that some of the clusters survive the shock wave. The reaction then can be described as follows:

$$(NO)_n + O \rightarrow NO_2^* + (NO)_{n-1} \qquad (28)$$
$$NO_2^* \rightarrow NO_2 + h\nu \qquad (29)$$

Here, the third body required in the normal three-body reaction (4) is carried along with the NO, so that basically every collision with atomic O could yield a stabilized NO_2^*. Therefore, NO clusters could account for the anomalously high photon emission rate provided there is sufficient concentration of clusters in the mixing zone. Subsequently, wind tunnel simulation experiments by Golomb and Good[154] demonstrated that such clusters (of which the dimer is the simplest form) indeed exist in expanding NO jets and that there is a direct relationship between cluster abundance in the free jet and photon emission rate in the headglow. Recently, it has also been shown that when the NO reservoir pressure, p_0, multiplied by the release orifice diameter, d, exceeds a value of 100 Torr cm, the headglow intensity becomes independent of the NO flow rate and the total pressure in the mixing zone.[150] The brightness remains dependent, however, on the atomic O flux impinging onto the bow shock:

$$B = q \dot{n}_0 \text{ photons cm}^{-2} \text{ sec}^{-1} \qquad (30)$$

where $\dot{n}_0$ is the O flux (free stream O density multiplied by the stream velocity), and q is the efficiency factor by which photons are emitted per impinging O atom. This efficiency was determined to be $q = 0.005$ with an uncertainty factor of 2.

Once the anomalous headglow intensity was explained, and the relevant efficiency factor determined in wind-tunnel calibrations, the NO release technique became a useful tool for upper atmosphere [O] determination.[150] In the rocket release, as in the wind-tunnel calibrations, the photometer is situated near the release orifice, viewing the headglow from its center. The measured brightness, B, is telemetered to the ground. From known vehicle velocity (which equals the stream velocity), ambient O densities are readily determined via Eq. (30). Current results are in good agreement with those derived from mass spectrometry.[155]

2. **Trimethyl Aluminum (TMA)**. Woodbridge observed that aluminized grenades produce a long-lasting luminescence when exploded in the 100-125 km altitude region.[156] Rosenberg and Golomb proved that an aluminum compound must be involved.[157] They detonated two charges of high explosive at 114 km altitude, one containing Al powder as an additive, the other none. Only the aluminized charge produced luminescence. Rosenberg et al. developed a more convenient method to release an aluminized compound in the upper atmosphere—TMA.[158] TMA is a pyrophoric liquid and its handling and loading require special care. Typically, the liquid is dispensed from a pressurized container along the rocket trajectory, starting at about 90 km. A part of the liquid flash-vaporizes upon release, the remainder freezes. The frozen particles follow the ballistic trajectory and evaporate upon re-entry into the denser regions of the atmosphere at 100-90 km. The flash-vaporized molecules react with some constituent of the atmosphere, presumably atomic O, since the glow is most intense where atomic O is most abundant. TMA produces no headglow, only an afterglow which takes several seconds to develop. The afterglow spectrum appears to be a continuum in the visible.[159] This glowing trail follows the ambient wind motions. It grows radially by eddy or molecular diffusion, respectively.

The following sequence of reactions appears most probable:

$$\text{TMA} + \text{O} \rightarrow \text{AlO} + \text{Radicals} \qquad (31)$$

$$AlO + O + M \rightarrow AlO_2 + M + h\nu \text{ (chemiluminescence)} \qquad (32)$$

$$AlO_2 + O \rightarrow AlO + O_2 \text{ (recycling)} \qquad (33)$$

$$AlO + O \rightarrow Al + O_2 \text{ (destruction)} \qquad (34)$$

This scheme is in accord with the following observations. In the TMA clouds, about 0.3 to 0.6 photons are emitted per vaporized TMA molecule.[158] The high photon yield implies a cyclic process such as provided by Reaction (33) in which the intermediary molecule is regenerated. Fluorescence of the blue-green AlO bands ($B^2\Sigma^+ - X^2\Sigma^+$) has been observed in twilight releases of TMA;[159] thus, the presence of AlO is established. From the fluorescence spectral intensity distribution atmospheric temperatures are now routinely deduced (see e.g. Ref. 160). The bond energy of AlO is about 5.1 eV,[161] about equal to that of O_2, explaining the fact that AlO is not readily reduced to Al by atomic O. Most metal monoxides have bond energies smaller than D_o O_2 and as a result do not persist in the O-rich region of the atmosphere, cf. e.g. FeO below.

3. <u>Acetylene</u>. C_2H_2 has been released to simulate the luminous and radio-frequency effects observed following the passage through the upper atmosphere of hydrocarbon/liquid oxygen burning missiles.[157] C_2H_2 probably comes closest to the hydrocarbon fragments and radicals found in missile exhausts. These fragments are believed to be responsible for the CH and C_2 band emission in missile plumes.[162] The highly energetic radicals may also enter into chemi-ionization reactions, causing the enhanced RF reflectivities of missile plumes at E-layer altitudes. The glow requires one to several seconds after release to develop. The spectrum is similar to that observed in the laboratory[81] and shows[163] an apparent weak continuum upon which are superimposed the CH 431.5 nm band (strong), the CH 389.0 nm band (weak), and the C_2 516.5 nm band (weak). In addition there is a pronounced emission at about 577.0 nm $[O(^1S) \rightarrow O(^1D)?]$. Laboratory studies of the C_2H_2/O luminescence indicate that the CH bands become relatively more intense than the C_2 bands as pressure decreases.[164] The same effect has been observed in the upper atmosphere.

4. <u>Iron Pentacarbonyl</u>. $Fe(CO)_5$ has been released both at night and in twilight, in the 100-150 km region.[165] In twilight, when the sun still illuminates the ascending rocket, the fluorescence of the FeO orange bands is clearly observed in the headglow surrounding the rocket. The headglow is of short persistence and evolves into an afterglow of different spectral characteristics. In a subsequent, as yet unpublished, experiment it has been proven that the twilight afterglow consists of resonance scattering by FeI atoms. At night, $Fe(CO)_5$ produces a bright glow streaking through the sky, with no afterglow at all. Because of the reddish tint of the headglow, it is probable that some FeO molecules are produced in the emitting state, leading to chemiluminescence. An important observation is that in twilight the FeO fluorescence quickly disappears, indicating that

this molecule is consumed in a fast reaction such as $FeO + O \rightarrow Fe + O_2$. This reaction is exothermic by ≈ 0.9 eV, and laboratory experiments indicate that it is very fast.[166]

5. Other Chemical Releases. Several chemicals cause weak chemiluminescence in the upper atmosphere which is insufficient to obtain good wind, diffusion or composition measurements; however, the resulting glows might be interesting for the elucidation of the reaction mechanisms involved. These chemicals are CS_2, NO_2 and $Pb(CH_3)_4$. CS_2 shows intense SO_2 afterglow emission in the laboratory,[167] but in the upper atmosphere it is barely photographable.[157] This is attributable to the two steps necessary to produce the glow, $CS_2 + O \rightarrow CS + SO$, followed by Reaction (11). Apparently the steady-state concentration of SO following the release of CS_2 is insufficient for high photon fluxes. Similarly, NO_2 requires a two-step mechanism (Reactions 5 and 4) to produce light. However, not all multistep reactions lead to weak emission in the atmosphere as demonstrated by the TMA and C_2H_2 reactions. In the case of TMA, the strong chemiluminescence can be attributed to recycling of AlO; for C_2H_2 the actual emission steps are mainly[81] atom stripping and exchange reactions which tend to be much faster[168] than association reactions, such as (4). The weak chemiluminescence produced[169] by the release of $Pb(CH_3)_4$ may be attributed to a reaction such as $PbO + O \xrightarrow{M} PbO_2 + h\nu$.

B. Natural Chemiluminescence (Airglow)

The many chemically highly active species (atoms, ions, free radicals and excited species) present in the undisturbed upper atmosphere give rise to a weak luminescent phenomenon, the airglow. The airglow can be observed from the ground, away from the interference of moon and star light and aurorae, as a diffuse omnipresent glow. Rocket and satellite measurements have considerably increased our knowledge of these glows and their origins. Chemiluminescent reactions are by far the dominant process of the night glow; in day glows chemiluminescence is much less important than radiative processes produced by solar irradiation. For general reviews of airglows and the excitation mechanisms involved, see e.g. Refs. 170-174. Known emission features which are thought to be mainly or in part due to chemiluminescence are[170a] the $O_2(A\ ^3\Sigma_u^+ - X^3\Sigma_g^-)$ Herzberg bands and $(b\ ^1\Sigma_g^+ - X^3\Sigma_g^-)$ atmospheric bands due to Reaction (17), the NO_2 continuum due to the O/NO reaction (4), the $O(^1S-^1D)$ 557.7 nm green line, the $Na(^2P-^2S)$ 589.3 nm D line and the $OH(v' \leq 9)$ vibration-rotation Meinel bands.

Some of the airglow measurement methods used or proposed are very similar to those employed in conjunction with the more intense artificial glows. A major difference from the latter is that in the natural glow one no longer has one known reactant concentration (i.e. that of the release agent). However observations of natural glows can be far more extensive in time and space which constitutes an important advantage.

Application to spatial isolation of the emitting layers and atmospheric dynamics is similar to the artificial glows and has been discussed (Sears,[170b] see also Silverman[173]). Temperature can be derived from several emission features. The Meinel OH bands are due to the reaction

$$H + O_3 \rightarrow OH(v \leq 9) + O_2 \qquad (35)$$

Rotational equilibration requires only a few collisions and the rotational temperature of the Meinel bands is used as a measure of ambient temperature below about 90 km where collisions are sufficiently frequent.[170c,171a] The $O(^1S)$ state has a radiative lifetime of 0.74 sec, orders of magnitude higher than the collision time at around 100 km, where a maximum in the intensity of the $O(^1S-^1D)$ emission occurs, and the emitter at that altitude may therefore be considered to be thermalized; the Doppler line profile of this line is used as a measure of temperature.[171a] The $O(^1S)$ formation near 100 km is due to a reaction involving three $O(^3P)$ atoms (possibly via intermediate formation of an excited O_2 molecule). However, a second O-green line intensity maximum is observed near 180 km to which the reaction

$$O_2^+ + e^- \rightarrow O(^1S) + O(^3P) \qquad (36)$$

contributes in a major way. Reaction (36) is 2.8 eV exothermic; since the only products are atomic species this energy must go into translational energy and as a result the emission profile is nonthermal at this higher altitude.[171a] Dandekar and Turtle[175] have measured the intensity of the oxygen green line with a rocket-borne photometer to determine the distribution of $O(^3P)$ as a function of altitude near 100 km. The results are in reasonable agreement with those measured by other means (mass spectrometry and NO releases).

In closing, it must be remarked that in view of the highly complex environment in which the natural airglows occur, measurements of these glows should be considered as very useful adjuncts to other upper atmospheric measurements rather than as supplying uniquely correct data. Agreement of results from various methods and their use in model atmospheres is leading to an ever more accurate knowledge of the upper atmosphere. It is likely that chemiluminescence is also going to play an important role in studying the upper atmosphere of other planets, once night glow observations of those environments are made.

V. EPILOGUE

It may be concluded from this article that measurement techniques based on gas-phase chemiluminescence have found widespread use in the laboratory for reaction kinetics, quantitative analysis and gas dynamic studies. While there are several gaseous environments of major technological interest, applications of basic knowledge and laboratory techniques have thus far been mainly restricted to upper atmospheric studies and air pollutant monitoring. Recent rapid developments and acceptance of chemiluminescence techniques in this latter area suggest that similar developments may be possible elsewhere. Direct observations of practical combustion sources and monitoring and control of industrial process streams are examples of areas for which chemiluminescence techniques would appear to hold promise. A beginning along these lines is the study of NO formation in internal combustion engines via observation of the O/NO and O/CO continua by Lavoie, Heywood and Keck.[176,177] The advantages of chemiluminescence for monitoring process streams were pointed out a number of years ago[178,179] but no actual work along these lines has come to our attention, suggesting that this is still largely virgin territory.

Acknowledgment. We have benefited from the comments and information supplied by a number of people. Particularly, Drs. K.L. Wray, E.M. Bulewicz, M. Steinberg, J.C. Keck, R.A. Young, W.J. Miller and H.S. Pergament have been most helpful. Thanks are also due to Ms. Helen Rothschild and Ms. Evangeline Stokes for the careful editing and typing of the manuscript, respectively.

REFERENCES[†]

1. G. Agricola, *De Re Metallica* (First Latin Edition 1556; translated into English by H.C. and L.H. Hoover), Dover Publications, New York, 1950, Book VII.
2. F. Kaufman, Progr. React. Kin. **1**, 1 (1961).
3. B.A. Thrush, Science **156**, 470 (1967).
4. A. Fontijn, Progr. React. Kin. **6**, 75 (1972).
5. F. Kaufman, Proc. Roy. Soc. **A247**, 123 (1958).
6. J.T. Herron and H.I. Schiff, Can. J. Chem. **36**, 1159 (1958).
7. A.A. Westenberg and N. deHaas, J. Chem. Phys. **40**, 3087 (1964).
8. F.A. Morse and F. Kaufman, J. Chem. Phys. **42**, 1785 (1965).
9. F. Kaufman and J.R. Kelso, J. Chem. Phys. **46**, 4541 (1967).
10. M.A.A. Clyne, C.J. Halstead and B.A. Thrush, Proc. Roy. Soc. **A295**, 355 (1966).

[†] References for which an NTIS number has been cited may be obtained from Dept. A, National Technical Information Service, Springfield, Va. 22151.

11. C.J. Halstead and B.A. Thrush, Proc. Roy. Soc. A295, 363 (1966).
12. L. Elias, J. Chem. Phys. 38, 989 (1963).
13. L. Elias and H.I. Schiff, Can. J. Chem. 38, 1657 (1960).
14. J.T. Herron and R.E. Huie, Progr. React. Kin. (in press).
15. P.B. Davies and B.A. Thrush, Trans. Faraday Soc. 64, 1836 (1968).
16. P.A. Leighton, Photochemistry of Air Pollution, Academic Press, New York, 1961, Chap. 8.
17. F. Stuhl and H. Niki, Chem. Phys. Lett. 7, 197 (1970).
18. R.A. Young, G. Black and T.G. Slanger, J. Chem. Phys. 49, 4758 (1968).
19. A. Fontijn, C.B. Meyer and H.I. Schiff, J. Chem. Phys. 40, 64 (1964); A. Fontijn and H.I. Schiff in Chemical Reactions in the Lower and Upper Atmosphere, Interscience, New York, 1961. p. 239.
20. A. Fontijn and J. Lee, J. Opt. Soc. Am. 62, 1095 (1972).
21. M. Vanpee, K.D. Hill and R. Kineyko, AIAA J. 9, 135 (1971).
22. D.E. Paulsen, W.F. Sheridan and R.E. Huffman, J. Chem. Phys. 53, 647 (1970).
23. A.T. Stair and J.P. Kennealy, J. Chim. Phys. (Paris) 64, 124 (1967).
24. A. Sharma, J.P. Padur and P. Warneck, J. Chem. Phys. 43, 2155 (1965); J. Phys. Chem. 71, 1602 (1967).
25. F. Kaufman, "The Air Afterglow Revisited," this symposium.
26. M. Mandelman, T. Carrington, R.A. Young, York U., Toronto, to be submitted.
27. M.A.A. Clyne and B.A. Thrush, Proc. Roy. Soc. A269, 404 (1962).
28. F. Kaufman and J.R. Kelso, in Preprints of Papers, Symposium on Chemiluminescence, Durham, 31 Mar.-2 Apr., 1965, p. 65.
29. R.P. Hartunian, W.P. Thompson and E.W. Hewitt, J. Chem. Phys. 44, 1765 (1966).
30. M.A.A. Clyne and B.A. Thrush, Disc. Faraday Soc. 33, 139 (1962).
31. L. Elias, J. Chem. Phys. 44, 3810 (1966).
32. J.E. Morgan, L. Elias and H.I. Schiff, J. Chem. Phys. 33, 930 (1960).
33. P. Harteck, R.R. Reeves and G. Mannella, J. Chem. Phys. 29, 608 (1958).
34. I.M. Campbell and B.A. Thrush, Proc. Roy. Soc. A296, 222 (1967).
35. R.A. Young and R.L. Sharpless, Disc. Faraday Soc. 33, 228 (1962).
36. R.A. Young and R.L. Sharpless, J. Chem. Phys. 39, 1071 (1963).
37. R.A. Young and G. Black, J. Chem. Phys. 44, 3741 (1966).
38. R.J. McNeal and S.C. Durana, J. Chem. Phys. 51, 2955 (1969).
39. R.A. Young and R.L. Sharpless, J. Geophys. Res. 67, 3871 (1962).
40. R.L. Brown, J. Chem. Phys. 52, 4604 (1970).

41. W. Brennen and R.L. Brown, J. Chem. Phys. 52, 4910 (1970).
42. W. Brennen and E.C. Shane, Chem. Phys. Lett. 2, 143 (1968).
43. E.C. Shane and W.Brennen, Chem. Phys. Lett. 4, 31 (1969).
44. A.N. Wright and C.A. Winkler, Active Nitrogen, Academic Press, New York, 1968.
45. R.W.F. Gross, J. Chem. Phys. 48, 1302 (1968).
46. R.W.F. Gross, and N. Cohen, J. Chem. Phys. 48, 2582 (1968).
47. T. Carrington and D. Garvin, Comprehensive Chemical Kinetics, Vol., 3, Formation and Decay of Excited Species, C.H. Bamford and C.F.H. Tipper, Eds., Elsevier, Amsterdam, 1969, Chap. 3.
48. B.A. Thrush, Ann. Rev. Phys. Chem. 19, 371 (1968).
49. L.F. Phillips, Can. J. Chem. 46, 1450 (1968).
50. A. Fontijn and R. Ellison, J. Phys. Chem. 72, 3701 (1968).
51. I.M. Campbell, S.B. Neal, M.F. Golde and B.A. Thrush, Chem. Phys. Lett. 8, 612 (1971).
52. J.A. Meyer, D.W. Setser and W.G. Clarke, J. Phys. Chem. 76, 1 (1972).
53. G.W. Taylor and D.W. Setser, Chem. Phys. Lett. 8, 51 (1971); J. Am. Chem. Soc. 93, 4930 (1971).
54. K.H. Becker and K.D. Bayes, J. Chem. Phys. 48, 653 (1968).
55. E.L. Milne, M. Steinberg and H.P. Broida, J. Chem. Phys. 42, 2615 (1965).
56. A.G. Gaydon, Dissociation Energies and Spectra of Diatomic Molecules, 3rd Edition, Chapman and Hall, London, 1968. a. Chaps. 3-6; b. Chap. 7.
57. C.D. Jonah, R.N. Zare and Ch. Ottinger, J. Chem. Phys. 55, 263 (1972).
58. R.A. Young, J. Chem. Phys. 33, 1044 (1960); 34, 1295 (1961).
59. R.E. Lund and H.J. Oskam, J. Chem. Phys. 48, 109 (1968).
60. K. Loomis, A. Bergendahl, R.R. Reeves, Jr. and P. Harteck, J. Am. Chem. Soc. 91, 7709 (1969).
61. J.E. Morgan and H.I. Schiff, J. Chem. Phys. 38, 2631 (1963).
62. J.E. Morgan and H.I. Schiff, Can. J. Chem. 42, 2300 (1964).
63. B. Khouw, J.E. Morgan and H.I. Schiff, J. Chem. Phys. 50, 66 (1969).
64. K.M. Sancier, W.J. Fredericks, J.L. Hatchett and H. Wise, J. Chem. Phys. 37, 860 (1962).
65. K.M. Sancier, D.J. Schott and H. Wise, J. Chem. Phys. 42, 1233 (1965).
66. K.M. Sancier, J. Chem. Phys. 42, 1240 (1965).
67. A.G. Gaydon and H.G. Wolfhard, Flames, Their Structure, Radiation and Temperature, 3rd Edition, Chapman and Hall, London, 1970.
68. J.D. Winefordner, Ed., Spectrochemical Methods of Analysis, Wiley-Interscience, New York, 1971. a. C. Th.J. Alkemade and P.J.Th. Zeegers, Chap. 1; b. J. Ramirez-Munoz, Chap. 2.
69. T.M. Sugden, Ann. Rev. Phys. Chem. 13, 369 (1962).
70. R. Mavrodineanu, Ed., Analytical Flame Spectroscopy. Selected Topics, Macmillan, London, 1970. a. C.Th.J. Alkemade, Chap. 1; b. P.T. Gilbert, Chap. 5.

71. J.A. Dean and T.C. Rains, Eds., Flame Emission and Atomic Absorption Spectrometry, Vol. 1. Theory; Vol. 2. Components and Techniques, Marcel Dekker, New York, 1969, 1971.
72. R. Mavrodineanu and H. Boiteux, Flame Spectroscopy, Wiley, New York, 1965. a. p. 553; b. p. 549.
73. A.G. Gaydon, The Spectroscopy of Flames, Wiley, New York, 1957, Chap. 7.
74. T.M. Sugden, E.M. Bulewicz and A. Demerdache in Chemical Reactions in the Lower and Upper Atmosphere, Interscience, New York, 1961, p. 89.
75. E.M. Bulewicz in Twelfth Symposium (International) on Combustion, The Combustion Institute, Pittsburgh, 1969, p. 957.
76. P.J. Padley and T.M. Sugden, Proc. Roy. Soc. A248, 248 (1958).
77. E.M. Bulewicz and T.M. Sugden, Trans. Faraday Soc. 52, 1475 (1956).
78. J.L. Rosenfeld, PhD. Thesis, Cambridge, 1961.
79. R. Carabetta and W. Kaskan in Eleventh Symposium (International) on Combustion, The Combustion Institute, Pittsburgh, 1967, p. 321.
80. P.T. Gilbert in Proc. Xth Colloq. Spectrosc. Internat., E.R. Lippincott and M. Margoshes, Eds., Spartan Books, Washington, 1963, p. 171.
81. No recent comprehensive review of this subject exists. For pertinent discussions see e.g., the Biannual International Combustion Symposium volumes, The Combustion Institute, Pittsburgh and Refs. 72b and 73.
82. R.W. Reid and T.M. Sugden, Disc. Faraday Soc. 33, 213 (1962).
83. R.W.B. Pearse and A.G. Gaydon, The Identification of Molecular Spectra, 3rd Edition, Chapman and Hall, London, 1963.
84. R.M. Fristrom and A.A. Westenberg, Flame Structure, McGraw-Hill, 1965, p. 215.
85. A. Fontijn, D.E. Rosner and S.C. Kurzius, Can. J. Chem. 42, 2440 (1964).
86. B.F. Myers and E.R. Bartle, J. Chem. Phys. 48, 3935 (1968).
87. R.D. Sharma and K.L. Wray, J. Chem. Phys. 54, 4578 (1971).
88. K.L. Wray and E.V. Feldman in Fourteenth Symposium (International) on Combustion, The Combustion Institute, Pittsburgh, in press.
89. H.B. Palmer, J. Chem. Phys. 47, 2116 (1967).
90. K.L. Wray and S.S. Fried, J. Q. S. R. T. 11, 1171 (1971).
91. J.F. Bott and T.A. Jacobs, J. Chem. Phys. 52, 3545 (1970).
92. R.A. Carabetta and H.B. Palmer, J. Chem. Phys. 46, 1325 (1967).
93. R.A. Carabetta and H.B. Palmer, J. Chem. Phys. 46, 1333 (1967).
94. B.F. Myers and E.R. Bartle, J. Chem. Phys. 47, 1783 (1967).
95. D. Gutman, R.W. Lutz, N.F. Jacobs, E.A. Hardwidge and G.L. Schott, J. Chem. Phys. 48, 5689 (1968).
96. N. Cohen and R.W.F. Gross, J. Chem. Phys. 50, 3119 (1969).
97. R.W. Ladenburg, Ed., Physical Measurements in Gas Dynamics and Combustion, Princeton University Press, Princeton, 1954. a. Articles A1-A3; b. E. Winkler, Article A4.

98. M. Steinberg, K.S. Wen, T. Chen and C.C. Yang, AIAA Preprint No. 70-729, 1970.
99. G.W. Sutton and M. Camac, AIAA J. 6, 2402 (1968).
100. R.L. Schapker and M. Camac, AIAA J. 7, 2254 (1969).
101. R.A. Hartunian and D.J. Spencer, AIAA J. 5, 1397 (1967).
102. R.A. Hartunian and D.J. Spencer, AIAA J. 4, 1305 (1966).
103. F.P. DelGreco, D. Golomb, J.A. van der Bliek and R.A. Cassanova, J. Chem. Phys. 44, 4349 (1966); D. Golomb and R.E. Good, J. Chem. Phys. 49, 4176 (1968); the headglow extinction method is as yet unpublished.
104. C.R. Philbrick, AFCRL, Bedford, Mass., unpublished results.
105. V.H. Regener, J. Geophys. Res. 65, 3975 (1960); 69, 3795 (1964).
106. J.R. Smith, H.G. Richter and L.A. Ripperton, Final Report, EPA Contract No. PH 27-68-26, 1967.
107. J.A. Hodgeson, K.J. Krost, A.E.O'Keeffe, and R.K. Stevens, Anal. Chem. 42, 1795 (1970).
108. L.F. Ballard, J.B. Tommerdahl, C.E. Decker, T.M. Royal, and D.R. Nifong, Research Triangle Inst., NTIS PB 204 444, April 1971.
109. J.A. Hodgeson, B.E. Martin and R.E. Baumgardner, Paper No. 77, Eastern Analytical Symposium, New York, 1970.
110. G.W. Nederbragt, A. Van der Horst and J. Van Duijn, Nature 206, 87 (1965).
111. B.J. Finlayson, J.N. Pitts, and H. Akimoto, Chem. Phys. Lett. 12, 495 (1972).
112. G.J. Warren and G. Babcock, Rev. Sci. Instr. 41, 280 (1970).
113. Environmental Protection Agency, Federal Register 36 (228), 22384 (Nov. 25, 1971).
114. L.F. Ballard, J.B. Tommerdahl, C.E. Decker, T.M. Royal, and L.K. Matus, Interim Report, Phase I and II and Final Report, EPA Contract No. CPA 70-101, 1971.
115. R.K. Stevens, J.A. Hodgeson, L.F. Ballard, and C.E. Decker in Determination of Air Quality, G. Mamantov and W.D. Shults, Eds. Plenum, New York, 1970.
116. R.K. Stevens, T.A. Clark, C.E. Decker, and L.F. Ballard, Paper No. 72-13, 65th Annual Meeting, Air Pollution Control Association, Miami, June 1972.
117. A. Fontijn, A.J. Sabadell and R.J. Ronco, Anal. Chem. 42, 575 (1970).
118. R.J. Ronco and A. Fontijn, AeroChem Research Labs., Inc., NTIS PB 209 837, May 1971.
119. D.H. Stedman, E.E. Daby, F. Stuhl, and H. Niki, J. Air Poll. Control Assoc. 22, 260 (1972).
120. F. Stuhl and H. Niki, Ford Motor Co., Dearborn, Report No. SR-70-42, 1970.
121. J.E Sigsby, F.M. Black, T.A. Bellar, and D.L. Klosterman, Publication Preprint, EPA, 1972.
122. H. Niki, A. Warnick and R.R. Lord, Paper No. 710072, Society of Automotive Engineers, Detroit, January 1971.
123. Environmental Protection Agency, Federal Register 136 (128), 12652 (July 2, 1971).

124. J.A. Hodgeson, K.A. Rehme, B.E. Martin, and R.K. Stevens, Paper No. 72-12, 65th Annual Meeting Air Poll. Contr. Assoc., Miami, June 1972.
125. L.P. Breitenbach and M. Shelef, Ford Motor Co., Dearborn, Technical Report No. SR 71-130, 1971.
126. W.A. McClenny, J.A. Hodgeson and J.P. Bell, Paper No. Watr-60, 164th National Meeting, Am. Chem. Soc., New York, August 1972.
127. A.D. Snyder and G.W. Wooten, Monsanto Research Corp., NTIS PB 188 103, August 1969.
128. F.M. Black and J.E. Sigsby, Publication Preprint, EPA, 1972.
129. K.J. Krost, J.A. Hodgeson and R.K. Stevens, Publication Preprint, EPA, 1972.
130. A. Syty and J.A. Dean, Appl. Opt. $\underline{7}$, 1331 (1968).
131. A. Tewarson and H.B. Palmer in Thirteenth Symposium (International) on Combustion, The Combustion Institute, Pittsburgh, 1971, p.99.
132. B. Draeger, Heinrich Draegerwerk, West Germany Patent 1,133,918, July 26, 1962.
133. S.S. Brody and J.E. Chaney, J. Gas Chromatogr. $\underline{4}$, 42 (1966).
134. R.K. Stevens, A.E. O'Keeffe and G.C. Ortman, Environ. Sci. Tech. $\underline{3}$, 652 (1969).
135. A. Horning, EPA Contract No. EHSD 71-50, 1972.
136. J.W. Shiller, Bendix Tech. J. $\underline{4}$, 56 (1971).
137. R.K. Stevens, J.D. Mulik, A.E. O'Keeffe, and K.J. Krost, Anal. Chem. $\underline{43}$, 827 (1971).
138. M.F.R. Mulcahy and O.J. Williams, Chem. Phys. Lett. $\underline{7}$, 455 (1970).
139. W.A. Kummer, J.N. Pitts, Jr., and R.P. Steer, Environ. Sci. Tech. $\underline{5}$, 1045 (1971).
140. M.A.A. Clyne and B.A. Thrush in Ninth Symposium (International) on Combustion, Academic Press, New York, 1963, p. 177.
141. R.S. Braman and E.S. Gordon, IEEE Trans. $\underline{IM-14}$, 11(1965).
142. P.T. Gilbert, Anal. Chem. $\underline{38}$, 1920 (1966).
143. International Reference Atmosphere (CIRA 72), North Holland Publ. Co., Amsterdam, 1972 (in press).
144. I.S. Shklovskii in Artificial Earth Satellites, Vol. 4, Plenum Press, New York, 1961, p. 445.
145. Sky and Telescope, December 1971, p. 382 (unsigned article).
146. N.W. Rosenberg and H.D. Edwards, J. Geophys. Res. $\underline{69}$, 2819 (1964).
147. S.P. Zimmerman and K.S.W. Champion, J. Geophys. Res. $\underline{68}$, 3049 (1963).
148. D. Golomb and M.A. MacLeod, J. Geophys. Res. $\underline{71}$, 2299 (1966).
149. D. Golomb, D.F. Kitrosser and R.H. Johnson, in Space Res. XII, Akademie Verlag, Berlin, 1972, p. 733.
150. D. Golomb and R.E. Good, ibid., p. 675.
151. J. Pressman, L.M. Aschenbrand, F.F. Marmo, A. Jursa, and M. Zelikoff, J. Chem. Phys. $\underline{25}$, 187 (1956).
152. D. Golomb, N.W. Rosenberg, C. Aharonian, J.A.F. Hill, and H.L. Alden, J. Geophys. Res. $\underline{70}$, 1155 (1965).
153. A. Fontijn and D.E. Rosner, J. Chem. Phys. $\underline{46}$, 3275 (1967).

154. D. Golomb and R.E. Good, J. Chem. Phys. **49**, 4176 (1968).
155. C.R. Philbrick, G.A. Faucher and E. Trzcinski, in *Space Research XIII*, Akademie Verlag, Berlin, to be published, 1973.
156. D.D. Woodbridge in *Chemical Reactions in the Lower and Upper Atmosphere*, Interscience Publ. Co., New York, 1961, p. 373.
157. N.W. Rosenberg and D. Golomb in "Project Firefly, 1962-63," N.W. Rosenberg, Ed., AFCRL Environmental Research Papers, No. 15, AFCRL-64-364, 1964, Chap. 1.
158. N.W. Rosenberg, D. Golomb and E.F. Allen, J. Geophys. Res. **68**, 5895 (1963).
159. N.W. Rosenberg, D. Golomb and E.F. Allen, J. Geophys. Res. **69**, 1451 (1964).
160. D. Golomb, F.P. DelGreco, O. Harang, R.H. Johnson, and M.A. MacLeod in *Space Research VIII*, North Holland Publ. Co., Amsterdam, 1968, p. 705.
161. O.M. Uy and J. Drowart, Trans Faraday Soc. **67**, 1293 (1971).
162. N.W. Rosenberg, W.H. Hamilton and D.J. Lovell, Appl. Optics. **1**, 115 (1962).
163. C.D. Cooper in Ref. 157, p. 161.
164. N. Jonathan and G. Doherty in Ref. 157, p. 393.
165. G.T. Best, C.A. Forsberg, D. Golomb, N.W. Rosenberg, and W.K. Vickery, J. Geophys. Res. **77**, 1677 (1972).
166. M.J. Linevsky as quoted by A. Fontijn and S.C. Kurzius, Chem. Phys. Lett. **13**, 507 (1972).
167. P. Harteck and R. Reeves, Bull. Soc. Chim. Belg. **71**, 682 (1962).
168. E.M. Bulewicz, P.J. Padley and R.E. Smith, Proc. Roy. Soc. **A315**, 129 (1970).
169. R.A. Hord and H.B. Tolefson, Virginia J. Sci. **16**, 105 (1965).
170. B.M. McCormac, Ed., *The Radiating Atmosphere*, D. Reidel Publishing Co., Dordrecht-Holland, 1971. a; D.M. Hunten, p. 1, b; R.D. Sears, p. 116, c; G. Visconti, F. Congeduti and G. Fiocco, p.82.
171. B.M. McCormac and A. Omholt, Eds., *Atmospheric Emissions*, Van Nostrand, New York, 1969. a; G.C. Shepherd, p. 411.
172. J.F. Noxon, Space Sci. Rev. **8**, 92 (1968).
173. S.M. Silverman, Space Sci. Rev. **11**, 341 (1970).
174. J.W. Chamberlain, *Physics of the Aurora and Airglow*, Academic Press, New York, 1961.
175. B.S. Dandekar and J.P. Turtle, Planet Space Sci. **19**, 949 (1971).
176. G.A. Lavoie, Combust. Flame **15**, 907 (1970).
177. G.A. Lavoie, J.B. Heywood and J.C. Keck, Comb. Sci. and Techn. **1**, 313 (1970).
178. V.Ya. Shlyapintokh, O.N. Karpukhin, L.M. Postnikov, V.F. Tsepalov, A.A. Vichutinskii, and I.V. Zakharov, *Chemiluminescence Techniques in Chemical Reactions*, Consultants Bureau, New York, 1968, p. 181.
179. V.Ya. Shlyapintokh, R.F. Vassil'ev and O.N. Karpukhin, U.S.S.R. Patent 127779, Byul. Izobret, No. 8 (1960) as quoted in Ref. 178 and by R.F. Vassil'ev, Progr. React. Kin. **4**, 305 (1967).

WAYNE: I was rather mystified to hear you say that the NO + O_3 emission is the last "modification" of the air afterglow that we are going to hear about. I haven't heard this emission called the air afterglow since Lord Rayleigh's time, I think, although I'm not really that old. The point that I'd like to make concerns your statement that the glow is shifted by the bond energy of ozone. If I recollect correctly it isn't actually shifted by exactly that amount because the activation energy of the NO + O_3 reaction also contributes to the amount of energy available for light emission. It seems to me that this is probably true of any process involving activation energy, because, after all, the product molecules that are actually going to emit are the ones that got over the barrier. You have the exothermicity plus the activation energy available for emission, a fact that is not always recognized.

FONTIJN: If it isn't shifted by exactly the amount of the bond energy it is certainly close to it. I agree in principle with the last part of your comment.

THRUSH: Could I make a purely nonscientific comment. In these days when some people say that basic University research should be aimed at solving specific problems, it is interesting to reflect that the work which Dr. Clyne and Dr. Wayne and I did on the chemiluminescent NO + O_3 reaction was carried out entirely for the inherent interest of the problem. We realized that it was a highly reproducible reaction, but did not foresee its importance in measuring pollution at ground level and from supersonic aircraft.

FONTIJN: Actually I was as pleased as you were that we could justify in this manner the things we all have been playing with over the years. Of course, when I went to an agency to try to get funds for this idea, I used your two papers on the subject to demonstrate how well the method could work.

THORINGTON: I was interested to see the application of chemiluminescence yesterday. Everyone expostulated at the sight of those very fine chemiluminescent tubes which represent a possible practical application. And we are talking this morning about practical applications involving measurements. You mentioned high temperature chemiluminescence, and this is something I've been interested in for years and I've not heard much about it at this conference. Up in the ceiling we are getting light: about 90% of the power that goes into those light bulbs comes out as heat and only 10% is light. You have a high temperature filament in there that gets to anywhere from about 2800 to 3200°K. One has the possibility of producing all kinds of atomic and molecular species and totally recombining them either on a solid luminescent surface or in the gas phase to generate practical chemiluminescence. So I would like to invite anybody who has any interest in this to give us some ideas. We have funded research programs in this area for a while, but so far have drawn a complete blank. It is potentially a very useful application.

WAYNE: I would just ask in regard to that question whether anybody has looked for excitation of I or I_2 in the quartz-iodine lamp? Should there be any?

STEDMAN: A spectrum that we have taken looked just like the tungsten filament emission shifted to higher temperatures.

CHEMILUMINESCENCE ANALYSIS FOR TRACE ELEMENTS

W. Rudolf Seitz
NERC, Corvallis, EPA, Southeast Environmental Research
Laboratory, Athens, Georgia 30601
and
David M. Hercules
Department of Chemistry, University of Georgia
Athens, Georgia 30601

Introduction

Several metal ions catalyze the oxidation of luminol (5-amino-2,3-dihydrophthalazine-1,4-dione) by hydrogen peroxide in basic aqueous solution. This extensively studied reaction is one of the most efficient chemiluminescent reactions known (1-3). A few catalysts are effective even in the absence of hydrogen peroxide.

In the presence of excess reagents, the intensity of light emission is proportional to catalyst concentration, a relationship that can be used for trace analysis. Babko and co-workers (4-7) reported methods for cobalt, copper and iron, all based on catalysis of the luminol reaction. By simply mixing the reactants while exposing the container to a photographic film and measuring exposure as a function of concentration, they found detection limits of 1, 3 and 10 ppb for cobalt, copper and iron, respectively.

Chemiluminescence methods are more sensitive than most other analytical techniques without requiring expensive instrumentation. They can measure one particular chemical form of an element rather than total element, e.g. Fe(II) rather than total iron, a capability that can be very useful in determining water quality. Because of this potential the Southeast Environmental Research Laboratory and the University of Georgia have been collaborating on a project to develop chemiluminescence methods for analyzing natural waters. The direction of our research has been to design instrumentation adapted to the needs of the analytical chemist and to develop techniques for analyzing specifically for one catalyst in a complex mixture. We have developed specific methods for Cr(III) (8) and Fe(II) (9) that do not require separation. Prior ion exchange separation promises to extend chemiluminescence analysis to other catalysts, Cu(II), Co(II), Ni(II), VO^{+2}, etc. Non-catalysts can be determined indirectly by titration with catalysts, e.g. non-catalyst As(III) can be titrated with catalyst I_2.

In this paper we will discuss in detail the properties of various luminol-catalyst systems and then show some analytical applications of the luminol reaction.

Experimental

Apparatus

The chemiluminescence apparatus has been described in detail elsewhere (8, 9).

Chemicals

Luminol from Eastman Organic Chemicals was converted to the sodium salt and was purified by recrystallization from basic aqueous solution.

The purified sodium luminol was dissolved in 0.1 M $KOH-H_3BO_3$ buffer to control the pH in the reaction cell. The H_3BO_3 concentration was maintained constant while the amount of KOH was varied to achieve the desired pH.

All reagents were prepared using water from a Continental Water Conditioning Company deionization system.

Standard 0.100 M catalyst solutions were prepared by weighing. Other standards were prepared by dilution.

Analytical Characteristics of the Luminol Reaction

Catalysts

Table I lists metal ions that catalyze the luminol reaction only in the presence of peroxide, along with some of their analytical characteristics. These metals all have oxidation states separated by one electronic charge; Fe(II)-Fe(III), V(IV)-V(V), Ni(II)-Ni(III), etc. Except for Cu(II) the lower oxidation state is the catalyst for the metals listed in Table I. It may also be true of Cu(II) since Cu(III) can exist in aqueous solution (10). In the case of Cr(III), the transition to Cr(VI) involves Cr(IV) and Cr(V) (11) so this reaction includes one-electron transfers. Assuming that the ability to undergo a one-electron transfer causes a metal ion to be a catalyst, it can be predicted that many other metals besides those in the table will react with luminol and peroxide to give light; Ti, Mo, Ce, Ru, etc.

The limiting factor determining detection limits is light emission from H_2O_2 and luminol-buffer in the absence of added catalysts. We believe this background light is catalyzed by trace metal contaminants in the reagents. If this explanation of the background light is correct, it should be possible to reduce detection limits by purifying the H_2O_2 and luminol-buffer before making measurements. This has not yet been attempted.

Table I

Catalysts in the Presence of Peroxide

Catalyst	Detection Limit (M)	Linear Range (M)	Remarks
Co(II)	10^{-11}	10^{-11} to 10^{-7}	--
Cu(II)	10^{-9}	10^{-9} to 10^{-6}	in NH_3 for linearity
Ni(II)	10^{-8}	10^{-8} to 10^{-5}	--
Cr(III)	10^{-9}	10^{-9} to 10^{-6}	--
V(IV)	---	--	quite sensitive
Mn(II)	10^{-8}	--	requires amines
Fe(II)	10^{-9}	--	--

Cu(II) and Mn(II) behave differently from other catalysts. In a non-complexing medium, chemiluminescence vs. copper curves are non-linear at higher copper concentrations (10^{-7} M Cu). Chemiluminescence per standard addition of Cu(II) increases as Cu(II) concentration increases. Peaks on our apparatus are poorly defined and less reproducible than for other catalysts. In the presence of ammonia (and amines), the response becomes linear at higher concentrations and peaks are well-defined. Mn(II) does not catalyze chemiluminescence in water without complexing agents, but in the presence of amines it is activated as a catalyst. Complexation stabilizes Mn(III), thereby promoting catalysis.

Table II lists catalysts of the luminol reaction in the absence of peroxide. Without peroxide background levels are approximately a factor of 100 lower than they are with peroxide. This results in low detection limits even though the efficiencies of the catalysts listed in Table II are not as great as the efficiencies of the catalysis in Table I.

The only catalysts to be studied in detail thus far are Cr(III), Fe(II) and I_2. The rest have only been surveyed, so the data given in Tables I and II should be regarded as preliminary. In most cases, optimization of concentrations is likely to improve detection limits. In the case of Co(II) the listed detection limit of 10^{-11} M extrapolated from data at higher concentrations may be overly optimistic because of the difficulty in handling solutions at these concentrations.

Effect of pH

For most catalysts the optimum pH for chemiluminescence is around 11. Figure 1 shows chemiluminescence as a function of pH for Co(II), Cr(III), Fe(II) and MnO_4^-. The MnO_4^- data show that it is possible for a particular catalyst to have very different pH behavior from the other catalysts.

Effect of Peroxide (Oxygen Concentration)

Light emission from luminol systems involving peroxide is proportional to the peroxide concentration. This has been used to develop analytical methods for peroxide (12).

In the Fe(II)-oxygen-luminol system, the sensitivity of response is independent of oxygen concentrations at low Fe(II) concentrations. The upper concentration limit of linear response to Fe(II) increased as the rate of oxygen supplied to the cell increased (9).

Effect of Luminol Concentration

Each catalyst has its own characteristic dependence on luminol concentration. The effect of luminol concentration on several different catalysts is

Table II

Catalysts in the Absence of Peroxide

Catalyst	Approximate Detection Limit (M)	Linear Range (M)	Remarks
OCl^-	10^{-9}	---	requires O_2
Br_2	---	non-linear	not useful
I_2	10^{-9}	10^{-9} to 3×10^{-7}	squared and cubed response also observed
MnO_4^-	10^{-10}	10^{-10} to 10^{-7}	no O_2 necessary
Fe(II)	10^{-10}	10^{-10} to 5×10^{-7}	requires O_2
V(II)	---	---	requires O_2

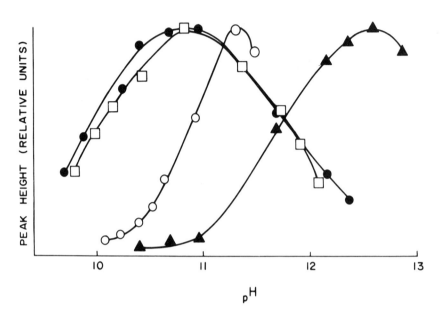

Figure 1. Effect of pH on chemiluminescence intensity for Fe(II) (●-●-●), MnO_4^- (▲-▲-▲), Cr(III) (□-□-□) and Co(II) O-O-O). Relative intensities have been disregarded so four catalysts can be shown on one graph.

Conditions:

Fe(II): 2×10^{-7}M Fe(II), 10^{-3}M Luminol, 80 cc O_2/min.

MnO_4^-: 5×10^{-8}M MnO_4^-, 10^{-3}M Luminol

Cr(III): 2×10^{-8}M Cr(III), 10^{-2}M H_2O_2, 10^{-3}M Luminol

Co(II): 2×10^{-8}M Co(II), 10^{-2}M H_2O_2, 10^{-3}M Luminol shown in Figure 2. Characteristically there is a decrease in efficiency at high luminol concentrations, which may be due to luminol (or its aminophthalate oxidation product) acting as an organic complex to reduce the "availability" of the metal ions for catalysis.

Effect of Flow Rate

Increasing the flow rate through the cell has two effects. The amount of catalyst entering the cell per unit time is proportional to flow rate. Thus if the light emitting reaction goes to completion before the catalyst leaves the cell, the intensity of emitted light should be proportional to flow rate.

Increasing the flow rate also reduces the residence time in the cell for the catalyst. If the light-emitting reaction does not go to completion before the catalyst leaves the cell, then reducing the residence time results in less of the light emission occurring in the cell and more light being lost in the exit tube.

If the light-emitting reaction is slow relative to the residence time in the cell, the concentration of catalyst leaving the cell will equal the concentration of catalyst entering the cell at all flow rates and peak height will be independent of flow rate.

In the presence of base (and peroxide) most of ions listed in Tables I and II react to produce forms that do not catalyze luminol chemiluminescence, e.g. $Fe(II)-Fe(III)$, $Cr(III) - Cr(VI)$, and $I_2 - I^- + IO_3^-$. The chemiluminescent reaction of luminol terminates when these species have reacted to an inert form. These species are not true "catalysts" because they change during the course of the reaction. We refer to them as "catalysts" only because this has been done by previous workers in the field.

There are a few "true" catalysts such as $Ni(II)$. Any $Ni(III)$ formed during the course of the chemiluminescent reaction will react with water and go back to $Ni(II)$. For $Ni(II)$ the concentration of active catalyst in the cell at steady state is independent of flow rate.

Figure 3 shows peak height vs. flow rate for several catalysts. For $Fe(II)$ and $Cr(III)$ peak height is proportional to flow rate at slower flow rates because they both rapidly react to inert forms in the reaction cell. For $Ni(II)$ peak height is independent of flow rate while $Cu(II)$ shows an intermediate behavior.

Varying flow rate can provide useful information about a particular system. For example from the information in Fig. 3 it would be possible to tell whether a particular peak was due to $Ni(II)$ in a sample that could contain $Cu(II)$, $Cr(III)$ and $Fe(II)$ by measuring peak height vs. flow rate.

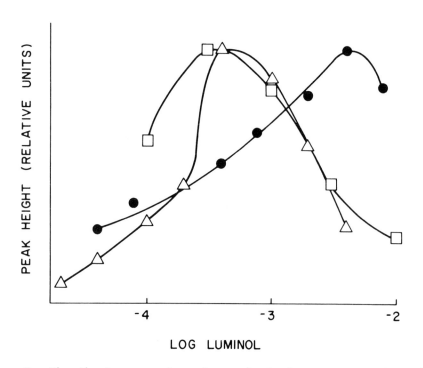

Figure 2. Chemiluminescence intensity vs. luminol concentration for Fe(II) (●-●-●), Cr(III) (□-□-□) and MnO_4^- (△-△-△). Relative intensities have been disregarded so the data for three catalysts can be shown on one graph.

Conditions:

- Fe(II): 10^{-7}M Fe(II), 80 cc O_2/minute, Cell pH 10.5-11.0
- Cr(III): 10^{-7}M Cr(III), 10^{-2}M H_2O_2, Cell pH 10.5
- MnO_4^-: 10^{-7}M MnO_4^-, Cell pH 10.5

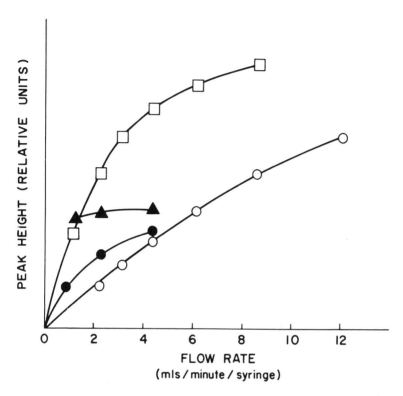

Figure 3. Chemiluminescence Intensity vs. Flow Rate for Ni(▲-▲-▲), Cu (●-●-●), Cr(III) (□-□-□) and Fe(II) (O-O-O). Relative intensities have been disregarded so that all four curves can be shown on one graph.

Conditions:

Ni(II): 10^{-6}M Ni(II), 10^{-2}M H_2O_2, 10^{-3}M Luminol

Cu(II): 10^{-6}M Cu(II), 10^{-2}M H_2O_2, 10^{-3}M Luminol, 10^{-2}M NH_3 in sample and background (No KOH-H_3BO_3 buffer)

Cr(III): 4×10^{-7}M Cr(III), 10^{-2}M H_2O_2, 10^{-3}M Luminol, 2×10^{-2}M EDTA

Fe(II): 2×10^{-8}M Fe(II), 80 cc O_2/minute

Effect of Organic Complexation

Organic complexation reduces the effectiveness of metal ion catalysis. Figure 4 shows the effect of four different complexes on Fe(II)-catalyzed light. The stronger the complex, the more effectively it reduces light. Response is still linear in the presence of organic ligands. This effect occurs for all metal ion catalysts except Mn(II), which is activated by the presence of amines, and may make it possible to determine not only the concentration of a metal ion but also its chemical form in an aqueous sample.

Sample Chemistry

The only restriction on the sample (and background) medium that flows into the reaction cell is that it does not interfere with the luminol reaction. The pH can vary from 2-12. Extra base or acid can be added to the 0.1 M $KOH-H_3BO_3$ buffer to compensate for any effect that the sample medium may have on the pH of the reaction. The pH range can be extended to 1 to 13 by working with more concentrated $KOH-H_3BO_3$ buffer. This will lead to higher background light emission from catalyst impurities in the reagents. This will also increase the heat generated by neutralization reactions in the cell, which may affect an analysis adversely.

Reducing agents present in equal or greater concentration than the H_2O_2 in the cell are unacceptable. Also concentrations of primary amines greater than 10^{-3} M react with peroxide and luminol to produce light.

Specific Analysis for Chromium

Cr(III) can be determined specifically by adding EDTA to complex metal ions that would otherwise interfere (8). The Cr(III)-EDTA complex does not catalyze chemiluminescence either, but it is kinetically slow to form (13). The only metal ions to interfere at less than a 1000-fold excess are Co(II), Fe(II) and Fe(III). These interferences can be accounted for by running a blank. This is done by heating the sample loop of the injection valve in water at 80 to 90°C followed by cooling to room temperature for 6-8 minutes. This treatment causes all the Cr(III) in the sample loop to form the non-catalyzing Cr(III)-EDTA complex. The light emission catalyzed by Fe(II), Fe(III) and Co(II) is unchanged by heating and cooling the sample may be subtracted from the total.

The conditions used for Cr(III) analysis were 2×10^{-2} M EDTA in the sample bottle and background, 10^{-3} M EDTA in the H_2O_2 and luminol-buffer solution to complex trace catalyst contaminants in these reagents before they enter the cell, and a pH of 4.4 in the sample bottle. The sample bottle pH is important because the rate of Cr(III)-EDTA formation increases as pH increases (13) while at very low pH's EDTA solubility decreases.

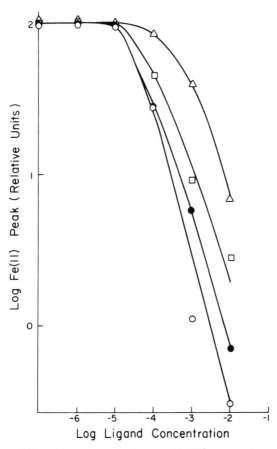

Figure 4. Effect of ligand concentration on Fe(II) - catalyzed chemiluminescence.

△-△-△   ethylene diamine

□-□-□   glycine

●-●-●   tartrate

O-O-O   2,4-pentane dione

Conditions:

pH 10.5-11.0, O_2 80 cc/minute, 10^{-7}M Fe(II), 4×10^{-4}M Luminol

A detection limit of 5×10^{-10}M Cr(III) is imposed by low level light emission from the reagents. Although the EDTA complexes trace metal impurities in the reagents to reduce background significantly, there is still light emission that may be catalyzed by Fe(III) impurities in the EDTA itself. Response is linear up to 10^{-6}M. Above 10^{-6}M, Cr(III) peaks rise to an initial spike and decay to a lower value. This is probably due to precipitation of Cr(OH)$_3$ ($K_{sp} = 10^{-30}$) in the cell after initial supersaturation.

Table III shows some values for total Cr in NBS orchard leaves (Standard Reference Material 1571) determined by chemiluminescence after wet ashing (14). Table IV shows some Cr(III) concentrations measured in natural waters (8). The values are reproducible and in the range expected for natural waters (15). Because chemiluminescence measures only Cr in the uncomplexed Cr(III) form, it cannot be directly compared to other methods for chromium.

Table III
Chemiluminescence Analysis of Orchard Leaves (13)

2.1, 2.1, 2.6 µg/gm

NBS value   2.3 µg/gm

The orchard leaves (NBS Standard Reference Material (1571)) were wet ashed with perchloric and nitric acids with a maximum temperature of 265°C.

Table IV
Chromium(III) Analysis of Natural Water Samples

Sample	Cr(III) Concentration Found (M)	Average	Equivalent Cr(III) Concn of Blank, (M)
Oconee River	2.5×10^{-8}		2×10^{-7}
	3.0×10^{-8}	3.2×10^{-8}	
	4.0×10^{-8}	(1.7 ppb)	
Lake Lanier	2.0×10^{-8}		
	2.2×10^{-8}	2.1×10^{-8}	4×10^{-8}
	2.0×10^{-8}	(1.1 ppb)	
Tap Water	1.8×10^{-7}		
	1.9×10^{-7}	1.9×10^{-7}	3×10^{-8}
	2.0×10^{-7}	(10 ppb)	
Tara Pond	0.8×10^{-7}		
	1.0×10^{-7}	1.0×10^{-7}	2×10^{-7}
	1.1×10^{-7}	(5 ppb)	

The ability to distinguish different forms of chromium by chemiluminescence is of value in studying the relationship between the chemical form of chromium and its impact on the environment.

Specific Analysis for Iron

Most of the catalysts not requiring peroxide are either strong oxidizing or reducing agents. By adding a reducing agent, such as sulfite, to destroy all the oxidants and oxidizing catalysts and to convert iron to Fe(II), chemiluminescence analysis is specific for iron (9). Sulfite is not a sufficiently strong reductant to generate V(II), or Cr(II), the other reducing agents that catalyze luminol chemiluminescence in the absence of peroxide.

The chemiluminescence detection limit for Fe(II) is 10^{-10}M; response is linear up to 5×10^{-7} M. In addition to the effect of organics discussed earlier, several metal ion--Co(II), Cr(III), Ni(II), Mn(II) and Cu(II)--interfere to reduce chemiluminescence. Co(II), Cr(III) and Ni(II) must be present in large excess to interfere. Response in the presence of Cu(II) remains linear with concentration. Mn(II) is the most serious interference causing a curvature in calibrations. These interferences are discussed in detail in reference (9).

Table V compares some iron concentrations measured by chemiluminescence with values obtained by other methods. The natural water samples were analyzed for total iron by the method of standard additions using sulfite as a reducing agent. To attain the optimum range for chemiluminescence analysis, they were diluted by a factor of 100 for tap water, 250 for river water and 500 for pond water.

No color developed for any of the three water samples when a conventional phenanthroline method for total iron (16) was applied to them without prior solvent extraction or ashing. This suggests that if Fe(II) were present in these samples, the colorimetric method would not specifically determine it in the presence of Fe(III), since the ashing or solvent extraction procedure would alter the Fe(II)/Fe(III) ratio in the sample.

Ion Exchange

To determine those metals that catalyze luminol chemiluminescence but aren't readily analyzed specifically, ion exchange separation is required. This can be done by inserting an ion exchange column directly in the flow system. We have recently started working on such a system using anion exchange in concentrated chloride for the separation. In strong chloride metal ions form anionic chloride complexes that preferentially stay on the resin. As the chloride concentration is reduced the metal ions come off the resin. The metals that don't readily form chloride complexes come off the column first

Table V

Sample	CL Analysis	Atomic Absorption	Colorimetry
Orchard Leaves 1* NBS SRM 1571	240, 220 µg/g	—	282 µg/gm
Orchard Leaves 2* NBS SRM 1571	237, 251 µg/g	—	282 µg/gm
Pond Water	0.72, 0.83 µg/ml	0.74 µg/ml	—
River Water	0.39, 0.38 µg/ml	0.34 µg/ml	—
Tap Water	0.18, 0.15 µg/ml	0.30 µg/ml	—

* Orchard leaves 1 and 2 are separate ashings with nitric and perchloric acids of the same material.

while the metals that form stronger chloride complexes stay on the column among the various metal ions. Metal ions differ more in their tendency to form anionic chloride complexes than in their relative affinities to cation exchangers. Most of the work in this area has been by Krause and associates (17-19) using HCl. Because of the pH requirements of the luminol reaction, it is necessary to work in concentrated LiCl. Although we have just recently started this work, we have already obtained separations with chemiluminescence detection. Figure 5 shows one of our separations (20).

Chemiluminescence Titrations

Because some of the catalysts in Table II are common titrants, it is possible to do chemiluminescence titrations where the amount of light catalyzed by a system is measured as a function of titrant added. For example, in the titration of arsenic (III) with iodine (20), no light will be catalyzed by the titration mixture until an excess of iodine is present. Beyond the endpoint, light intensity will be proportional to excess iodine. The endpoint can be determined by extrapolation.

Iodine-Luminol Reaction

Because iodine is the most versatile titrant of the species in Tables I and II, a study of the iodine-luminol reaction was undertaken to optimize conditions for using iodine analytically. Babko et al. (21) reported linear response to iodine concentration with a detection limit of 1×10^{-5} M. Using our apparatus, it was discovered that response was first order, second order, third order or a combination thereof depending on the pH of the chemiluminescing reaction. Figures 15 and 16 show log peak height vs. log iodine concentration at cell pH's 10.60 and 11.60 respectively. The response at pH 10.60 was computer fitted to a combination of first and second order components. Peak heights calculated from the computer least squares fit are included in Fig. 6. The response at cell pH 11.60 is similarly resolved into first and third order components, shown in Fig. 7.

Experiments in the presence of oxygen showed that the first order response required oxygen while the second and third order processes did not. In another experiment it was shown that triiodide did not catalyze the reaction.

In base iodine disproportionates
$$I_2 + 2OH^- \longrightarrow I^- + OI^- + H_2O \quad (22),$$
Hypoiodite then disproportionates
$$3OI^- \longrightarrow 2I^- + IO_3^- \quad (22)$$

in a complex reaction, the mechanism of which is not known. The first order response to iodine concentration must be the reaction of hypoiodite with oxygen and luminol. The second and third order responses must be due to the reaction of one of the intermediates in the disproportionation of hypoiodite

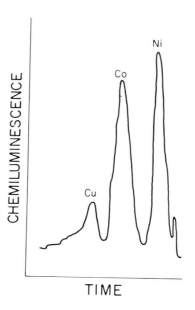

Figure 5. Ion exchange separation of Ni, Co, and Cu in LiCl with chemiluminescence detection.

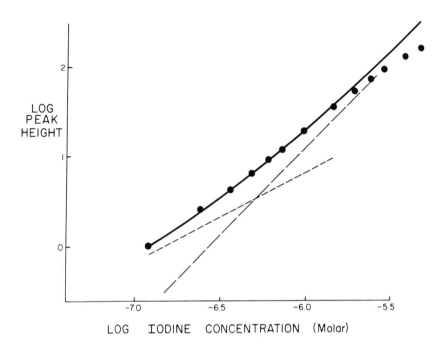

Figure 6. Log peak height vs. log iodine concentration at cell pH 10.60

●-●-●   Experimental points
————   Calculated least squares response
- - - - - - -   First order component of calculated response
— — — —   Second order component of calculated response

Conditions:

    1.5×10^{-4} Luminol

    80 cc O_2/minute

    10^{-2}M HCl in sample background

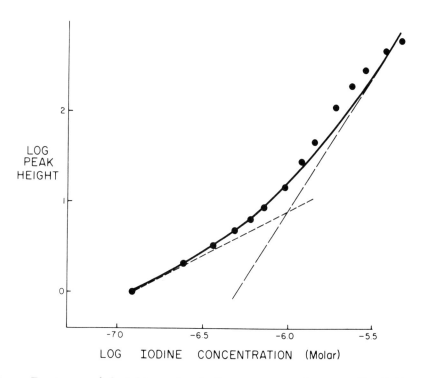

Figure 7. Log peak height vs. log iodine concentration at cell pH 11.60

● ● ●	Experimental points
———	Calculated least squares response
- - - - - -	First order component of calculated response
— — — —	Third order component of calculated response

Conditions:

    1.5×10^{-4}M Luminol

    80 cc O_2/minute

    10^{-2}M HCl in sample background

to iodide and iodate. The possible intermediates are IO_2^-, $I_2O_2^=$, $HI_2O_2^-$, and $I_3O_2^=$ (23). All of them contain two oxygens which explains why molecular oxygen isn't required for the reaction; all of them can undergo a 4-electron reduction to form stable iodine species, e.g. IO_2^- + Luminol $\rightarrow$ Aminophthalate + I^-.

For analytical use the greatest range of first order response is obtained at higher pH's where the third order process occurs. The third order process becomes significant at higher concentrations than the second order process (see Figs. 6 and 7). Response is first order with respect to iodine concentration from 5×10^{-10} to 3×10^{-7} M.

Titrations

The titration of arsenic is of interest because of its toxicity. Arsenic can be separated by distillation from 6 M HCl as $AsCl_3$ (24). Table VI shows the results of 12 chemiluminescence titrations of 2×10^{-7} M As(III) with I_2 in a 10^{-2} M phosphate buffer (pH 7), and compares them to the classical titration (to a starch endpoint) of the 0.05 M stock solutions. The titrations were done by adding 50-microliter aliquots of 10^{-4} M I_2 to 500 mls. of sample 2×10^{-7} M in arsenic. Three percent more iodine was necessary to reach the chemiluminescence endpoint than the classical endpoint. An actual titration is shown in Figure 8.

Table VI

Chemiluminescence Titration of 100.0 Nanmoles As(III) with Iodine

Set	Titration	Nanomoles I_2 To Reach End Point	Set Average
I	1	102.5	103.5
	2	102.5	
	3	105.0	
II	1	103.0	
	2	103.0	103.0
	3	103.0	
III	1	100.0	
	2	101.0	100.5
	3	100.5	
IV	1	105.5	
	2	107.0	105.0
	3	103.0	

Each set represents separate 10^{-3} M As and 10^{-4} M I_2 standards prepared by dilution from 0.1 M As and 0.05 M I_2 standards. The 0.105 M I_2 standard was standardized vs. the 0.100 M As (primary standard).

The fact that precision within sets is greater than precision among different sets indicates that precision is lost in preparing standards by dilution.

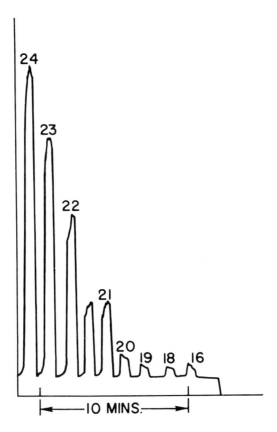

Figure 8. Data showing chemiluminescence titration of As(III) with I_2.

The numbers over the peaks indicate how many 5-nanmole aliquots of iodine were added before running the peak.

Iodine titration has been used to determine SO_2 in air (25). Table VII shows the results for chemiluminescence titrations of 2×10^{-7} M $SO_3^=$ with iodine in 500 ml of 10^{-2} M phosphate buffer at pH 7 (26). A theoretical detection limit of less than 0.01 ppm SO_2 in air was calculated, based on a 500 cc air sample collected in 25 mls. of solution. The time for a titration is 15-20 minutes.

Table VII

Chemiluminescence Titration of 94.0 Nanomoles of $SO_3^=$ with I_2

Titration	Nanomoles I_2 to End Point
1	97.0
2	97.0
3	97.0
4	97.7
5	97.5
6	98.5
7	98.8
8	99.0
9	98.8
10	98.3
11	100.0
Average	98.1

Instead of a many-point titration, a known spike of catalyst can be added to the sample and the change in peak height due to the sample and the change in peak height due to the sample concentration. This reduces the time and the sample volume required for an analysis. For analyses where sample is collected in a solution (e.g. SO_2 bubbled into a scrubber) the reduction in sample volume will increase the sensitivity of the method. Another advantage of this approach is to eliminate the need for a volume correction as required by the titration. The fact that a certain amount of solution is consumed in getting each point of a chemiluminescence titration has to be corrected for if it is not possible to work under conditions where the solution needed per measurement is very small relative to total volume being titrated. At present we are doing titrations of SO_2 using an initial sample volume of 25 mls. and 3/4 ml. per measurement.

Fe(II), V(II), MnO_4^- and OCl^- are other possible chemiluminescence titrants.

Conclusion

The work reported here is just a start in the area of chemiluminescence analysis. We hope we have demonstrated the potential of this approach and have stimulated interest in applying chemiluminescence analysis to a variety of problems.

Disclaimer

Use of trade names does not imply endorsement by the Environmental Protection Agency or the Southeast Environmental Research Laboratory.

Acknowledgments

This work was supported in part through funds provided the University of Georgia by PHS, NIH Research Grant No. GM17913-01 from the National Institute of General Medical Sciences.

References

1. H. O. Albrecht, Z. Phys. Chem., 136, 321 (1928).
2. E. H. White, "A Symposium on Light and Life", W. D. McElroy and B. Glass, Ed., The John Hopkins Press, Baltimore, Md., 1961, p. 183.
3. F. McCapra, Quart. Rev., 20, 485 (1966).
4. A. K. Babko and N. M. Lukovskaya, Zh. Anal. Khim., 17, 50 (1962).
5. A. K. Babko and L. I. Dubovenko, Z. Anal. Chem., 200, 428 (1964).
6. A. K. Babko and N. M. Lukovskaya, Zavod. Lab., 29, 404 (1963).
7. A. K. Babko and I. E. Kalinichenko, Ukr. Khim. Zh., 31, 1316 (1965).
8. W. Rudolf Seitz, W. W. Suydam and D. M. Hercules, Anal. Chem., 44, 957 (1972).
9. W. R. Seitz and D. M. Hercules, Anal. Chem., 44, 2143 (1972).
10. H. A. Laitinen, Chemical Analysis, p. 446, McGraw-Hill, New York (1960).
11. F. H. Westheimer, Chem. Revs. 45, 419 (1949).
12. O. Ojima and R. Iwaki, Nippon Kagaka Zasshi, 78, 1632-5 (1957).
13. R. E. Hamm, J. Amer. Chem. Soc., 75, 5670 (1953).
14. R. Li, unpublished results, University of Georgia, 1972.
15. W. Merts, Physiological Reviews, 49, 168 (1969).
16. Standard Methods, APHA, 13th ed. (1971).
17. K. A. Krause, Trace Analysis, pp 34-101, ed. J. H. Yoe and H. J. Koch, John Wiley and Sons, Inc., New York (1957).
18. K. A. Krause and G. E. Moore, J. Amer. Chem. Soc., 75, 1460 (1953).
19. G. E. Moore and K. A. Krause, J. Amer. Chem. Soc., 74, 843 (1952).
20. M. Neary, unpublished results, University of Georgia, 1972.
21. Willard, Furman and Bricker, Elements of Quantitative Analysis, pp 256-258, 4th edition, D. Van Nostrand Co., Inc., Princeton, New Jersey (1956).

22. A. K. Babko, L. V. Markova and N. M. Lukovskaya, Zh. Anal. Khim. 23, 401-6 (1968).
23. F. A. Cotton and G. Wilkinson, Advanced Inorganic Chemistry, Chapter 22, Interscience (1962).
24. J. A. Morgan, Quart. Rev. Chem. Soc., 8, 123 (1954).
25. D. Liederman, J. E. Bowen and O. I. Milner, Anal. Chem., 30, 1543 (1958).
26. F. P. Terraglio and R. M. Manganelli, Anal. Chem., 34, 675 (1962).
27. W. Hardy, unpublished results, University of Georgia, 1972.

LEE: As everyone knows, hemoglobin is one of the most effective catalysts of the luminol reaction. Since it is very likely that you will get heme pigments as a contaminant from sewerage in water supplies would this tend to interfere with the use of this reaction as a tool for analyzing trace metals?

SEITZ: If it is there in high enough concentration then in principle it is possible. In practice I doubt that the heme concentrations are anywhere near the iron concentration. From what I understand it biodegrades pretty readily. So you have a selection factor of several orders of magnitude right there.

KEARNS: You didn't mention magnesium as a possible element you could analyze.

SEITZ: Magnesium is an alkali metal and the alkali metals don't work. It seems that at least two valence states separated by one electronic charge for metals are required to be effective as catalysts. And yet most of the metals in the first transition state are catalysts. Probably quite a few in the higher transition series are effective, but we haven't really made a survey or attempted to determine how many actually make the luminol reaction go.

CHEMICAL LIGHT PRODUCT RESEARCH AND DEVELOPMENT[1]

Michael M. Rauhut

Organic Chemicals Division

American Cyanamid Company

Chemiluminescence research was initiated in the Author's laboratory in 1961 with the objective of developing efficient chemical lighting systems for use in practical lighting applications. The preliminary phase of the program consisted of market research, feasibility analysis and selection of an optimum research approach.

The preliminary market study[2] indicated a reasonable probability for markets in the emergency and portable lighting fields. Emergency lighting would benefit from the cold, flameless characteristics of chemiluminescence, while portable lighting would benefit from the high light-energy storage density possible in theory from chemiluminescence. Thus, chemical light could be used safely at accident scenes involving automobiles, aircraft, boats, etc., where spilled gasoline would prohibit flare use. Similarly, it could be used safely in inherently hazardous situations such as coal mines and the repair of gas transmission lines. In the portable lighting field a high efficiency chemiluminescent system could replace flashlights and lanterns for marking and illumination applications, and again safety and reliability would be additional advantages. Application examples in this area would include use in hiking, camping, small boats, life rafts, life jackets and use in office buildings, factories and homes during power failures. Market Research also helped define the general characteristics a practical system would require, and these are indicated in Chart I.

Feasibility analysis began with a calculation, summarized in Chart II, of the maximum light output possible from chemiluminescence. We defined the visual light output from one liter of a chemiluminescent system as the "LIGHT CAPACITY". As seen in the chart, light capacity is the integral of luminous intensity with respect to reaction

CHART I

PRODUCT REQUIREMENTS

- HIGH BRIGHTNESS
- LONG LIFETIME
- CONVENIENT UTILIZATION
- STORAGE STABILITY
- SAFETY
- LOW COST

CHART II

POTENTIAL CHEMILUMINESCENCE CAPABILITY

LIGHT CAPACITY = VISIBLE LIGHT OUTPUT PER UNIT VOLUME

$$\text{LIGHT CAPACITY (LUMEN HOURS/LITER)} = \int IdT/\text{VOLUME}$$

	LIGHT CAPACITY	=	LUMINANT CONC.	x	QUANTUM YIELD	x	PHOTOPIC FACTOR	x CONST.
THEORETICAL	173,000	=	5 $\underline{M}$	x	1.00	x	0.85	
REASONABLE	300	=	0.1 $\underline{M}$	x	0.10	x	0.75	

PRACTICAL EQUIVALENTS

CHEMILUMINESCENCE	40 W TUNGSTEN BULB BURNING FOR
173,000	2 WEEKS
300	40 MINUTES

time for one liter of reacting system. It is clear that brightness and lifetime are limited by the light capacity, and that light capacity must be high for brightness and lifetime to be satisfactory. It can be shown, as indicated in Chart II, that the light capacity is determined by the concentration of the chemiluminescent compound in the system, the chemiluminescence quantum yield and the photopic factor. All three factors must be high to achieve high light capacity. The photopic factor defines the sensitivity of the human eye to the color of the emitted light and is determined by the spectral distribution. As a theoretically optimum system, we can imagine a luminant concentration of 5 $\underline{M}$, a quantum yield of 100% and a photopic factor of 0.85. (A photopic factor of one would require narrow band emission in the yellow at 555 NM, and this is not possible for organic fluorescers. A photopic factor of 0.85 corresponds to the yellow fluorescence of fluorescein. Blue and red colors are substantially less effective in terms of the human eye response.) A perfect system would generate 173,000 lumen hours liter^{-1} of light. This is equivalent to the light produced by a 40 watt incandescent bulb burning continuously for two weeks. It seemed reasonable that a system operating at only 0.2% of the theoretical efficiency (300 lumen hours liter^{-1}) would have practical utility.

We were also encouraged by the knowledge that the feasibility of high quantum yield had been demonstrated by the firefly. According to Seliger and McElroy, the firefly quantum yield is 88%.[3] Other literature information, however, was less encouraging. The firefly reaction would not be useful for our purposes because it requires an enzyme which only the firefly knows how to make. Many simpler chemiluminescent reactions were known, but the quantum yields were far too low to be useful. We calculated the light capacity of the luminol reaction, for example, to be about 0.2 lumen hours liter^{-1} or only about 0.06% of the 300 lumen hours liter^{-1} we felt would be necessary for a practical lighting system. We concluded after a few preliminary experiments that the required three to four orders of magnitude increase in light capacity would probably not be achieved through structural modifications or formulation changes of the then known chemiluminescent reactions. As the only alternative, it appeared that achievement of our goal would require the discovery of a new chemiluminescent reaction incorporating the factors required for high quantum yield. Since these fundamental factors were not known, we began with a series of mechanism studies to determine criteria for efficient chemiluminescent reaction design.

This was an ambitious undertaking. At that time (in 1961) all of the known chemiluminescent reactions had been discovered by accident, and the mechanisms were unknown. Fortunately, we were able to obtain support from the Office of Naval Research, and later from the Naval Ordnance Laboratory and the Naval Weapons Center, which provided the program with the scope and momentum which were essential for its success.

The overall program is summarized in Chart III. The fundamental phase of the program provided oxalic ester chemiluminescence, which was found to have the inherent efficiency required for practical chemical light.[4] This reaction provided quantum yields as high as 28%. We have published this work and I need not go into it in any detail. At the end of that stage, oxalic ester chemiluminescence was only a reaction, and a substantial amount of applied research was required to develop practical formulations and devices to meet the product requirement summarized in Chart I. Following that, manufacturing procedures and markets had to be developed leading to manufacturing and sales.

CHART III

CHEMICAL LIGHT RESEARCH AND DEVELOPMENT

PHASE I

MARKET RESEARCH

FUNDAMENTAL RESEARCH

PHASE II

PERFORMANCE OPTIMIZATION

PROCESS RESEARCH

DEVISE DESIGN RESEARCH

MARKET RESEARCH

PHASE III

PROCESS DEVELOPMENT

DEVICE MANUFACTURING DEVELOPMENT

MARKET DEVELOPMENT

PHASE IV

MANUFACTURING

MARKETING

Chart IV summarizes the formulation research. In oxalic ester chemiluminescence, an oxalic ester reacts with hydrogen peroxide in the presence of a fluorescer and a catalyst to generate the first singlet excited state of the fluorescer, which is the source of emitted light. The color can be varied by changing the fluorescer, because it is independent of the energy-releasing reactants. Moreover, the brightness and lifetime can be varied by modifying the

CHART IV

SYSTEM VARIABLES

ROCCOR + FLUORESCER + H_2O_2 + CATALYST $\longrightarrow$ LIGHT

$$LT. \; CAP. = \int_{300} IdT = K \cdot QY \cdot CONC. \cdot P \; FACTOR$$
$$\phantom{LT. \; CAP. = \int_{300} IdT = K \cdot } 0.10 \quad 0.10 \; \underline{M} \quad 0.74$$

CRITICAL DESIGN CRITERIA

ESTER: • QY   • SOLUBILITY   • STABILITY   • REACTIVITY

FLUORESCER: • QY   • SOLUBILITY   • STABILITY   • COLOR

CATALYST: • QY   • SOLUBILITY   • STABILITY   • ACTIVITY

SOLVENTS: • QY   • SOLUBILITY   • INERT   • NON-TOXIC
              • HIGH FLASH POINT AND BOILING POINT   • MILD ODOR
              • LOW FREEZING POINT   • INEXPENSIVE • PURE

catalyst. An electronegatively-substituted ester is required for high quantum yield, but an effective ester must also be soluble, stable and reactive. The stability requirement was the most difficult to achieve. While a variety of polycyclic aromatic hydrocarbons, such as 9,10-diphenylanthracene and rubrene, are effective fluorescers, an optimum fluorescer had to be selected in terms of stability and photopic factor, as well as chemiluminescent efficiency.

Bases strongly catalyze the reaction but also tend to decrease the quantum yield. Catalyst selection was a major part of the formulation problem. Solvent selection, in spite of the numerous requirements the solvent must meet, was relatively easy. Almost all known liquids fail to meet one or more of the requirements listed in Chart IV, and elimination of the obvious misfits allowed us to concentrate on phthalic ester solvents early in the game. In particular, dimethyl and dibutyl phthalates are non-toxic and have high flash points as required for safety. This effort was done primarily on an empirical basis and ultimately provided a two-component system having the ester and fluorescer combined in one solution and the hydrogen peroxide and catalyst combined in a second solution.

You will recall that "ease of utilization" was a prime product requirement. There are a variety of ways the two component solutions can be mixed conveniently. An especially easy-to-use device is one we call a lightstick. It is a sealed polyethylene tube containing the hydrogen peroxide-catalyst solution in a thin-walled glass pod which floats in the ester-fluorescer solution. The lightstick is activated simply by bending to break the glass pod. I will demonstrate two lightsticks of this design. One provides green light with a light capacity of 280 lumen hours liter^{-1}; the second is a new formulation which gives yellow light with a light capacity of 900 lumen hours liter^{-1}. The brightness-time curves for these formulations are shown in Chart V. Note that the lower luminosity of the green system is primarily useful for marking applications, while the brighter yellow system can be used for illumination as well as for marking. The yellow system can also be formulated to provide a good marking brightness level (greater than two foot lamberts cm^{-1}) for more than 14 hours.

The development phase of the program required the establishment of practical manufacturing procedures for the chemicals, formulations, and lightsticks. A substantial process development effort was required to scale-up laboratory synthetic procedures for the oxalic ester and fluorescer to plant size, in part because of the novel chemistry involved, but primarily because of the need for exceptional product purity. Both the performance and the stability of the system is highly sensitive to water and trace contaminants so that rigorous manufacturing specifications were required. Novel mass-production equipment for lightstick fabrication was designed for us by an independent company. The lightsticks are fabricated for us

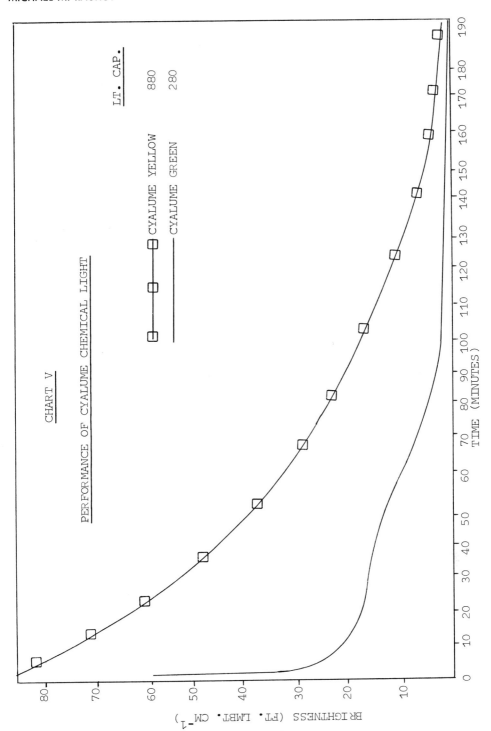

CHART V

PERFORMANCE OF CYALUME CHEMICAL LIGHT

by the Fenwal Corporation of Ashland, Massachusetts. Other development activities included finding suppliers for non-commercial raw materials, the establishment of analytical procedures for quality control and extensive shelf-life testing.

Additional market research and preliminary market development led Cyanamid to a decision to market lightsticks primarily through independent distributors. Cyanamid's marketing is done through Dr. R. R. Miron, Manager of the New Ventures Department of Cyanamid's Organic Chemicals Division, and through the Glendale Optical Company, a Cyanamid subsidiary. Most sales to customers, however, are handled by independent companies.

Although we have made a start at exploiting the practical value of chemical light, it is clear that the major opportunities still lie before us. The most efficient system now known operates at only about 0.5% of the theoretical maximum light capacity, and we feel that an order of magnitude increase in light capacity is a reasonable expectation from future research.

REFERENCES

1. Principle contributors to Cyanamid's Chemical Light Research and Development Program include: Dr. M. T. Beachem, Dr. L. J. Bollyky, Dr. T. Brownlee, Dr. D. Maulding, Dr. A. G. Mohan, Mr. A. M. Semsel, Mr. R. L. Narburgh, Mr. B. G. Roberts, Mr. J. F. Russ, Dr. D. Sheehan, Mrs. M. L. Vega and Mr. R. Whitman. We appreciate support of this program by the Advanced Research Projects Agency under contract to the Office of Naval Research and by the U. S. Naval Ordnance Systems Command under contracts to the U. S. Naval Ordnance Laboratory and to the U. S. Naval Weapons Center.

2. Market research was carried out principally by Dr. A. M. Swift and Dr. R. R. Miron.

3. H. H. Seliger and W. D. McElroy, Arch. Biochem. Biophys., 88, 136 (1960).

4. M. M. Rauhut, Accounts Chem. Res., 2, 80 (1969). M. M. Rauhut, L. J. Bollyky, B. G. Roberts, M. Loy, R. H. Whitman, A. V. Iannotta, A. M. Semsel and R. H. Clarke, J. Am. Chem. Soc., 89, 6515 (1967).

UNIDENTIFIED: How much does a stick cost?

RAUHUT: I think what you probably mean is how much does it cost to buy one in a retail store. I should have mentioned that these are sold by Coollight Corporation, an independent company, to the consumer market and the retail price is one dollar.

HELLER: What does your market research indicate about developing new colors, in addition to a green and a yellow. Does your market research tell you how much better you are going to do with other colors?

RAUHUT: Not really very quantitatively. It is really difficult to know what the market potential is for a product that really doesn't exist or have any close relative. We do think a red color would have an important use because people associate red with danger. Particularly for highway markings and that sort of thing a red system ought to have some advantages.

HERCULES: How many man years went into the development of the light stick?

RAUHUT: All I can say is it was a lot, but I've never really added it up. It does extend over an 11 year period of time and we had as many as six people at one time working on it but it was at a lower rate most of the time. A very substantial number of man years has been devoted to it.

THORINGTON: Since you have elected to use photometric terms in evaluating the light and since you have a difference in the color between the first generation and the second generation, this approach obscures any real difference in the quantum efficiency. In other words you are recording luminous units, and just visually it looks to me like your second generation has a spectral distribution which is closer to the photopic sensitivity of the eye.

RAUHUT: I would have thought that too. We associate yellow with high eye sensitivity. Actually it doesn't turn out that way when you make the calculations. The photopic factor for the yellow system is actually about 7 or 8% lower. The reason primarily is that the spectral envelopes are quite broad so that a substantial amount of the light is emitted toward the red end where the eye sensitivity is low. But actually the yellow system in quantum yield terms is more efficient and operates at about 16% whereas the green system is about 10 or 11% efficient.

THORINGTON: In a comparison with the incandescent lamp there would appear to be a very big difference in the spectral distribution. If one takes the power efficiency of the incandescent lamp I think you'll have about 1/5th of the burning time instead of 40 minutes, to correlate with the chemiluminescence. In other words the difference in the spectral distribution there is a very large one.

RAUHUT: We have just determined the luminous output of incandescent lamps using standard handbook values which give you the luminosity per watt. Using this conversion factor, we have converted the incandescent lamp power to lumens.

THORINGTON: If you convert that same power into this same spectral distribution you'd have much more brightness and many more lumens.

RAUHUT: Yes; of course the incandescent lamp is not really efficient because its distribution is based on the color temperature. A lot of the radiation from it is in the red, which you don't see. On the other hand, the light that does come out of this device is visible light, chemiluminescence.

REYNOLDS: Since the matter has been brought up involving the eye response it might be interesting to keep in mind the possibility a class of light source where the scotopic eye is the one to be matched because you need it most when you are dark adapted, and the device might look better on that basis than when compared to the eye response at a higher light level.

McELROY: Assuming you could get cheap luciferin you could bypass the enzyme by using the ester of luciferin which gives extremely bright light in the chemiluminescent reaction. I don't have any idea, though, how high a concentration is required to sustain it over a long period of time. Perhaps some of the other workers could tell us this. It certainly is extremely brilliant.

RAUHUT: Yes, we've wondered about that too, we don't have any knowledge of it ourselves but you have emphasized an important practical point: that one needs not only a high quantum yield but a high concentration as well.

McCAPRA: We have measured the quantum yield for the luciferin simple phenyl ester and its just under 10% - it's not bad. The difficulty is in concentration terms you lost terribly. Anything like a useful concentration quenches the reaction.

APPLICATIONS OF BIOLUMINESCENCE AND CHEMILUMINESCENCE[1]

H. H. Seliger

McCollum-Pratt Institute and Department of Biology
The Johns Hopkins University, Baltimore, Maryland 21218

A. INTRODUCTION

Chemiluminescence (bioluminescence) arises from the excited states of products of an exothermic reaction. It is analogous to photoluminescence in that the individual excited molecular species are not in thermal equilibrium with their neighbors nor with the solvent molecules. Since relaxation to stable energy levels occurs at rates corresponding to vibrational collision frequencies (ca. 10^{12} sec^{-1}) the negative free energy change in the chemical reaction leading to the electronically excited product must occur as a single step. During the lifetime of the excited state (ca. 10^{-9} sec) of the product molecule, non-luminescent pathways such as quenching (formation of non-fluorescent excimers), energy transfer to non-fluorescent acceptors and intramolecular radiationless deactivation may compete with the luminescent transition to the ground state. In general, three criteria must be satisfied simultaneously to achieve a sensible chemiluminescence. These are

I. $$-\Delta F \geq \frac{hc}{\lambda_m}$$

where λ_m is the long wavelength limit for excitation of the product molecule and ΔF is the free energy change for the reaction.

II. Efficient chemical pathway leading to the excited state of the

---

[1] Work supported under AEC Contract AT(11-1)3277.

product molecule. This can be represented as a fraction ϕ_{ex} of the total chemical pathways.

III. Efficient fluorescence quantum yield ϕ_f of the excited state of the product molecule.

Thus the overall quantum yield of the chemical reaction is

$$\phi_c = \phi_{ex}\phi_f \tag{1a}$$

and is defined operationally as

$$\phi_c(N) = \frac{\text{total number (or rate) of quanta emitted by } P^*}{\text{total number (or rate) of molecules of N reacted}} \tag{1b}$$

where N is one of the reactant molecular species and P^* is the excited state of the product molecule P.

The definitions in (1a) and (1b) are sufficient for a major class of chemiluminescent reactions, <u>direct</u> <u>chemiluminescence</u>. However there are many cases where ϕ_f due to competing non-luminescent pathways is extremely low. In these cases (Vassil'ev 1962; 1967; Vassil'ev <u>et al</u>., 1963) suitable acceptor molecules A with high fluorescent quantum yields may be introduced such that

$$P^* + A \longrightarrow P + A^* \quad (\text{efficiency } \phi_{tr}) \tag{2}$$

occurs with high probability. In these cases of <u>sensitized chemiluminescence</u>, equation (1a) takes the form.

$$\phi_c(N) = \phi_{ex}(P^*) \, \phi_{tr} \, (P^* \longrightarrow A^*) \, \phi_f \, (A^*) \tag{3}$$

The general forms of the equations for most oxygen-requiring chemiluminescent (bioluminescent) reactions in solution are:

$$N^- + O_2 \longrightarrow N\text{-}O\text{-}O^- \longrightarrow P^* + V \tag{4}$$

where N^- is the anion of the reactant molecule, O_2 is the dissolved oxygen molecule, $N\text{-}O\text{-}O^-$ is the intermediate peroxy addition complex and V is a collective symbol for the remaining components in the exothermic reaction which may also carry away vibrational energy.

$$N + O_2^- \xrightarrow{H^+} N\text{-}O\text{-}OH \longrightarrow P^* + V \tag{5}$$

where O_2^- is the superoxide radical ion.

$$\dot{N} + O_2 \longrightarrow \dot{N}O_2 \xrightarrow{H} N\text{-}O\text{-}OH \longrightarrow P^* + V \tag{6}$$

$$\overset{\bullet}{NO_2} + \overset{\bullet}{NH} \longrightarrow N\text{-}O\text{-}OH + \overset{\bullet}{N} \qquad (7)$$
$$\downarrow$$
$$P^* + V$$

$$\overset{\bullet}{NO_2} + \overset{\bullet}{NO_2} \longrightarrow NO_2 + {}^{1,3}NO_2^* \xrightarrow{A} A^* \qquad (8)$$

$$\overset{\bullet}{NO_2} + \overset{\bullet}{NO_2} \longrightarrow O_2^* + \text{products} \qquad (9)$$

$$N + O_2^* \longrightarrow N\overset{O}{\underset{O}{\diagdown}} \longrightarrow P^* + V \qquad (10)$$

Weak light emissions in the oxidation of a large variety of organic and inorganic compounds have been known for many hundreds of years (see Harvey, 1957 for a complete historical review through 1900; Trautz, 1905; Trautz and Shorygin, 1905; Audubert, 1939; Vassil'ev et al., 1959; Shlyapintokh et al., 1960; Ahnström, 1961; Ashby, 1961; Stauff and Rümmler, 1962; Barenboim et al., 1969). The number of direct chemiluminescence photons emitted per molecule reacted can range from 10^{-14}- 10^{-15} (Audubert, 1939) through 10^{-2} for Luminol (Lee and Seliger, 1965; 1970; 1972) to as high as 1 in the enzyme-catalyzed oxidation of firefly luciferin (Seliger and McElroy, 1959; 1960). The high intrinsic efficiency of modern phototubes for the detection of chemically generated light quanta at rates below 100 $\sec^{-1}$ presents us with a unique external probe for monitoring chemical (or biochemical) reactions with a sensitivity such that specific enzymatic oxidations can be examined in single cells. In principle, one could introduce into a complex chemical process requiring precise regulation of either conditions or reactants, a trace impurity in parts per billion whose resultant chemiluminescence would be proportional to the concentration of some critical intermediate in the chemical process. Thus the externally observed light intensity could be incorporated into the overall control system (Shlyapintokh et al., 1968).

It is the purpose of this review to describe a range of applications of bioluminescent and chemiluminescent reactions and to possibly stimulate the further application of these luminescent reactions to other areas.

B. PHOTOELECTRIC DETECTION OF LIGHT QUANTA

The initial impetus for the development of phototubes and electronic instrumentation for the efficient detection of photons came from nuclear physics and the techniques of scintillation counting (for reviews see Birks, 1953; 1964; Wang and Willis, 1965). Subsequently the requirements for low-noise phototubes in such diverse fields as liquid scintillation counting and astronomy resulted in new photocathode surfaces with low thermionic yields even at room temperature and even an ingenious continuous dynode geometry for electron multiplication giving saturation pulses for initial photoelectrons analogous to the Geiger-Muller counter, developed by Bendix as the Channeltron phototube.

1. Pulse Detection versus Current Detection of Low Intensity Light Levels

The limit of photon detection is set by the fluctuations in the dark noise of the phototube. This may be the result of (a) leakage across elements of the internal structures of the phototube elements, (b) leakage across the external connectors or tube base, (c) shot noise due to random thermal emission from the photocathode and dynode surfaces, (d) Johnson noise associated with the load resistor, (e) electrons produced by secondary ionization of residual gas molecules within the phototube, (f) secondary photoelectrons produced by recombination radiation emitted upon neutralization of ionized residual gas molecules on dynode surfaces, (g) field emission from the glass envelope or other insulator surfaces due to the high potentials applied to the electrode structures (up to 3000 volts), (h) beta particles or beta particle-produced luminescence emitted into the electron multiplication region due to natural radioactivity of the glass envelope (mainly ^{40}K) and the electrode structures, external radioactivity or cosmic ray background.

In the electron multiplier phototube an incident photon will have a photoelectric conversion efficiency ε_λ and for each photoelectron produced, a probability f of reaching the first dynode, giving rise to an avalanche of secondary electrons. The total charge collected at the anode of the phototube due to single photoelectrons produced at the photocathode and going through m stages of secondary amplification, will exhibit Poisson statistics (see Breitenberger, 1955; Birks, 1964). The multiplication factor of a dynode is the product of the secondary emission coefficient and the fraction of secondary electrons collected at the <u>succeeding</u> dynode. The overall gain for m dynodes reduces to

$$M = R^m \qquad (11)$$

for equivalent stages of gain. The variance, Var (X), for a Poisson distribution is defined as

$$\text{Var}(X) = \langle X \rangle \tag{12}$$

The relative variance

$$v(X) = \frac{\text{Var}(X)}{\langle X \rangle^2} \tag{13}$$

is, for a Poisson distribution, equal to

$$v(X) = \frac{1}{\langle X \rangle} \tag{14}$$

For an electron cascade due to m dynode stages, the relative variance of the gain is given (Birks, 1964 p. 150) as

$$v(M) = \frac{1}{R - 1} \tag{15}$$

where it is assumed that all gain stages R are equal. The total charge q collected at the anode will be given by

$$q = N_\lambda \mathcal{E}_\lambda fM \tag{16}$$

where $N_\lambda \mathcal{E}_\lambda$ represents the mean photoelectron production due to the photon spectral intensity incident on the photocathode.

In the case of single photon counting, as a separate case from scintillation counting, $v(q) = v(M)$.

From the definition of Var(q)

$$\frac{\langle q^2 \rangle - \langle q \rangle^2}{\langle q \rangle^2} = \frac{1}{R - 1}$$

$$\frac{\langle q^2 \rangle}{\langle q \rangle^2} = \frac{R}{R - 1} \tag{17}$$

The following approximate treatment is based on Robben (1971). We can examine the characteristics of a phototube for either pulse counting or current measurement of very weak light sources in terms of a photoelectron noise factor, S. Let us consider first the case of pulse counting. Since there is always an arbitrary discriminator setting in pulse counting we do not always count all of the single photoelectron pulses. We count only those pulses whose charge is greater than q_d, the discrimination level. Thus the counting rate r, due to ℓ photons per second incident on the photocathode will be

$$r = \ell \, \mathcal{E} \, f \, \alpha(q_d) = n \, \alpha(q_d) \tag{18}$$

where $\alpha(q_d)$ is a function of the discriminator level q_d, and n is the counting rate if all photoelectrons reaching dynode 1 were detected. Ideally then from (14), if X = nt counts are observed over a time t, $v(X) = 1/nt$. However, as a result of the characteristics of the phototube and electronics there will be an additional component in the observed variance. The total effect is included in S_p, the photoelectron noise factor for pulse counting as

$$v(X) = \frac{S_p}{nt} \tag{19}$$

If r is the observed counting rate, $v(X) = \frac{1}{rt}$, so that

$$S_p = \frac{1}{\alpha(q_d)}. \tag{20}$$

S_p is a function of discriminator setting q_d.

Under the same conditions the rate of n photoelectrons per second produces a current $i = n \langle q \rangle$. Assuming a Poisson distribution for equal stages of dynode multiplication,

$$v(i) = \frac{n(\langle q^2 \rangle - \langle q \rangle^2)}{n^2 \langle q \rangle^2} = \frac{1}{n} \left(\frac{\langle q^2 \rangle}{\langle q \rangle^2} - 1 \right) \tag{21}$$

Since from (19), $S_p - 1$ is the fractional increase in noise-to-signal ratio squared $S_c = \langle q^2 \rangle / \langle q \rangle^2$.

From (17),

$$S_c = \frac{R}{R-1}. \tag{22}$$

In practice a light source which gives of the order of 10^5 photoelectron pulses per second is used for testing the phototube. This counting rate is very much larger than the dark noise counting rate and the latter can therefore be neglected. A plot of log r versus integral discriminator setting at various phototube voltages should yield curves which rise linearly at high values of q_d, approach a plateau as q_d decreases and then rise sharply as $q_d \to 0$. Extrapolation of the plateau portion of the optimum curve to $q_d = 0$ will give n, the true photoelectron counting rate, and from (20)

$$S_p = \frac{n}{r} \quad (23)$$

In a typical phototube $R \simeq 5$ so that from (22) $S_c \simeq 1.25$. We should expect therefore that S_p for a good phototube should also be equal to or less than 1.25.

Without changing the conditions of illumination we now measure the mean anode current $\langle i \rangle$. The mean charge $\langle q \rangle$ per photoelectron pulse is thus

$$\langle q \rangle = \frac{\langle i \rangle}{n} \quad (24)$$

from which the phototube gain can be calculated. The light source is then removed and the dark counting rate r_d and dark current i_d are measured. It is usually found that the dark current i_d is significantly larger than the product of $r_d S_p \langle q \rangle$, the latter two previously measured with the external light source. Since most of the contributions to dark noise listed above will give rise to electron cascades at the anode which do not go through all dynode multiplication stages, the anode current will be the sum of true photocathode-emitted thermionic electron current plus this extraneous current.

From (19), for pulse counting of dark noise

$$v(X_d) = v(r_d t) = \frac{S_p}{n_d t} = \frac{S_p}{n_d t} \times \frac{r_d S_p}{n_d} = \frac{1}{n_d t}\left(\frac{D_p}{n_d}\right) \quad (25)$$

where

$$D_p = R_d S_p^2 \quad (26)$$

Is the effective dark counting rate for pulse measurement.

The effective dark counting rate for current measurement is

$$D_c = D_p + \frac{i_d - D_p \langle q \rangle}{\langle q \rangle} \quad (27)$$

The figure of merit of a phototube in terms of its potential to be used for current measurement of low intensity light sources is

$$F_c = \frac{D_c}{D_p} = \frac{i_d}{r_d S_p^2 \langle q \rangle}. \quad (28)$$

Ideally F = 1. Thus we have a criterion for judging whether any particular phototube should be used for low intensity light measurements. Since it is possible to obtain "good" commercially available phototubes, and since Robben (1971) in a detailed review paper has shown that under conditions where $F \simeq 1$ there is no significant difference in observed signal to noise ratios by pulse counting techniques as compared with current measurements, there is no <u>a priori reason to use the more complex pulse counting techniques over the much simpler D.C. amplifier current measurements for optimum detection of very weak chemiluminescent reactions</u>.

C. APPLICATIONS

1. Illumination

At the International Exposition in Paris in 1900, R. Dubois illuminated a vast hall of the Palace of Optics by means of large glass jugs coated over their interiors with a gelatin nutrient medium previously inoculated with luminous bacteria, producing light as bright as a clear moon (Dubois, 1914). At that time it was proposed, because of the absence of heat associated with the light, that luminous bacteria be used in powder magazines and in mining, where there was always the serious danger of explosions. More recently application has been made of the efficient chemiluminescence of Luminol, tetrakis dimethylamino ethylene (TMAE) and sensitized chemiluminescence in the oxalic acid ester oxidation. The latter has been developed into a practical, commercially available light "wand" for temporary illumination (Rauhut, this symposium).

2. Detection of Specific Reactants in Bioluminescence and Chemiluminescence

a) <u>ATP in Firefly Reaction</u>. In the firefly bioluminescent reaction the first step is the production of an activated enzyme-substrate complex, specific for ATP.

$$LH_2 + E + ATP \xrightleftharpoons{Mg^{++}} E \cdot LH_2 AMP + PP \qquad (29)$$

The subsequent reaction with dissolved molecular oxygen results in an excited state enzyme-decarboxy keto luciferin product, emitting yellow-green light (562 nm) with a quantum yield close to unity (McElroy et al., 1969). Thus with all reactants except ATP in excess, the bioluminescent light intensity or total light emitted should be directly proportional to ATP concentration or total ATP, respectively.

i) <u>Extraterrestrial Life Detection</u>. Because of the extreme sensitivity of the photon detection technique and owing to the ubiquitous presence of ATP in all life forms, a method proposed in the NASA program for detection of extraterrestrial life involved the detection of bioluminescence of a firefly *in vitro* reaction mixture containing excess luciferin, luciferase and Mg^{++} ions in buffered oxygenated solution to which an extraterrestrial grab sample could be added. Light emission would be direct evidence for ATP and presumably of the previous synthesis of ATP by living organisms. As the result of work by Chappelle and others at NASA, it was found that purified luciferase (Sephadex columns) at approximately 1 mg/ml in a solution of 0.05 $\underline{M}$ Tris buffer at pH 7.4 containing 0.01 $\underline{M}$ $MgSO_4$ and 1 mg/ml of the luciferin substrate could be lyophilized, stored indefinitely at -60°C and reactivated by addition of H_2O prior to assay for ATP (Chappelle and Levin, 1968). With the most sensitive instrumentation the assay technique is reported to be able to detect as little as 0.01 pgm or 2×10^{-5} pmole of ATP. With partially purified luciferase the presence of "residual light" limits the technique to 1 pgm ATP. The kinetics of light emission show an initial rise in light intensity, the peak height of which depends upon the rapidity of injection and subsequent mixing, followed by a complex decay, depending on luciferase purity and concentration and on ATP concentration. Some investigators (Seliger and McElroy, 1965 Appendix VII; Chappelle and Levin, 1968) prefer to use the flash height, while others (Holm-Hansen and Booth, 1966; Holm-Hansen, 1971) prefer to integrate the light intensity over a fixed time.

ii) <u>Determination of Bacterial Contamination</u>. With present techniques it is possible to detect as few as 10^4 cells of <u>Escherichia coli</u>, which contains 10^{-16} g ATP per bacterium. In a study of 19 species of bacteria Chappelle and Levin found a mean of 2×10^{-16} g ATP/bacterium with a coefficient of variation of 100 percent, due mainly to three species with "abnormally" high ATP/cell; <u>Bacillus globigii</u> (5.4×10^{-16} g), <u>Klebsiella pneumoniae</u> (5.0×10^{-16} g) and <u>Mycobacterium smegmatis</u> (8.9×10^{-16} g). In seven species of marine bacteria representing five genera, Hamilton and Holm-Hansen (1967) have reported values ranging from 0.5×10^{-15} to 6.5×10^{-15} g ATP per bacterium; a mean of 2.3×10^{-15} g ATP/bacterium with a coefficient of variation of approximately 100 percent, due mainly to <u>Chromobacterium marinum</u> (6.5×10^{-16} g).

The apparent disagreement between these two independent series of measurements amounts to a factor of 10 and is particularly significant since this same factor of 10 is present for three <u>Pseudomonas</u> species measured by Hamilton and Holm-Hansen and two <u>Pseudomonas</u> species measured by Chappelle and Levin. It is obvious therefore that a concerted intercomparison be instituted among various laboratories using the ATP technique for bacterial assays. One of the possible sources of this discrepancy is discussed by Hamilton and Holm-Hansen (1967). They find variations of an order of magnitude in ATP/cell depending on the stage of the bacterial culture, i.e. high during lag growth and low at stationary or senescent phase and in starved cells. The data of Hamilton and Holm-Hansen are presented for chemostat cultures containing $\sim 6 \times 10^8$ cells/ml, similar to the concentrations ranging from $6\text{-}33 \times 10^8$/ml reported for the batch cultures of Chappelle and Levin. No mention is made of stage of growth of the latter investigators. D'Eustachio and Levin (1967) reported a constancy of ATP/cell during growth phases and Levin et al. (1968) separately confirmed the data of Chappelle and Levin (1968), so that the question is still unresolved.

iii) <u>Microbial Biomass in Marine Biology</u>. The use of the ATP assay in marine biology for determining the biomass of heterotrophically growing cells at ocean depths below the euphotic zone and of phytoplankton in the euphotic zone has been pioneered by Holm-Hansen and his associates (Holm-Hansen and Booth, 1966; Hamiltion and Holm-Hansen, 1967; Coombs et al., 1967; Holm-Hansen, 1969; 1970; 1972). It has been estimated that over 99 percent of all organic material in the ocean is found in the dissolved state or in nonliving particulate material (Hamilton and Holm-Hansen, 1967). In the euphotic zone, depending on whether samples are taken from ocean water, more turbid inshore water or highly turbid estuarine water the biomass can vary from 20 to close to 100 percent of the total organic matter. It has been demonstrated that ATP is present in living cells, diffuses rapidly out of dead cells and in general is not adsorbed appreciably on sediment or detrital particles. This latter adsorption appears to be a problem however if large volumes are filtered from waters containing high sediment loads (Sutcliffe et al., 1972). The ratio of cellular organic carbon to ATP has been found to be close to 286 for marine and freshwater algae (Coombs et al., 1967; Syrett, 1958; Holm-Hansen, 1969; 1970) bacteria (Hamilton and Holm-Hansen, 1967; Kelly and Syrett, 1966; Cole et al., 1967; Forest, 1965) molds (Dawes and Large, 1970) tumor cells (Jones, 1969) and the zooplankters <u>Calanus helgolandicus</u> and <u>C. hyperbarius</u> (Holm-Hansen, 1972). For comparison carbon to chlorophyll <u>a</u> ratios of around 100 have been reported by Loftus et al. (1972); Eppley and coworkers (Anon, 1971); Holm-Hansen, 1969; and Parsons et al., 1961. The ATP assay thus furnishes a corroborative check with chlorophyll <u>a</u> assays of phytoplankton for the estimation of biomass in surface waters (where phytoplankton are the major source of total organic carbon). It has the further advantage of generating total biomass numbers and therefore in principal

Carbon(from ATP) - Carbon(from chlorophyll <u>a</u>) = Heterotrophic Carbon

(30)

The phytoplankton carbon numbers obtained in these cases divided into primary production rates obtained by ^{14}C uptake studies give generation rate constants for photosynthesis in the water volume. These rate constants, together with carbon:nitrogen:phosphorus ratios in phytoplankton and measurements of N and P soluble nutrient concentrations in the sampled waters permit estimates to be made of turnover rates which are essential in any systematic approach to an ecosystem study.

b) <u>Reactants in Firefly Bioluminescence and in ATP Formation.</u> The firefly reactants, mixed anaerobically, can be used for extremely sensitive detection of oxygen concentrations. This has been applied in the detection of oxygen evolution in photosynthesizing algae. In an analogous manner the firefly system minus ATP can be used in chloroplast extracts for the detection of rates of photophosphorylation. Stanley and Williams (1969) have used the ATP generated in the reaction

Adenosine 5' phosphosulfate + PP + ATP sulphurylase $\rightleftharpoons$ ATP + sulfate   (31)

to determine as little as 10^{-11} moles of inorganic pyrophosphate. In general, therefore, any biochemical reaction generating ATP (transphosphorylase and kinase reactions) can be assayed by measuring firefly bioluminescence.

c) <u>FMN Specificity in Bacterial Bioluminescence.</u> In the <u>in vitro</u> bacterial bioluminescent reaction the reactants are (Hastings <u>et al.</u>, this symposium).

$FMNH_2 + E + RCHO + O_2 \longrightarrow$ blue bioluminescence   (32)

with quantum yields per $FMNH_2$ molecule of 0.05 (Lee, 1972).

i) Extraterrestrial Life Detection. In view of the ubiquitous occurrence of flavin compounds in living organisms, another of the methods proposed in the NASA program for detection of extraterrestrial life has been the bioluminescence of a bacterial in vitro reaction mixture for the specific detection of FMN as $FMNH_2$. The technique has a distinct advantage over fluorometry in terms of the specificity for flavin mononucleotide (Chappelle <u>et al.</u>, 1967). The bacterial light reaction utilizes a crude extract of <u>Achromobacter fischeri</u>, dissolved in 0.05 M Tris buffer at pH 7.4 at a concentration of 1 mg/ml. This solution, mixed 2:1 with a saturated solution of the bisulfite addition complex of dodecylaldehyde in the same buffer, can be lyophilized and stored at -20°C. The extraction of FMN is in 6% butanol in 0.01 $\underline{M}$ Tris buffer containing 10^{-3} $\underline{M}$ EDTA, reduced by addition of dry sodium borohydride $NaBH_4$ to make 1 mg/ml of solution and by addition of a minute aliquot of $PdCl_2$ catalyst solution (>1 ppm $PdCl_2$). Chappelle has established a linear relationship between the peak intensity of a bioluminescence emission versus time curve and FMN concentration down to 10^{-11} g FMN in the reaction cuvette. The limit of detection for Escherichia coli (7 x 10^{-17} g FMN/cell) would be approximately 10^5 cells, to be compared with the ATP technique discussed above.

d) <u>Other Reactants in Bacterial Bioluminescence</u>. In crude extracts of luminous bacteria a DPNH dehydrogenase will react with DPNH and FMN to reduce the latter to $FMNH_2$, subsequent to which the coupled luciferase enzyme, in the presence of aldehyde (longer than C_6), will produce bioluminescence. The crude <u>in vitro</u> bacterial enzyme extract can therefore be used to detect concentrations of DPNH, and in any preceding enzyme reaction which results in DPNH, the concentration of substrate being oxidized. Malate or oxalacetate can be detected using the reaction (Stanley, 1971)

$$\text{oxalacetate} + \text{DPNH} \xrightleftharpoons{\text{malate dehydrogenase}} \text{Malate} + \text{DPN} \quad (33)$$

For detection of malate excess DPN is added to push the equilibrium to the left. Oxalacetate is assayed by the rate of DPNH oxidation (the rate of decrease in the DPNH dependent bacterial bioluminescent intensity). Biejerinck (1902) used the oxygen specificity of luminous marine bacteria to measure oxygen evolution due to photosynthesis in a suspension of crushed clover leaves previously made anaerobic (see Harvey, 1952 pp. 34-38 for other early experiments).

e) <u>Calcium Ion Specificity in Aequorea Bioluminescence</u>. The specificity of Aequorin bioluminescence for Ca^{++} (and to a much smaller extent Sr^{++}) ions provides the basis for a quantitative micro method which can easily detect 10^{-8} g Ca^{++} in 2 ml of the reaction solution (<u>ca</u>. 10^{-7} <u>M</u>) (Shimomura <u>et al</u>., 1962; 1963a; 1963b). The photoprotein aequorin is fairly stable in saturated ammonium solution containing EDTA at dry ice temperatures. The technique has been used to examine the nature of intracellular Ca^{++} ion transients in electrically stimulated single muscle fibers of the acorn barnacle <u>Balanus nubilus</u> (Ridgway and Ashley, 1967). The authors were able to demonstrate that Ca^{++} release after simulation was rapid, and could be observed prior to the onset of tension, which began 5 msec after the beginning of stimulation. The peak Ca^{++} concentration occurs before the muscle has reached 10% of its maximum tension, in qualitative agreement with the previous data of Jobsis and O'Conner (1966). The conclusion appears to be that there is no linear relationship between Ca^{++} concentration and tension. The kinetic limitations of this technique have been investigated by double stopped flow measurements (Hastings <u>et al</u>., 1969). The following kinetic scheme is proposed:

$$A + Ca^{++} \underset{k_{-1}}{\overset{k_1}{\rightleftharpoons}} A \cdot Ca^{++} \xrightarrow{k_2} X \xrightarrow{k_3} Y^* \xrightarrow{k_4} Y + h\nu \quad (34)$$

At optimum Ca^{++}, the formation of X ($k_2 \simeq 1\ sec^{-1}$) is the limiting reaction.

Since $k_3 = 100\ sec^{-1}$, the response time of the luminescence to initial rapid changes in Ca^{++} concentration will be rapid. However owing to the rate limiting constant k_2 it appears difficult to define absolute Ca^{++} concentrations except over periods of seconds.

f) Chemiluminescent Assay for Trace Elements. The aqueous chemiluminescence of Luminol can be catalyzed by trace amounts of a variety of ionic species (Seitz and Hercules, this symposium), such as Co(II), Cu(II), Ni(II), Cr(III), V(II), Mn(II), Fe(II), MnO_4^-, I_2 and Br_2. The latter authors suggest the feasibility of chemiluminescence titration, and use as an example the case of the titration of As(III) with I_2. The reaction produces $2\ I^- + As(V)$ and in the Luminol solution no chemiluminescence is observed until excess I_2 is present. In all of the above cases the detection limit ranges between 10^{-8} M and 10^{-11} M. The use of the Luminol chemiluminescent reaction in forensic medicine as a detector of heme is also well established.

3. Air Pollution

Dubois (1914) reported that general anesthetics produced changes in the bioluminescent intensity of luminous bacteria. Harvey (1952) lists a series of old researches dating from 1889 on the effects of alkaloids, vapors of ether, alcohols, tobacco smoke, metal salts, etc. on the brightness of cultures of luminous bacteria. These initial observations have recently been extended and strains of luminous bacteria have been isolated for use as nuclear radiation detectors, as air pollutant detectors (Serat et al., 1965) and most recently (Newsweek, October 9, 1972), as substitutes for German police dogs in the "sniffing out" of heroin. All of these applications make use of some chemical inhibition of one or more of the factors involved in the mechanism of in vivo bacterial bioluminescence.

Chemiluminescent detection of ozone, oxides of nitrogen and sulfur dioxide as homogeneous gas phase reactions can give sensitivities in the ppb (parts per billion) range (Fontijn et al., this symposium). Chemiluminescence is observed when an excess of a second reactant species is added to the molecule of interest in a static or an air flow system. Alternatively the molecule of interest can be converted to a reactant species which gives rise to chemiluminescence. In several practical cases chemiluminescence in cooled flames, provides a sensitive assay. The intense near UV emission of S_2 obtained when sulfur compounds are burned in a hydrogen-rich flame, is used as an assay for SO_2. Heterogeneous chemiluminescence such as that obtained when air containing ozone is bubbled through Luminol can also have a wide range of applicability. In all of these cases the detector is calibrated by adding known concentrations of gas samples.

4. Low Level Biological Chemiluminescence

The concept of using bioluminescence (or chemiluminescence) as a reaction rate tool (Johnson, 1948) was brought to its extreme fruition with the development of the new generation of electron multiplier phototubes. There are several areas of investigation which are accessible only to photon detection techniques. These are the very weak luminescence emissions from living matter, particularly rapidly growing (oxidizing) cells (Konev et al., 1961; Stauff et al., 1963; Stauff and Ostrowski, 1967; Barenboim et al., 1969).

The extreme sensitivity of photon detection has brought the original mitogenetic radiation of Gurvich (see discussion in Barenboim, 1969 pp. 121-125) onto a more quantitative level. It has been possible at least to verify that these weak chemiluminescences do in fact occur and are most likely radical recombinations with or without sensitization. It can be demonstrated that addition of H_2O_2 to whole cells or homogenates greatly stimulates the observed luminescence (Konev et al., 1961; Steele, 1963). That weak UV chemiluminescence has any biological function remains to be investigated. White et al. (1972) suggest that excited states can be produced chemically in biological molecules not normally exposed to photoexcitation, a "photobiology without light" (see Bun-Hoi and Sung, 1970; Pettrus and Moore, 1970 and Lamola, 1971). It is conceivable that the weak chemiluminescences observed in biological systems are tracers of reactions which involve excited electronic states which are normally quenched (proceed efficiently).

5. The Study of Excited States and Kinetics of Reactions

The chemiluminescence method for study of excited states has several advantages (Vassil'ev, 1967): a) Excitation is uniform throughout the mass of the sample. b) There is no background scatter from the source and therefore reactions with very low φ_{ex} may be detected. c) The absorption of exciting light by other components is eliminated; strongly fluorescent components are not excited. d) Since excitation is by chemical reaction the usual selection rules associated with electronic photoexcited transitions do not apply and population of ordinarily "forbidden" levels is possible.

Vassil'ev, for example, examined the mechanism of energy transfer from a ketone donor (weakly luminescent) to an anthracene acceptor. This could not have been studied by photoexcitation since the anthracene would have been excited along with the ketone. In the oxidation of ethylbenzene in benzene an enhancement of luminescence curve (analogous to a Stern-Volmer quenching curve) plotted as a function of acceptor concentration allowed the determination of luminescence yields as low as 5×10^{-4}. Introduction of high Z atoms into the acceptor molecules (bromine additions to anthracene) enhances the interaction between spin and orbital angular moments of the electrons and then results in a mixing of triplet and singlet states. The resulting enhancement of energy transfer between triplet ketone excited states populated by chemical oxidation and the halogen substituted anthracene as compared with alkyl or phenyl anthracene was greater by one to three orders of magnitude. These experiments furnished the basis for the conclusion that triplet levels are populated in the chemiluminescence of methylethylketone and ethylbenzene. Luminol (Nilsson, 1969) or pyrogallol (Nilsson, 1964) have been used as indicators to study the kinetics of peroxidase reactions.

McElroy et al. (1969) have analyzed the shape and shift in the bioluminescence emission spectrum in the firefly to infer the nature of the local dielectric environment at the active sites of excited-state, enzyme-bound product molecules.

It has been possible with low noise phototubes to study the endogenous rhythm of spontaneous bioluminescence in single dinoflagellate cells which emit of the order of 200 photons per second.

D. SUMMARY

Light emissions from bioluminescent and chemiluminescent reactions have a wide range of applicability. A major area which has not been fully exploited is that of process control, where the instrumentally observed rate of chemiluminescence can be incorporated into the controlling system. This would have direct application to polymerization reactions involving radical initiators, the scrubbing of SO_2 in power plant air pollution control equipment, bacterial and tissue growth chemostats--any reaction where photon emission is proportional to the concentration of one of the reactants.

REFERENCES

Ahnström, G. (1961) Acta Chem. Scand. 15:463.
Anonymous (1971) "Eutrophication in coastal waters:nitrogen as a controlling factor", Water Pollution Control Research Series, 16010 EHC, 12/71 67 pp.
Ashby, G. (1961) J. Polymer Sci. 50:99.
Aubert, R. (1939) Trans. Farad. Sco. 35:197-204.
Barenboim, G. M., A. N. Domanskii and K. K. Turoverov (1969) Luminescence of Biopolymers and Cells, Plenum Press, N.Y. 229 p.
Birks, J. B. (1953) Scintillation Counters, McGraw Hill, N. Y., 148 p.
Birks, J. B. (1964) The Theory and Practice of Scintillation Counting, MacMillan Co., N. Y. 662 p.
Breitenberger, E. (1955) in Prog. Nuclear Phys. (Ed., O. R. Frisch) 4:56.
Bun-Hoi, N. P. and S. S. Sung (1970) Naturwissen. 57:135.
Chappelle, E. W. and G. V. Levin (1968) Biochem. Medicine 2:41-52.
Chappelle, E. W., G. L. Picciolo and R. H. Altland (1967) Biochem. Medicine 1:252-260.
Cole, H. A., J. W. T. Wimpenny and D. E. Hughes (1967) Biochim. Biophys. Acta 143:445-453.
Coombs, J., P. J. Halicki, O. Holm-Hansen and B. E. Volcani (1967) Exper. Cell Research 47:315-328.
Dawes, E. A. and P. J. Large (1970) J. Gen. Microbiol. 60:31-40.
D'Eustachio, A. J. and G. V. Levin (1967) Bacteriolog. Proc. Abstract.
Dubois, R. (1914) La Vie at La Lumière, Libraire Felix Alcan, Paris 338 p.
Fontijn, A., D. Golomb and J. A. Hodgeson (1972) Int. Symp. Chemilum., Athens, Ga.
Forrest, W. W. (1965) J. Bacteriol. 90:1013-1016.
Hamilton, R. D. and O. Holm-Hansen (1967) Limnol. Oceanog. 12:319-324.
Harvey, E. N. (1952) Bioluminescence, Academic Press, N. Y. 649 p.
Harvey, E. N. (1957) A History of Luminescence, Amer. Phil. Soc., Phila., Pa. 692 p.
Hastings, J. W., G. Mitchell, P. H. Mattingtly, J. R. Blinks and M. Van Leeuwen (1969) Nature 222:1047-1050.
Hastings, J. W., T. Baldwin, A. Eberhard, M. N. Nicoli, T. Cline and K. Nealson (1972) Int. Conf. Chemilum., Athens, Ga.
Holm-Hansen, O. (1969) Limnol. Oceanogr. 14:740-747.
Holm-Hansen, O. (1970) Plant and Cell Physiol. 11:689-700.
Holm-Hansen, O. (1971) Symposium on Estuarine Microbiology, Columbia, S. C. (USCD 10P20-107).
Holm-Hansen, O. (1972) U. Calif. San Diego Report 10P20-124.
Holm-Hansen, O. and C. R. Booth (1966) Limnol. Oceanogr. 11:510-519.
Johnson, F. H. (1948) Scientific Monthly 67:225-235.
Jones, P. C. T. (1969) J. Cell Physiol. 73:37-42.

Kelly, D. P. and P. J. Syrett (1966) J. Gen. Microbiol. 43:109-118.
Konev, S. V., N. A. Troitsky and M. A. Katibnikov (1961) V. Internat. Biochem. Congr. Moscow.
Lamola, A.A.(1971) Biochem. Biophys. Res. Commun. 43:893.
Lee, J. (1972) Biochemistry 11:3350-3359.
Lee, J. and H. H. Seliger (1965) Photochem. Photobiol. 4:1015-1048.
Lee, J. and H. H. Seliger (1970) Photochem. Photobiol. 11:247-258.
Lee, J. and H. H. Seliger (1972) Photochem. Photobiol. 15:227.
Levin, G. V., E. Usdin and A. R. Slonim (1968) Aerospace Med. 39:14.
Loftus, M. E., S. V. Rao and H. H. Seliger (1972) Ches. Sci., in press.
McElroy, W. D., H. H. Seliger and E. White (1969) Photochem. Photobiol. 10:153-170.
Nilsson, R. (1964) Acta Chem. Scand. 18:389-401.
Nilsson, R. (1969) Biochim. Biophys. Acta 184:237-251.
Parsons, T. R., K. Stephens and J. D. H. Strickland (1961) "On the chemical composition of eleven species of marine phytoplankters", J. Fish. Res. Bd. Canada 18:1001-1016
Pettrus, J. A., Jr. and R. E. Moore (1970) Chem. Commun. 1093.
Rauhut, M. M. (1972) Int. Conf. Chemilum., Athens, Ga.
Ridgway, E. B. and C. C. Ashley (1967) Biochem. Biophys. Res. Commun. 29:229-234.
Robben, F. (1971) Appl. Optics 10:776-796.
Seitz, W. R. and D. M. Hercules (1972) Int. Symp. Chemilum., Athens, Ga.
Seliger, H. H. and W. D. McElroy (1959) Biochem. Biophys. Res. Commun. 1:21.
Seliger, H. H. and W. D. McElroy (1960) Arch. Biochem. Biophys. 88:136.
Seliger, H. H. and W. D. McElroy (1965) Light: Physical and Biological Action, Academic Press, N. Y. 417 p.
Serat, W. F., F. E. Budinger and P. K. Mueller (1965) AIHL Report No. 14, State of California Dept. of Public Health.
Shimomura, O., F. H. Johnson and Y. Saiga (1962) J. Cell Comp. Physiol. 59:223.
Shimomura, O., F. H. Johnson and Y. Saiga (1963a) J. Cell Comp. Physiol. 62:1.
Shimomura, O., F. H. Johnson and Y. Saiga (1963b) "Microdetermination of calcium by aequorin luminescence" Science 140:1339.
Shlyapintokh, V. Y., R. F. Vassil'ev, O. N. Karpukhin, L. M. Postnikov and L. A. Kibalko (1960) J. Chim. Phys. 57:1113.
Shlyapintokh, V. Y., O. N. Karpukhin, L. M. Postnikov, V. F. Tsepalov, A. A. Vichutinskii and I. V. Zakharov (1968) Chemiluminescence Techniques in Chemical Reactions, Consultants Bureau, N. Y. 222 p.
Stanley, P. E. (1971) Anal. Biochem. 39:441.
Stanley, P. E. and S. G. Williams (1969) Anal. Biochem. 29:381.
Stauff, J. and J. Ostrowski (1967) Z. fur Naturforsch. 22b:734-

Stauff, J. and G. Rümmler (1962) Z. Phys. Chem. N. F. 34:67.
Stauff, J., H. Schmidkunz and G. Hartman (1963) Nature 198:281-
Steele, R. H. (1963) Biochemistry 2:529.
Sutcliffe, W. H., E. A. Orr and O. Holm-Hansen (1972) U. Calif. San Diego Report 10P20-110.
Syrett, P. J. (1958) Arch. Biochem. Biophys. 75:117-124.
Trautz, M. (1905) Z. Phys. Chem. 53:1.
Trautz, M. and P. P. Shorygin (1905) Z. wissensch. Photogr. Photophys. Photochem. 3:121.
Vassil'ev, R. F. (1962) "Secondary processes in chemiluminescent solutions" Nature 196:668.
Vassil'ev, R. F. (1967) Prog. Reaction Kinetics 4:305-352.
Vassil'ev, R. F., O. N. Karpukhin and V. Y. Shlyapintokh (1959) Dokl. Acad. Nauk. SSSR 125:106-109.
Vassil'ev, R. F., A. A. Vichutupkii and A. S. Cherasov (1963) Dokl. Acad. Nauk. SSR 149:124.
Wang, C. H. and D. L. Willis (1965) Radiotracer Methodology in Biological Science, Prentice Hall, N. J. 382 p.
White, E. H., E. Rapaport, H. H. Seliger and T. A. Hopkins (1971) Bioorganic Chem. 1:92-122.

REYNOLDS: I'd like to emphasize a point made by Dr. Seliger, which is very useful. This concerns the possibility of using a current of integrating photodetection technique. I would like to ask if he would agree that a word of caution should be made in the following sense? One wouldn't want to achieve the low background counting rate that he quite properly recommends at the expense of spectral sensitivity. One way to get low dark current counting rates is to have a bialkali cathode. In certain applications this cathode might just not have the spectral sensitivity for the wavelengths that one is concerned about. I think a good trick here would be to try to be sure that one was using as small a cathode area as possible and still have the solid angle detection of the reaction cell or whatever else is involved. This would minimize dark current.

SELIGER: Something else should also be mentioned. It has been reported rather recently, and there's a lot of literature that bears on this, that if you take the simple precaution of painting the glass envelope with conducting paint and then connecting it through a high resistance to the photocathode potential, it is possible to reduce the noise by a factor of between 10 and 100. That simple arrangement with almost any phototube will allow you to use it for the count measurements rather than having to go through the difficulties of doing pulse techniques.

REYNOLDS: If you don't like to use paint there is a good rule of thumb. Extending a conducting shield to about 1 1/2 tube radii beyond the end of the tube will also reduce the dark current.

Abstracts of Short Contributions

INFLUENCE OF HALOGEN IONS ON THE ELECTROGENERATED CHEMILUMINESCENCE OF 9,10-DIPHENYLANTHRACENE. T. Kihara and K. Honda, Institute of Industrial Science, University of Tokyo.

The electrochemiluminescence (ECL) of 9,10-diphenylanthracene (DPA) has been intensively studied by Hercules, Bard and his co-workers and many other investigators. Concerning the mechanism of ECL of DPA, in addition to the well-known cation-anion annihilation process, the preannihilation ECL process due to oxidant or reductant in the electrolytic solution has been proposed.

In the present study, the preannihilation electrochemiluminescence of DPA was observed in the acetonitrile solution containing halide anions such as Cl^-, Br^-, and I^- as supporting electrolyte. The ECL was observable at the anodic potential corresponding to the anodic oxidation of halide anion which is more negative than the oxidation potential of DPA. Consequently, it is supposed that halogen formed (probably in the ionic state of X_3^-) plays the part of oxidant against DPA^- in the preannihilation process.

Experimental. A platinum wire of 12 cm long coiled around a glass tube was served as a working electrode. A platinum counter-electrode was set under the working electrode. The double-step controlled-potential of working electrode was controlled against a saturated calomel electrode connected through a glass filter. The intensity of the emission was measured by a photomultiplier.

Results. (1) Effect of Cl^-. Octadecyldimethylbenzylammonium chloride 1mM was used as supporting electrolyte to study the effect of Cl^-. The ECL intensity of DPA 1mM solution in acetonitrile was measured as a function of the potential of anodic step, while the potential of the cathodic step was always fixed at -2.0 V which is a little negative than the potential of the first reduction of DPA. Fig. 1 shows the relation between the ECL intensity (I_{max}) and the potential (E) of anodic step where o denotes the emission observable in the period of anodic step and Δ denotes the emission in the period of cathodic step. ECL begins to appear at about +1.0 V which is nearly the oxidation potential of Cl^-, but more negative than the oxidation potential of DPA. (2) Effect of Br^-. The measurements under the same experimental conditions were carried out in the case of acetonitrile solution containing tetrabutylammonium bromide 1mM. (Fig. 2) The ECL in the period of anodic step begins to appear at +0.7 V corresponding to the oxidation potential of Br^- which suggests the preannihilation process between DPA^- and Br_3^-. On the other hand, ECL behavior in the period of cathodic step suggests the ordinary cation-anion annihilation process. (3) Effect of I^-. Tetrabutylammonium iodide 1mM was used as supporting electrolyte. (Fig. 3) In the anodic step, the emission begins to appear at +0.3 V corresponding to the oxidation potential of I^- and in the cathodic step, it appears from +1.2 V.

The former will be attributed to the preannihilation and the latter to the cation-anion annihilation.

Conclusion. In the acetonitrile solution containing halide anion (X^-), the ECL of DPA due to the preannihilation was observed. It is suggested that the oxidation product of X^- (probably in the ionic state of X_3^-) will take the part of acceptor against DPA^-.

The energy balance of the present ECL system is shown in the next Table.

Table. Enthalpy difference of DPA-ECL system.

System	$E_{ox} - E_{red}$ (V)	ΔH (eV)*	Energy Balance
$DPA^- - DPA^+$	3.05	3.25	sufficient
$DPA^- - Cl_3^-$	2.82	3.02	sufficient
$DPA^- - Br_3^-$	2.52	2.72	deficient
$DPA^- - I_3^-$	2.15	2.35	deficient

* --- $\Delta H = E_{ox} - E_{red} + 0.2$ (V)

In the cases of Br^- and I^-, the triplet-triplet annihilation process should be taken into consideration.

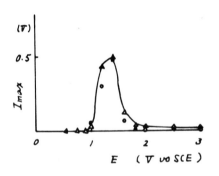

Fig. 1   Fig. 2

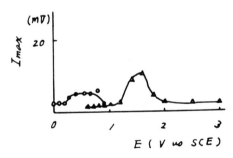

Fig. 3

CHEMILUMINESCENT FORMATION OF BrF AND IF $^3\Pi$ (0^+). M. A. A. Clyne, J. A. Coxon and L. W. Townsend, Queen Mary College, London

Combination of F $^2P_{3/2}$ with Br $^2P_{3/2}$ or I $^2P_{3/2}$ ground state atoms in the presence of singlet O_2 $^1\Delta_g$, $^1\Sigma_g^+$, leads to formation of BrF $B^3\Pi(0^+)$ or IF $B^3\Pi(0^+)$, which were detected by observation of their band emission spectra (975 $\geq \lambda >$ 483 nm), $B^3\Pi(+) \to X\ ^1\Sigma^+$. It was shown that formation of BrF $B^3\Pi(0^+)$ explicitly requires energy transfer from O_2 $^1\Delta_g$ or $^1\Sigma_g^+$. The most probable excitation process is envisaged to occur by combination of Br $^2P_{3/2}$ + F $^2P_{3/2}$ atoms (in the presence of a third body) to give BrF $A^3\Pi$ (1) [or possibly $^3\Pi$ (2,0$^-$)], followed by collisional energy transfer from singlet oxygen with BrF $A^3\Pi$ (1) to BrF $B^3\Pi(0^+)$. At least 38 new bands of the B→X system of BrF have been identified, and the (previously uncertain) vibrational numbering of the $B^3\Pi(0^+)$ state has been confirmed. Vibrational constants are reported for the first time for levels of the X $^1\Sigma^+$ ground state up to v" = 10, corresponding to an energy of 6262 cm^{-1} above v" = 0. The available data on the dissociation energy of BrF are briefly surveyed, and a value of D° (BrF) = 2.384 eV is favoured. For the B→X system of IF, many new bands have been observed, and vibrational constants for levels up to v" = 19 (10381 cm^{-1} above v" = 0) of the ground state are reported.

CHEMILUMINESCENCE IN THE OZONE PHOTOCHEMICAL SYSTEM. R. P. Wayne, Physical Chemistry Laboratory, Oxford.

Excited species may be produced in photochemical processes (i) as primary photolytic products, (ii) by energy transfer from such products, or (iii) by secondary chemical reactions. Radiation from these species is thus a form of chemiluminescence rather than fluorescence. This paper describes some observations of "photochemiluminescence" in ozone. Although a number of reactions known to occur in the ultraviolet photolysis of ozone are sufficiently exothermic to excite a variety of states of O and O_2, the only molecular species which have been observed by optical emission are $O_2(^1\Delta_g)$, a primary product[1,2] of photolysis at λ = 254 nm

$$O_3 + h\nu_{\lambda=254nm} \to O(^1D) + O_2(^1\Delta_g) \qquad (1)$$

and $O_2\ (^1\Sigma_g^+)$, formed by energy transfer from $O(^1D)$ to ground state molecular oxygen[1]

$$O(^1D) + O_2(^3\Sigma_g^-) \to O(^3P) + O_2(^1\Sigma_g^+) \qquad (2)$$

The long radiative lifetime[3] of $O(^1D)$ and its high reactivity[4] with respect to O_3 have hitherto precluded observation of emission from the excited atomic fragment of reaction (1), although recently $O(^1D)$ has been explicitly detected in O_3 photolysis[4]. Strong circumstantial evidence (see, for example reference 5) also indicates that the species is formed. The primary quantum yield, $\Phi_{O_2(^1\Delta_g)}$, for $O_2(^1\Delta_g)$ production in reaction (1) has been measured[2] by determining the absolute intensity of emission of the $O_2(^1\Delta_g \rightarrow {}^3\Sigma_g^-)$ emission band at $\lambda = 1270$ nm. A flow technique was employed, and the experiment divides into three parts. First, the detector is calibrated by determining the response to known concentrations of $O_2(^1\Delta_g)$. $O_2(^1\Delta_g)$ is produced, for this purpose, in a discharge, and its concentration established from the amount of ozone it destroys under conditions where the reactions

$$O_2(^1\Delta_g) + O_3 \rightarrow 2O_2 + O \qquad (3)$$

$$O + O_3 \rightarrow 2O_2 \qquad (4)$$

account for virtually all $O_2(^1\Delta_g)$ and O_3 loss. Secondly, the absolute absorbed radiation intensity is estimated from the extent of photochemical ozone decomposition for contact times where the secondary process

$$O(^1D) + O_3 \rightarrow 2O_2 \qquad (5)$$

is essentially the only reaction following (1). Finally, the intensity of the $\lambda = 1270$ nm emission band, and hence $[O_2(^1\Delta_g)]$, is measured as a function of $[O_3]$ to yield the value of $\phi_{O_2(^1\Delta_g)}$. Within experimental error, $\phi_{O_2(^1\Delta_g)}$ is unity at $\lambda=254$ nm, which leaves little room for $O_2(^1\Sigma_g^+)$ production - a species energetically accessible at this wavelength. Other results[4] confirm that $O_2(^1\Sigma_g^+)$ can only be a minor product. Kinetic evidence[5] indicates that $\phi_{O(^1D)}$ drops off from a presumed value of unity at $\lambda=254$ nm to near zero at $\lambda > 310$ nm. On the other hand, an excited primary molecular product, capable of reacting with O_3 to yield $O(^3P)$, is formed at wavelengths at least as long as 334 nm:

$$O_3 + h\nu_{\lambda=334nm} \rightarrow O(^3P) + O_2^\dagger \qquad (6)$$

where $O_2^\dagger$ could, on energetic grounds, be in either the $^1\Delta_g$ or $^1\Sigma_g^+$ states. The energy transfer process (2) leads to emission of the $\lambda=762$ nm $O_2(^1\Sigma_g^+ \rightarrow {}^3\Sigma_g^-)$ band in the photolysis of O_3/O_2 mixtures[1] or of pure molecular oxygen[6-8]. The efficiency, α, of transfer (2) compared to overall quenching of $O(^1D)$

$$O(^1D) + O_2(^3\Sigma_g^-) \rightarrow O(^3P) + O_2(^3\Sigma_g^- \underline{OR} \, ^1\Sigma_g^+) \qquad (7)$$

has recently been redetermined[9] in this laboratory by studies of the photolysis of ozone. $O_2(^1\Sigma_g^+)$ reacts with O_3 according to

$$O_2(^1\Sigma_g^+) + O_3 \rightarrow 2O_2 + O \qquad (8)$$

so that if conditions are chosen such that only reactions (1), (7) and (8) can occur, for each quantum of radiation absorbed, then $(1+\alpha)$ molecules of ozone are consumed and $(1+\alpha)$ ground state oxygen atoms are produced. Both ozone removal and O atom formation have been studied and give similar results for α. The value of α appears to lie between 0.5 and 0.6.

1. M. Gauthier and D. R. Snelling, *Chem. Phys. Letts.*, *5*, 93 (1970).
2. I. T. N. Jones and R. P. Wayne, *Proc. Roy. Soc. A*, *321*, 409 (1971).

3. J. W. Chamberlain, *Physics of the Aurora and Airglow*, Academic Press, N.Y. (1961).
4. R. Gilpin, H. I. Schiff and K. H. Welge, *J. Chem. Phys.*, *55*, 1087 (1971).
5. I. T. N. Jones and R. P. Wayne, *Proc. Roy. Soc. A.*, *319*, 273 (1970).
6. T. P. J. Izod and R. P. Wayne, *Proc. Roy. Soc. A.*, *308*, 81 (1968).
7. J. F. Noxon, *J. Chem. Phys.*, *52*, 1852 (1970).
8. R. A. Young and G. Black, *J. Chem. Phys.*, *47*, 2311 (1967).
9. D. J. Giachardi and R. P. Wayne, *Proc. Roy. Soc. A.*, in press (1972).

THE EXCITATION OF IODINE BY SINGLET OXYGEN. R. G. Derwent[†] and B. A. Thrush, University of Cambridge, Department of Physical Chemistry.

Intense chemiluminescent emission by I_2 and I is observed when traces of iodine are added to oxygen gas containing singlet molecular oxygen in a discharge flow system. The yellow I_2 emission comes from vibrational levels of the $B^3\Pi_0{}^+{}_u$ state (which is responsible for visible absorption by I_2) almost up to its dissociation limit into $I(^2P_{3/2}) + I(^2P_{1/2})$ and requires the pooling of the energy of three $O_2(^1\Delta_g)$ or one $O_2(^1\Delta_g)$ and one $O_2(^1\Sigma_g^+)$. Arnold, Finlayson and Ogryzlo (J. Chem. Phys., *44*, 2529, 1966) who discovered the I_2 emission attributed it to the recombination process.

$$I(^2P_{3/2}) + I(^2P_{1/2}) + M = I_2(B) + M \quad (1)$$

They also observed strong infra-red emission by $I(^2P_{1/2})$ and enhanced emission by $O_2(^1\Sigma)$ and suggested the mechanism

$$O_2(^1\Delta) + I(^2P_{3/2}) = O_2(^3\Sigma) + I(^2P_{1/2}) + 0.8 \text{ kcal/mol} \quad (2)$$

$$O_2(^1\Delta) + I(^2P_{1/2}) = O_2(^1\Sigma) + I(^2P_{3/2}) + 6.8 \text{ kcal/mol} \quad (3)$$

We have made a quantitative study of these processes in a discharge flow system where ca. 10^{-4} Torr of iodine vapour was added to discharged oxygen at 1-10 Torr pressure which had been passed over mercuric oxide to reduce the atomic oxygen concentration below one part per million. The gas contained 2-5% $[O_2{}^1\Delta]$ and there was a steady state concentration of 10^{-3}-10^{-2}% $(O_2{}^1\Sigma)$ produced by the energy pooling process

$$2O_2(^1\Delta) = O_2(^1\Sigma) + O_2(^3\Sigma) + 7.6 \text{ kcal/mol} \quad (4)$$

and removed by diffusion to the walls (where $\gamma \sim 10^{-2}$) or quenching agents (H_2O, H_2) added to distinguish the roles of $O_2(^1\Delta)$ and $O_2(^1\Sigma)$. All experiments were carried out at $22 \pm 2°C$. The I_2 emission comes predominantly from levels around $v' = 25$ some 5 kcal/mol below the dissociation limit. This emission is stronger from the gas near the walls of the flow tube particularly at high pressures and high ($O_2\,^1\Sigma$), its overall intensity decreases strongly at higher total pressures. Kinetic analysis shows that iodine is present predominantly as I atoms. I_2 is efficiently dissociated in collisions with $O_2(^1\Sigma)$, the iodine atoms diffuse to the wall where they recombine to give I_2 iodine molecules, then diffuse into the gas phase where they are redissociated

$$I_2 + O_2(^1\Sigma) = O_2(^3\Sigma) + I + I + 1.6 \text{ kcal/mol} \qquad (5)$$

or undergo stepwise electronic excitation via the metastable $A(^3\Pi_{1u})$ state:

$$I_2 + O_2(^1\Sigma) = I_2(A) + O_2(^3\Sigma) + 3.8 \text{ kcal/mol} \qquad (6)$$

$$I_2 + O_2(^1\Delta) = I_2(B) + O_2(^3\Sigma) \qquad (7)$$

an average of 55% of the energy of $O_2(^1\Delta)$ is converted into electronic plus vibrational energy of I_2 in (7). This stepwise excitation does not occur with Br_2 despite the similar energy of B state because the A state of Br_2 lies at too high an energy to be excited by $O_2(^1\Sigma)$. Transfer of energy from $O_2(^1\Sigma)$ to I_2 occurs at every collision - $(k_5 + k_6) = 1.4 \times 10^{14}$ cm^3 mol^{-1} s^{-1}; reactions (6) and (7) are both faster than one collision in ten. Reaction (1) is negligible as a source of excited iodine in our system. Very strong emission by $I(^2P_{1/2})$ at 1315 nm and the enhanced emission by $O_2(^1\Sigma)$ were also studied. Reaction (2) which is almost thermoneutral was shown to reach equilibrium in our system with $k_2 = (4.6 \pm 1.5) \times 10^{13}$ cm^3 mol^{-1} s^{-1} and $k_{-2} = (1.6 \pm 0.3) \times 10^{13}$ cm^3 mol^{-1} s^{-1}. The spin allowed energy transfer process (3) has a rate coefficient of $(1.6 \pm 0.2) \times 10^{10}$ cm^3 mol^{-1} s^{-1} which is 1000 times that of the closely related spin forbidden energy pooling process (7). Both processes gave 5% $O_2(^1\Sigma, v' = 1)$ about twice the proportion predicted by Franck-Condon factors. In general no single approach; - statistical theory, Franck-Condon factors or the importance of energy resonance gives a correct overall picture of the various processes studied in this work.

†Present Address: Air Pollution Division, Warren Springs Laboratory, Stevenage, Herts.

VACUUM-UV EMISSION BY ACTIVE NITROGEN, M.F. Golde[*] and B. A. Thrush, University of Cambridge

Vacuum ultraviolet emission from active nitrogen was first observed by Tanaka et al.,[1] who detected Lyman-Birge-Hopfield (LBH) bands, $N_2(a^1\pi_g - X^1\Sigma_g^+$, $v' \leq 6$). More recently, the v=0 level of the $N_2(a''^1\Sigma_u^- - X^1\Sigma_g^+)$ system has been shown to occur weakly[2] and we have reported detection of bands in the $N_2(B'^3\Sigma_u^- - X^1\Sigma_g^+)$ system. The modes of population of the emitting states following the recombination of N atoms have received considerably less study than that of $N_2(B^3\pi_g)$, which emits the well-known yellow afterglow. The study of emission from $N_2(a''^1\Sigma_u^-)$ revealed that this state is formed by reaction of N atoms with molecules in the $N_2(B^3\pi_g)$ state and removed by quenching by N and N_2. We have studied the vacuum-UV emission by active nitrogen in a discharge-flow apparatus at pressures between 1 and 6 Torr and find that the $(a^1\pi_g)$ state of N_2 is populated by two parallel processes: (a) A two-body inverse predissociation populates rotational levels J > 13 of $N_2(a^1\pi_g)$, $v'=6$, from which there is rapid rotational and vibrational relaxation; this gives an emission intensity proportional to $[N]^2$. (b) The reaction

$$N(^4S) + N_2(B^3\pi_g) = N_2(a^1\pi g) + N(^4S)$$

populates mainly lower levels of the $(a^1\pi g)$ state, giving emission proportional to $[N]^3$ which is enhanced by argon carriers. At these pressures, collisional vibrational relaxation dominates the removal of higher levels of the $(a^1\pi_g)$ state, the rate constant increasing with vibrational quantum number; level $v'=0$ is quenched principally to the $(a''^1\Sigma_u^-)$ state. The addition of small amounts of carbon monoxide to active nitrogen induces strong emission of the CO Fourth Positive bands $(A^1\pi - X^1\Sigma^+)$,[3] which extend up to $v'=8$. We find the kinetic behaviour of the emission to be closely similar to that of the LBH emission of N_2 and deduce that the $CO(A^1\pi)$ state is excited by energy transfer from $N_2(a^1\pi_g)$ and is removed predominatly by radiation. Comparison of the absolute intensities of the vacuum UV- emission by N_2 and CO yields the following rate constants:

$N_2(a^1\pi_g)$, v=0 + N_2 = quenched products, $k = (1.0 \pm 0.1) \times 10^{12} cm^3 mol^{-1} s^{-1}$.

$N_2(a^1\pi_g)$, v=0 + CO = $CO(A^1\pi)$, v=0-2 + N_2, $k=(9\pm 1) \times 10^{13} cm^3 mol^{-1} s^{-1}$.

[*]Present address: University of Pittsburgh, Pittsburgh, Pa. 15213
1. Y. Tanaka, A. Jursa and F. J. Leblanc, 1957, The Threshold of Space (ed M. Zelikoff) p 89, Oxford: Pergamon Press.
2. I.M. Campbell and B. A. Thrush, 1969, Trans. Faraday Soc. 65, 32.
3. H. P. Knauss, 1928, Phys. Rev. 32, 417.

MECHANISM OF PRODUCTION OF ELECTRONICALLY EXCITED BaO IN THE REACTION OF Ba VAPOR WITH O_2; DISTRIBUTION OF ELECTRONIC STATES IN THE PRODUCTS OF THE REACTIONS $Ba(g) + N_2O$ or $NO_2 \rightarrow BaO(A^1\Sigma$ or $X^1\Sigma) + N_2$ or NO. R. H. Obenauf, C. J. Hsu and H. B. Palmer, The Pennsylvania State University.

The reaction of Ba + O_2 was studied by a diffusion flame technique. Emission from the A → X system of BaO was found to be unaffected by the presence of atomic oxygen but greatly affected by addition of inert third body. Results are explained by the incorporation of the complex BaO_2 into the mechanism.

Preliminary measurements are reported of the distribution of electronic states in products of the highly exothermic atom-exchange reactions of barium vapor with N_2O and NO_2. The total photon yield for the N_2O reaction was 0.20 - 0.27 photon per molecule N_2O consumed while the yield for the NO_2 reaction was 0.015 ± .003.

CHEMICAL CO LASER PRODUCED FROM THE $O(^3P)$ + CH REACTION. M. C. Lin, Naval Research Laboratory, Washington, D.C. 20390.

Very strong 5 μm CO laser emission has been detected in the vacuum UV flash photolysis of mixtures of SO_2 and $CHBr_3$, in the presence of SF_6 or Ar. The emission consists of the vibrational-rotational lines of CO ranging from $\Delta v \sim 19 \rightarrow 18$ to $6 \rightarrow 5$ according to our preliminary frequency analysis. The primary pumping mechanism is proposed as follows:

$$SO_2 + h\nu \rightarrow O(^3P) + SO(^3\Sigma-), \lambda \geq 165 \text{ nm}$$

$$CHBr_3 + h\nu \rightarrow CH + 3Br$$

$$O + CH \rightarrow CO\dagger(X^1\Sigma^+) + H + 176 \text{ kcal/mole}$$

The contribution of other secondary reactions such as

$$CO(^3\pi) + M \rightarrow CO\dagger(X^1\Sigma^+) + M$$

$$CHO^+ + e \rightarrow CO\dagger(X^1\Sigma^+) + H,$$

however, cannot be ruled out. Both CO $(^3\pi)$ and CHO^+ + e are possible primary products of the O + CH reaction. Strong CO laser emission at 5 μm was also observed when $CHBr_3$ and O_2 were flashed in the presence of SF_6 or Ar. The emission is probably produced from the following exothermic reaction.

$$CH + O_2 \rightarrow CO + OH + 159 \text{ kcal/mole}$$

The results obtained for these two systems will be present.

EXCITATION OF CO IN THE NITROGEN AFTERGLOW. R.A. Young and William Morrow, Centre for Research in Experimental Space Science, York University.

Excitation of CO by energy transfer in discharged nitrogen has been verified. The Cameron and fourth positive bands have been observed. The dependence of the Cameron bands on [N] and [CO] imply that the precursor to their excitation is quenched by atomic nitrogen. The quenching of the $a^3\pi$ state of CO is not predominantly due to atomic nitrogen implying that its quenching rate coefficient is less than 10^{-11} cm^3/sec. The precursor is strongly quenched by H_2 and hence cannot be $N_2(A^3\Sigma)$. Atomic oxygen quenches the precursor much less than atomic nitrogen and quenches the $CO(a^3\pi)$ state less than CO itself.

PHOTO-EXCITATION AND PHOTO-DISSOCIATION LASERS. (I) 1-μm NO LASER EMISSIONS RESULTED FROM THE $C(^2\pi) \longrightarrow A(^2\Sigma^+)$ and $D(^2\Sigma^+) \longrightarrow A(^2\Sigma^+)$ TRANSITIONS. M.C. Lin, Naval Research Laboratory.

1-μm laser emission was detected when NO (with or without dilution) was flash-photolyzed in vacuum UV above 165 nm. The emission was identified as the (0,0) and (1,1) bands of both $C(^2\pi) \longrightarrow A(^2\Sigma^+)$ and $D(^2\Sigma^+) \longrightarrow A(^2\Sigma^+)$ transitions. The wavelengths of these transitions are 1.224, 1.228, 1.100 and 1.108 μm, respectively. The total laser emission was found to vary linearly with flash energy; this implies that the pumping mechanism is a one-photon excitation process. The effects of various additives (He, Ar, H_2, CO, N_2, O_2, N_2O, CO_2, CF_4 and SF_6) have been studied; H_2 was found to be the most effective quencher, whereas N_2O, the most efficient promoter of the total laser emission. When both CO_2 and an excess amount of He or Ar were added to NO, 10.6-μm CO_2 laser emission was observed concurrently with the 1-μm emission. The 10.6 μm emission can be attributed to the near resonant energy transfer process

$$NO^*(v) + CO_2(000) \longrightarrow NO^*(v-1) + CO_2(001)$$

where $NO^*(v)$ represents the NO molecule in the vth vibrational level of $A(^2\Sigma^+)$, $C(^2\pi)$ or $D(^2\Sigma^+)$ states.

THE CHEMILUMINESCENCE OF LITHIUM PHOSPHIDES. Richard A. Strecker, Jonathan L. Snead, and Gilbert P. Sollott, Pitman-Dunn Laboratory, Frankford Arsenal, Philadelphia.

The chemiluminescence of a series of lithium phosphides most of which were prepared for the first time has been characterized. Sensitivity of the emissions of the solid salts to change in the organic group, particularly the shift to longer wavelength obtained with the di-α-naphthylphosphide relative to phenyl, indicates that the chemiluminescence is of molecular rather than crystal origin. Factors affecting the relationship of phosphide chemiluminescence to molecular structure are discussed. Comparison of chemiluminescence and fluorescence data obtained on several phosphides and intermediate and final oxidation products indicates that electronically excited phosphide molecules are the major emitters. The lithium salts of dicyclohexyl- and diphenylphosphine oxides were discovered to be chemiluminescent, and a structure of these salts in the crystalline state is suggested based on a study of infrared spectra. In some instances a minor component was detected in the phosphide chemiluminescence spectra, and this emission is attributable to the lithium salt of the phosphine oxide in the case of cyclohexyl, but not in the case of phenyl. Mechanistic aspects of the oxidation and chemiluminescence of lithium phosphides are discussed.

PHOTO-OXIDATION AND DELAYED CHEMILUMINESCENCE IN POLY (ETHYLENE 2,6 NAPHTHALENE DICARBOXYLATE). R.R. Richards and R.S. Rogowski, Langley Research Center, NASA, Virginia.

The thermal decomposition of photo-oxidation products in poly (ethylene 2,6 naphthalene dicarboxylate) (PEN 2,6) was studied to elucidate the mechanism of the interaction with oxygen. PEN 2,6 film exposed to ultraviolet light and oxygen at room temperature exhibited chemiluminescence when heated above $80°C$. The decay rate of luminescent intensity was measured between $80° - 140°C$ and found to fit a first-order rate law with activation energy of 26 k cal/mole. The chemiluminescent spectrum and the fluorescent spectrum of the polymer film were identical with maxima at 430 nm. Probable mechanisms are proposed for the photo-oxidation and subsequent chemiluminescence.

A NOVEL SENSITIZED CHEMILUMINESCENCE. AN INFRARED EMISSION FROM METHYLENE BLUE SENSITIZED BY SINGLET OXYGEN. Dale E. Brabham and Michael Kasha, Florida State University.

Singlet oxygen may sensitize chemiluminescence in chromophores by several different proposed mechanisms, [1,2,3,4] in which an excited singlet state of the chromophore is the emitting state. These mechanisms have been proposed on the basis of energetic criteria, kinetic data, and comparison of the chemiluminescence emission with the photoluminescence of the chromophore. Against the background of the previous studies, we were surprised to make the observation of a possible triplet state "chemiluminescence" emission in fluid solution sensitized by singlet oxygen. When singlet molecular oxygen sensitizes methylene blue, a strong emission band at 830 nm is recorded (Steinheil 3 prism f/4 spectrograph, E. K. I-N plates). The singlet oxygen was produced by the reaction of 30% hydrogen peroxide with sodium hypochlorite (commercial bleach) in water at 20°C. The observed emission does not correspond to the methylene blue fluorescence (red), but is spectroscopically identical with the low temperature rigid glass solution (EPA, 77°K) phosphorescence spectrum of the dye. Stauff and Fuhr[5] sensitized methylene blue luminescence in aqueous solution by bubbling $^1\Delta_g$ oxygen through the cell. The spectral characteristic of the observed emission was not reported. We propose that the $^1\Sigma_g^+$ molecular oxygen produced in the reaction (indirectly from the $^1\Delta_g$ oxygen[6]) sensitized dye excitation by transferring its energy directly to the triplet state of the dye, which in turn emits luminescence in a radiative transition from this triplet state directly to the ground state, in the fluid aqueous solution system at 20°C. Triplet state emissions in fluid solutions are observed under photoexcitation conditions under favorable conditions of short intrinsic lifetime and favorable intersystem crossing ratio. Consideration of the excited state dynamics for methylene blue suggest that oxygen-sensitized chemiluminescence from the triplet state could be produced by the proposed mechanism, even though simple direct photoexcitation of methylene blue does not reveal a triplet state emission.

(1) A. U. Khan and M. Kasha, J. Am. Chem. Soc. 88, 1574 (1966).
(2) E. A. Ogryzlo and A. E. Pearson, J. Phys. Chem. 72, 2913 (1968).
(3) T. Wilson, J. Am. Chem. Soc. 91, 2387 (1969).
(4) S. R. Abbott, S. Ness and D. M. Hercules, J. Am. Chem. Soc. 92, 1128 (1970).
(5) J. Stauff and H. Fuhr. Ber. Bunsengesellschaft Phys. Chem. 73, 245 (1969).
(6) S. J. Arnold, M. Kubo and E. A. Ogryzlo, Adv. Chem. 77, 133 (1968).

CHEMILUMINESCENCE IN THE OXIDATION REACTIONS OF SOME QUINONES AND THEIR POLYMERS I. CHARACTERISTICS OF REACTIONS AND LUMINESCENCE, Danuta Slawinska, Institute of Physics Institute of Technology, Szczecin, Poland

An important role of quinones /Q/ and their polymers /PQ/ in biochemical and soil processes was the reason for which the investigation of Q and PQ chemiluminescence /CL/ was carried out. The water and water-methanol solutions of Q and PQ in concentration $10^{-5} - 10^{-3}$M were oxidized by means of O_2 or HWO_2 in the pH range 6 - 11 at temperatures 15 - 40°C. The intensity of CL was measured with the help of sensitive, quantometric or current photoelectric system. Simultaneously, EPR and optical emission and absorption spectra were measured. Moreover, light-induced CL and EPR signals from PQ were investigated. It was found that the strong CL with quantum yield /γ/ of the order 10^{-6} give purpurogallin-Q, polymers of o-benzene-Q, and a weak one with $\gamma = 10^{-9}$ adrenochrome, tetra-Cl-o-benzo-Q, polymers of p-benzoQ, melanins and humic acids. The spectral range of CL was 400 -720 nm with maxima at 470, 510, 570, 635 and 670 nm depending on Q or PQ, pH of solution and the presence of catalysts. Fluorescence spectra of solution after the reaction were found to be similar to that of CL only in the case of pur-purogalline-Q, adrenochrome and tetra-Cl-o-benzo-Q. In other cases fluorescence spectra were shifted in the blue region in comparison with CL ones. During the oxidation of Q and PQ or action of visible light on them, EPR signal appears. An amplitude of EPR signal is higher in argon atmosphere. EPR spectra of Q show hyperfine structure and those of PQ a singlet only. 1,4-benzo-Q, 2-methyl-1,4-benzo-Q, UQ_6, menadione and polymers of 1,2-benzo-Q give broad absorption band in the region 500 - 700 nm. When O_2 or H_2O_2 is added, the band and EPR signal disappears and CL appears. The absorption bands of purpurogallin-Q, α-tocopherol-o-Q, tetra-Cl-o-benzo-Q and adrenochrome at 450-500 nm under the same conditions diminish or disappear. Reduction of purpurogallin-Q, adrenochrome or polymers of o-benzo-Q with ascorbate in anaerobic condition gives no CL. It has been found that the action of visible light on Q and PQ results in singlet EPR signal formation and ultra-weak, short-lived CL with $t_{1/2}$ = 100 sec and maximal intensity of the order $5 \cdot 10^4 - 10^6$ quanta/ sec mg substrate for some Q and PQ, respectively. Two important facts have been established: 1. unsubstituted Q give more intense CL than substituted /CH_3-, CH_3O-/ ones, 2. the action of H_2O_2 on PQ results in stronger CL than that on Q, and this reaction leads to the decrease of absorption in 300 - 700 nm and decrease of pH solution of about 0.2 - 0.3. From the time- and temperature-dependences of CL intensities, the rate constants of the drop of luminescence and activation energies, respectively, were calculated.

CHEMILUMINESCENCE IN THE OXIDATION REACTIONS OF SOME QUINONES AND THEIR POLYMERS II. MECHANISM OF GENERATION OF EXCITED MOLECULES AND LIGHT EMISSION. Janusz Slawinski, Laboratory of Physics Agricultural University, Szczecin, Poland.

On the basis of experimental data concerning the oxidation reactions of quinones /Q/ and their polymers /PQ/ as well as characteristics of luminescence, the general mechanism of Q and PQ chemiluminescence /CL/ is proposed. Since CL occurs in alkaline solution /without catalysts/ and intermediate semiquinone are formed, the first process may be: $Q + OH^- \longrightarrow SQ^{\cdot-} + \cdot OH$. In dependence on: 1. the structure of Q and PQ and 2. the presence of O_2 or H_2O_2 in the reacting solution, the following reactions can take place: 1/ if Q are substituted and O_2 is present: $SQ^{\cdot-} + O_2 \longrightarrow Q + \cdot O_2^-$. $\cdot O_2^-$ might give in further step an excited $^1\Lambda_g$ and $^1\Sigma_g^+$ or their combination molecules, according to mechanism proposed by Stauff, and Khan and Kasha. This "oxygen path" of CL might explain emission spectra of 1,4-benzo-Q and 2-methyl-1,4-benzo-Q with maxima at 570, 635 and 670 nm, as well as weak CL from UQ_6, menadione e.t.c. 2/ in the case of unsubstituted Q, especially p-isomers in the presence of O_2 the polymerisation takes place. Due to the high values of free valency in α, β -positions of the $-C \cdots C-$ bond, high molecular, brown, alkali-dissolved products are formed. They exhibit fluorescence in the range 400 - 650 nm. Such reactions with low $E_{akt} \approx 20$ kJ/mole liberate the energy $-\Delta H = 50 - 100$ kJ/mole i.e. too small for an excitation of electronic levels, corresponding to the observable blue and green CL. However, it is possible to some degree that polymers might accept energy from O_2^*, $(O_2)_2^*$ or $\supset C = O^*$ and act as activators of CL, according to Vassil'ev and Kasha. 3/ When unsubstituted Q or PQ, especially o-isomers and OOH^- or O_2^- ions are present, the processes of oxidation degradation are privileged. The nucleophilic reaction:

$$H - O - O^- + \supset C^{+\delta} = O^{-\delta} \longrightarrow \supset C \begin{matrix} O^- \\ O - O - H \end{matrix}$$

leads via peroxide intermediates to opening of the quinone ring and formation of carbonyl compounds, fluorescing at 460 - 510nm, as for example, carboxylic acids of α -tropolone from purpuro - gallin-Q. The energy liberated $-\Delta H = 210 - 950$ KJ/mole + $E_{akt} \approx 35$ kJ/mole is sufficient for excitation of the products to their lowest electronic levels $n \longrightarrow \pi^*$ or $\pi \longrightarrow \pi^*$. Such mechanism is supported by the observed decrease in optical density of o-Q and PQ and pH decrease, which correlate with CL intensity. Calculations of electron densities, free valency and bond-order values of Q by means of Huckel's MO method additionally confirm the proposed mechanism. The participation of Q CL in processes such as electron transport in respiration, melanin-dyes formation, radiation injury and oxidative degradation of humic acids in soils and coals, as well as the possibilities of application of CL for studying some processes, are discussed.

A COMMENT ON THE MECHANISM OF LUMINOL CHEMILUMINESCENCE, T. Goto, M. Isobe and K. Ienaga, Department of Agricultural Chemistry, Nagoya University

Most chemiluminescence mechanisms have been suggested to involve a dioxetane intermediate which decomposes to produce carbonyl compound(s) in an excited state. Formation of the excited state molecule has been interpreted theoretically. Mechanism of luminol chemiluminescence, however, differs from others. Among several mechanisms which have been proposed, one involving a concerted decomposition of a 6-membered endoperoxide intermediate would be most plausible, but no experimental support has been reported. To see whether such an endoperoxide decomposes smoothly to the expected products, photooxygenation of phthalazine derivatives has been investigated. 1,4-Dimethoxyphthalazines (I) having a substituent X (X=H, OMe, NH_2, etc.) in methanol were irradiated at r.t. by high pressure Hg lamp. Although I (X=NH_2) produced a blue pigment, I (X=H) and I (X=OMe) afforded the expected ester (III) as a sole product. This result indicates that the endoperoxide intermediate (II) can be smoothly decomposed to the ester (III) and nitrogen molecule. The product III (X=H) shows no fluorescence, but strong fluorescence is observed with III(X=OMe) and hence the peroxide II (X=OMe) might produce chemiluminescence during its decomposition. Accordingly, I (X=OMe) in methanol was irradiated at $-78°$ and the solution brought to r.t. in a photoelectric recording light integrator. Unfortunately no light was observed. This result might suggest that decomposition of the 6-membered cyclic peroxide (II) does not product an excited state molecule, but further confirmation is necessary. No sensitizer is necessary for photooxygenation of the phthalazines (I). This is explained by a self-sensitized photooxygenation mechanism as has been suggested in the case of substituted naphthalenes since the phthalazines (I) show similar phosphorescence spectrum to those of naphthalenes.

α-PEROXYLACTONES: INTERMEDIATES IN BIOLUMINESCENCE.

W. Adam, J.C. Liu, G. Simpson, and H.C. Steinmetzer, University of Puerto Rico.

Authentic α-peroxylactones 2, which were postulated as the active intermediates in the luciferin-luciferase bioluminescence of the sea pansy, the firefly, and the cypridina, confirmed in the latter case by oxygen-18 labeling, were prepared by subambient cyclization of the respective α-hydroperoxy acids 1 and characterized by spectral methods. At ambient conditions the α-peroxylactones 2 decarboxylate smoothly, leading to the Eyring parameters $\Delta H^{\ddagger}$ = +18.8 kcal/mol, $\Delta S^{\ddagger}$ = 8.9 gibbs/mol, and $\Delta G^{\ddagger}$ (300°K) = 21.5 kcal/mol in carbon tetrachloride for 2a; but these data are strongly solvent dependent. In this decarboxylation the carbonyl product 3 is formed essentially quantitatively under light emission. The self-luminescence peaks at 405-420 nm, depending on the substitution of the carbonyl product 3, and exhibits very low quantum yields. However, energy transfer to photo-acceptors such as rubrene, 9,10-diphenylanthracene, etc. is efficient, resulting in significant enhancement (ca. 10^3 fold) of luminescence yield, the fluorescence being characteristic of the photo-acceptor. Stern-Volmer quenching of the luminescence by the fluorescer suggest that high yields of triplet rather than singlet state carbonyl product 3 are formed. The mechanism of this chemelectronic decarboxylation is discussed in terms of a concerted process leading directly to triplet state carbonyl product 3, analogous to the related 1,2-dioxetanes. The relevance of α-peroxylactones 2 in bioluminescence is emphasized.

$$\underset{1}{\overset{R^1}{\underset{R^2}{\nearrow}}\overset{O}{\underset{OOH}{\searrow}}\overset{OH}{}} \xrightarrow[-20°]{DCC} \underset{2}{\overset{R^2}{\underset{R^1}{\nearrow}}\overset{O}{\underset{O}{\searrow}}\overset{O}{}} \xrightarrow[-CO_2]{r.t.} \underset{3}{\overset{R^2}{\underset{R^1}{\nearrow}}=O} + h\nu$$

a: R^1 = H, R^2 = tert-Butyl; b: R^1 = H, R^2 = 1-Adamantyl; c: R^1 = R^2 = Methyl; d: R^1 + R^2 = Adamantylidene

BIOLUMINESCENCE METHODS FOR DETERMINING PYROPHOSPHATE, MALATE AND OXALACETATE IN BIOLOGICAL MATERIAL. Philip E. Stanley, Department of Agricultural Biochemistry, Waite Agricultural Research Institute, The University of Adelaide, Glen Osmond, South Australia 5064.

These techniques determine as little as 10 picomoles of inorganic pyrophosphate, malate or oxalacetate by utilizing enzyme reactions where either adenosine triphosphate (ATP) or reduced nicotinamide adenine dinucleotide (NADH) is involved. These two compounds are determined by coupling them to either the bioluminescence reaction of the firefly (ATP) (1,2) or those of the marine bacterium Photobacterium fischeri NADH (2,3). Thus in measuring inorganic pyrophosphate (PP_i) use is made of ATP-sulphurylase enzyme from yeast (4) which mediates the following reaction:-

$$APS + PP_i \rightleftharpoons ATP + sulphate$$

adenosine 5'-phosphosulphate

The equilibrium constant, 10^{-8}, favours ATP formation which is readily measured by coupling to the luciferin-luciferase system of the firefly (1,2). Malate and oxalacetate in the reaction catalysed by the enzyme malate dehydrogenase (Reaction I) may be determined by coupling to the NADH-dependent Photobacterium light system (2,3).

$$Oxalacetate + NADH \rightleftharpoons Malate + NAD \quad \ldots (I)$$

The equilibrium constant for this reaction, 10^{-12}, favours the formation of malate. The rate of NADH oxidation is followed when oxalacetate is to be determined, but for the reverse reaction to assay malate, a large excess of NAD is required to overcome the very unfavourable equilibrium so that sufficient NADH is produced. The rates of these light reactions, as determined by photon output, are monitored continuously in a liquid scintillation spectrometer. The data are ordinarily produced in digital form but an analogue presentation, usually favoured by biochemists, can be made with either a ratemeter and chart recorder or a multichannel analyzer operated in the multiscale mode. The advantages of these bioluminescence methods over more conventional procedures include specificity sensitivity, speed and economy. Moreover, these methods may be applied to other compounds, e.g. ammonia (5) and APS (4), which are involved in biochemical systems where ATP or NADH are reactants.

1. Stanley, P.E. and Williams, S.G., Anal. Biochem. 29, 381 (1969)
2. Stanley, P.E., in "Organic Scintillators and Liquid Scintillation Counting" (D.L. Horrocks and C.-T. Peng. eds) pp. 607-620. Academic Press, New York and London, 1971.
3. Stanley, P.E., Anal. Biochem. 39, 441 (1971).
4. Balharry, G.J.E., and Nicholas, D.J.D., Anal. Biochem. 40, 1 (1970).
5. Nicholas, D.J.D., and Clarke, G.R., Anal. Biochem. 42, 560 (1971).

LIGHT INHIBITION OF MECHANICALLY STIMULABLE LIGHT IN BIOLUMINESCENT DINOFLAGELLATES (1), H.H. Seliger, W. H. Biggley and B. Shur, Department of Biology, Johns Hopkins University

Under a photoperiodically entrained LD 12:12 cycle photosynthetic, bioluminescent, marine dinoflagellates exhibit extremely rapid increases in total photons emitted upon mechanical stimulation (TMSL) subsequent to D_o(3). The ratio of TMSL during scotophase (dark) to TMSL during photophase (light) is usually around 100. If however the light intensity during scotophase is set not to zero but to some intensity between zero and the regular photophase intensity the organisms exhibit intermediate values of TMSL. The range of light intensities over which this factor of 100 occurs however is extremely narrow. For Pyrodinium bahamense this range lies between 50-200 ft. c.(4) and the total mechanically stimulable light has the form $TMSL = 10^6 \left(\frac{I}{200}\right)^{-3.33}$, while below 50 ft. c. the exponent is lower by more than a factor of 10. Data will be presented for other dinoflagellate species.

(1) Research supported under AEC contract AT(11-1)3277.
(2) Supported on National Institutes of Health Training Grant GM-57.
(3) Biggley, W. H., E. Swift, P. J. Buchanan and H. H. Seliger 1969. J. Gen. Physiol. 54:96-122.
(4) Illumination during photophase was by cool white fluorescent lamps at 1200 ft. c. Reading represent only relative values.

PURIFICATION AND PROPERTIES OF CTENOPHORE PHOTOPROTEINS, W. W. Ward and H. H. Seliger, Department of Biology, The Johns Hopkins University

Two calcium-activated photoproteins have been extracted from the ctenophore, Mnemiopsis sp., and purified several thousand fold to homogeneity. The purification, involving ammonium sulfate and protamine sulfate fractionations and several ion-exchange and gel-filtration columns, yielded equal quantities of two photoproteins designated M-1 and M-2. These are completely resolved by DEAE cellulose chromatography and they differ slightly in molecular weight, electrophoretic mobility, and reaction kinetics. Their bioluminescence spectra are identical. The M-2 photoprotein can be activated by nine divalent cations, of which calcium and strontium are the most effective. A variety of alcohols including secondary, tertiary, unsaturated, and halogenated alcohols enhance total light yields from purified ctenophore photoproteins. Long chain aliphatic alcohols are strongly inhibitory. Purified ctenophore photoproteins are irreversibly inactivated by exposure to blue light. A preliminary photo-inactivation action spectrum indicates broad peaks near 260 nm and 285 nm. The bioluminescent system of Beroe ovata has also been investigated. This ctenophore, which feeds exclusively on Mnemiopsis, has a broader, red-shifted in vivo emission spectrum, but identical in vitro emission to that of Mnemiopsis. There is a single photoprotein from Beroe which differs from M-1 and M-2 in molecular weight and kinetics.

ACTIVE CENTER STUDIES ON BACTERIAL LUCIFERASE. M. Z. Nicoli, T. W. Cline, T. O. Baldwin, and J. W. Hastings. Harvard University, Cambridge, Mass. 02138

Bacterial luciferase catalyzes a bioluminescent oxidation of $FMNH_2$ by O_2 in the presence of a long-chain aliphatic aldehyde. The enzyme consists of two nonidentical subunits (α and β), neither of which has catalytic activity in the absence of the other. It has been shown that there is one $FMNH_2$ binding site per dimer (Meighen and Hastings, J. Biol. Chem. 246, 7666(1971)) and that succinylation of the α subunit affects intermediate stability and reduces the flavin binding affinity of the enzyme (Meighen et al., Biochemistry 10, 4062, 4069 (1971)). Furthermore, all mutant luciferases having altered kinetic parameters possess lesions in the α subunit (Cline and Hastings, Biochemistry, 11, 3359 (1972)). Modification of one cysteinyl residue with either N-ethylmaleimide ($k_{2(obs)}$ = 1670 M^{-1} min^{-1}) or iodoacetamide ($k_{2(obs)}$ = 7.0 M^{-1} min^{-1}) in 0.02 M PO_4, pH 7.0, inactivates luciferase, with a concomitant decrease in flavin binding. The rate of inactivation by N-alkylmaleimides is highly dependent on the chainlength of the alkyl group, longer chainlengths reacting more rapidly (N-n-butylmaleimide, $k_2(obs)$= 9300 M^{-1} min^{-1}; N-n-hexylmaleimide, $k_{2(obs)}$ = 49,000 M^{-2} min^{-1}), suggesting a hydrophobic environment for the thiol. Studies with ^{14}C-N-ethylmaleimide have shown that the reactive sulfhydryl is on the α subunit and is located in a tryptic peptide whose sequence is Phe-Gly-Ilu-Cys-Arg.

MICROSOMAL CHEMILUMINESCENCE. Antony R. Shoaf, Randolph M. Howes, Robert C. Allen, and Richard H. Steele, Tulane University School of Medicine.

Light emissions from non-bioluminescent biological preparations (tissues, tissue homogenates, mitochondria, and tissue extracts) have been interpreted widely to be due to the oxidation of cellular lipids and/or lipid peroxides mediated by free radical processes. Howes and Steele (1,2) described recently a chemiluminescence (CL) elicitable from rat liver microsomes by the addition of reduced diphosphopyridine nucleotide phosphate (NADPH) in the absence of an added oxidizable substrate. We present evidence that this CL correlates with the lipid peroxidation of the microsomes and that both processes are oxygen dependent. Having shown formerly that the addition of oxidizable substrates to these systems suppressed the CL and the lipid peroxidation while they were undergoing hydroxylations we suggested that these events were self-consistent with the known properties of singlet oxygen species. In this paper we extend these observations and demonstrate the influence of reduced diphosphopyridine nucleotide (NADH), and ascorbic acid (AAH_2) on these systems. The results reveal an ascorbic acid "catalyzed" enhancement of NADPH and NADH oxidations which display as an enhanced CL and which appear analogous to the activations described for plant systems by Beevers (3), by Kern and Racker (4), and by Nason et al. (5), and for pig adrenal cortical mitochondria by Kersten et al. (6). The addition of the hydroxylatable substrate acetanilide is shown to suppress the CL and the lipid peroxidation, particularly in the ascorbate system, while the substrate is hydroxylated. Bovine superoxide dismutase (SD) failed to suppress the NADH, NADPH, or AAH_2 initially inducible CL from these systems. When NADH or NADPH were added later, subsequent to marked lipid oxidation and loss of structural integrity, the re-inducible CL from the SD containing systems were considerably suppressed. Our data suggest that if the superoxide free radical anion, $\cdot O_2^-$, is generated in these systems it must be formed in a tightly coupled system, possibly within the intravesicular environment where it is inaccessible to SD. Alternately, the $\cdot O_2^-$ may be stabilized and/or transported by the quaternary group of phosphatidyl choline as suggested by the work of Strobel and Coon (7).

1. Howes, R.M., and Steele, R.H., Res. Commun. Chem. Path. Pharmacol., 2, 619 (1971).
2. Howes, R.M., and Steele, R.H., Res. Commun. Chem. Path. Pharmacol., 3, 349 (1972).
3. Beevers, H., Plant Physiol., 29, 265 (1954).
4. Kern, M., and Racker, E., Arch. Biochem. Biophys., 48, 235 (1954).
5. Nason, D., Wosilait, W.D., and Terrell, A.J., J. Biol. Chem., 210, 903 (1954).
6. Kersten, W., Schmidt, H., and Staudinger, H., Biochemische Z., 326, 469 (1955).
7. Strobel, H.W., and Coon, M.J., Am. Chem. Soc., 246, 7826 (1971).

FUNCTIONALITY OF ELECTRONIC EXCITATION STATES IN HUMAN MICROBICIDAL ACTIVITY. Robert C. Allen, Rune L. Stjernholm, Ruth R. Benerito, Richard H. Steele, Tulane University School of Medicine.

Recently a chemiluminescence (CL) has been observed when human polymorphonuclear leukocytes (PMN) phagocytize bacteria or particulate matter. This CL response correlates well with the stimulation of the hexose monophosphate shunt, which results in the generation of NADPH. The PMN possesses both CN^- -insensitive NADH and NADPH oxidases. Flavoprotein oxidases of this type are capable of univalent reduction of O_2. The reduced oxygen ($\cdot O_2, \cdot O_2H$) may then disproportionate yielding HOOH and singlet molecular oxygen $^1O_2$. The PMN also possesses a CN^--sensitive peroxidase, myeloperoxidase, which has microbicidal activity in the presence of HOOH and halide. In this reaction the HOOH is reduced to OH^- with the oxidation of the halide to the reactive halogonium species. In cases where the halogonium formed is Cl^+ or Br^+, there is the potential for further reaction with HOOH resulting in the generation of a haloperoxy anion. This unstable species can disintegrate to yield the original halide and $^1O_2$. $^1O_2$ has been demonstrated to be a potent microbicidal agent. Therefore, the biochemical generation of $^1O_2$ by the PMN might be closely associated with microbicidal activity. The CL response may be the result of the relaxation of excited carbonyl groups generated via $^1O_2$ mediated oxidations.

CHEMILUMINESCENCE ARISING FROM THE ACTION OF $^1\Delta_g$ - MOLECULAR OXYGEN ON CHLOROPHYLL-A. Hartmut Fuhr, University of California, Riverside.

Singlet-oxygen oxidations have been shown to occur when systems, containing a photosensitizer, ground state molecular oxygen and a substrate are irradiated with light of appropriate wavelength. Chlorophyll has been shown in in vitro experiments to be an efficient singlet oxygen sensitizer. Because of the great importance of chlorophyll in photosynthesis it appeared to be of interest to study how chlorophyll (here: chlorophyll-a) is itself possibly attacked by singlet molecular oxygen. Singlet oxygen was generated in a microwave discharge and was bubbled through solutions of chlorophyll-a in different solvents in a chemiluminescence detection unit. The disappearance of chlorophyll-a - as indicated by absorption spectroscopy - was followed by a chemiluminescent emission which was detected with a photomultiplier tube. The spectral distribution of the chemiluminescence was analyzed by cut-off filter spectroscopy. When the singlet oxygen source was shut off, a very long lasting afterglow was recorded. This eliminates a simple physical energy - transfer from singlet oxygen dimols to chlorophyll-a, thus generating electronically excited chlorophyll-a molecules. The afterglow decay curves indicate that a complicated mechanism is involved. A series of experiments dealing with quenching and enhancement of the observed chemiluminescence leads to a possible explanation of the mechanism of generation of the chemiluminescent species.

Epilogue

It is not lucky to dream such stuff
Dreaming men are haunted men.
Though Wingate's face looked lucky enough
To any eye that had seen him then,
Riding back through the Georgia Fall
To the white-pillared porch of Wingate Hall
Fall of the possum, fall of the coon,
And the lop-eared hound-dog baying the moon.
Fall that is neither bitter nor swift
But a brown girl bearing an idle gift
A brown seed-kernel that splits apart
And shows the Summer yet in its heart,
A smokiness so vague in the air
You feel it rather than see it there,
A brief, white rime on the red clay road
And slow mules creaking a lazy load
Through endless acres of afternoon,
A pine cone fire and a banjo-tune
And a julep mixed with a silver spoon.

Stephen Vincent Benet

Author Index

Adam, W., 228,281,283,322,323,356,493
Allen, R.C., 497,498
Anderson, J.M., 387
Baldwin, T.O., 369,496
Bard, A.J., 193,208
Berenito, R.R., 498
Biggley, W.H., 494
Brabham, D.E., 489
Broida, H.P., 101
Brundrett, R.B., 231
Carrington, T., 7,27,28,99
Chandross, E.A., 143
Cline, T.W., 369,496
Clyne, M.A., 481
Cormier, M.J., 1,344,361,378,380,387
Coxon, J.A., 481
Davidson, J.A., 111
DeLuca, M., 285,335,345,356,357,358,359
Dempsey, M.E., 345
Derwent, R.G., 483
Duthler, C.J., 101,109,110
Eberhard, A., 369
Eley, M., 377,378
Fontijn, A., 393,425,426
Fritsch, J.M., 249
Fuhr, H., 498
Golde, M.F., 73,485
Golomb, D., 393
Goto, T., 325,335,492
Gundermann, K.-D., 209,228,229,244
Hastings, J.W., 310,311,344,369,377,378,380,386,496
Heller, C.A., 249,263,281,282,459
Henry, R.A., 249
Hercules, D.M., 1,229,243,247,263,427,459
Herschbach, D.R., 29
Hodgeson, J.A., 393
Honda, K., 479
Hopkins, T., 311,312,356,357
Hori, K., 361
Howes, R.M., 497
Hoytink, G.J., 129,147
Hsu, C.J., 486
Hysert, D., 313
Ienaga, K., 492
Isobe, M., 492
Jaeschke, W., 131

Johnson, F.H., 337
Kasha, M., 489
Kaufman, F., 27,28,83,99,109
Kaufman, M., 130
Kearns, D.R., 128,129,141,449
Keszthelyi, C.P., 193
Kihara, T., 479
Kishi, Y., 325
Kubota, I., 325
Lee, D.C.-S, 265
Lee, E.K.C., 110
Lee, J., 1,129,141,228,245,247,248,281,309,310,356,381,386,449
Lin, M.C., 59,61,71,487
Liu, J.C., 493
Matheson, I.B.C., 245
McCapra, F., 283,313,322,323,358,460
McElroy, W.D., 243,285,309,310,311,312,322,344,380,460
Morrow, W., 487
Murphy, C.E., 381
Nealson, K.H., 369
Nicoli, M.Z., 369,496
Obenhauf, R.H., 486
Ogryzlo, E., 100, 111,127,128,129,130
Palmer, H.B., 486
Rauhut, M., 244,282,356,451,459,460
Reynolds, G., 460,478
Richards, R.R., 488
Rogowski, R.S., 488
Roth, M., 313
Sander, U., 131
Schaap, A.P., 263
Seitz, W.R., 427,449
Seliger, H.H., 242,243,323,335,377,461,478,495
Shimomura, O., 337,344
Shoaf, A.R., 497
Shur, B., 495
Simpson, G., 493
Slawinska, D., 490
Slawinski, J., 491
Smith, I.W.M., 43,59
Snead, J.L., 488
Sollott, G.P., 488
Stanley, P.E., 494
Stauff, J., 131,141
Stedman, D.H., 59,81,110,426
Steele, R.H., 497,498

Steinmetzer, H.C., 493
Stjernholm, R.L., 498
Strecker, R.A., 488
Suzuki, N., 325
Tachikawa, H., 193
Thorington, L., 426,459,460
Thrush, B.A., 27,71,73,81,99,110,141,425,483,485
Tokel, N.E., 193
Totter, J., 312
Townsend, L. W., 481
Ward, W.W., 495
Wayne, R.P., 81,108,127,128,129,425,426,481
Weller, A., 169,181,208
White, E., 228,231,242,243,244,247,248,310,323,344,358,359
Wilson, T., 265,281,282
Young, R.A., 27,28,487
Zachariasse, K., 169,181,208
Zaklika, K.A., 313

SUBJECT INDEX

A

Acanthoptilum,
  bioluminescence of, 362,367
2-acetamido-5-phenylpyrazine, 329
Acetylene, 415
Acridans,
  peroxides of, 315
Actinometric measurements,
  in ECL, 203
Actinometry apparatus,
  in ECL, 205
Activation energy,
  for reactions of CS+O, 51
  in dioxetane decomposition, 272
  in radiative recombination, 21
  in radical ion recombination, 190
Active center,
  bacterial luciferase, 378
Active site in firefly luciferase, 289,303
Acylaminopyrazine, 328,329,333
Adamantyliden dioxetane, 268
Adiabatic reaction,
  definition, 1
Aequorea, 362,367
Aequorin, 342,343
AF-350, 342,343,362,363,367
Afterglow, 141,487
  air, 83,395,406,413,416
  oxygen, 399
  nitrogen, 76,393
  of active nitrogen, 73
Airglow, 416
Air pollution, 405,473
Aldehyde,
  binding of, 375
  effect on bacterial bioluminescence, 370
  effect of chainlength on bacterial bioluminescence, 370,374,383

Aldehyde,
  oxidation of, 375
Alkali atoms,
  electronic excitation energy of, 104
Alkali reactions,
  by collisional excitation, 30
  by chemiluminescent atom exchange, 31
Amines,
  quenching by, 123
6'-aminoluciferin,
  effect on emission color in firefly luminescence, 296
3-aminophthalate, 210,242,246
  fluorescence of, 245
Aminophthalate ion,
  as light emitter in luminol reaction, 231
Aminopyrazine, 333
Ammonia, 407
AMP,
  effect on emission color in firefly luminescence, 301
Angular momentum, 14
Annihilation,
  cation-anion annihilation of naphthalene, anthracene and tetracene, 147
  energy considerations in cation-anion, 155
Anthozoa,
  bioluminescence of, 391
Anthracene,
  cation-anion annihilation rate constant, 160
  2,3 dicarboxylic acid, 215, 218
  2,3 dicarboxylic hydrazide, 214,217
  2,3 dicarboxylic hydrazide, relative quantum yield of, 220
Applications,
  of luminescence, 468
Arginine, 365

Arsenic, 445
Argon, 11
Ascorbic acid, 497
Atomic oxygen, 481,482,483,486, 487
ATP, 494
ε-ATP,
  effect on emission color in firefly luminescence, 297
ATP,
  detection of, 468
ATP binding site,
  in firefly luciferase, 300
Azodicarboxylates,
  chemiluminescence of, 224

B

Bacterial bioluminescence, 369, 341,494,495
Bacterial contamination,
  determination of, 469
Bacterial luciferase,
  as mixed function oxidase, 370
Barium, 486
Barium oxide, 486
Benzo(b)-phthalazine dione, 212
Bicyclic dioxetanes, 266
Bicyclic peroxide,
  in the chemiluminescence of hydrazides, 237
Biisoquinolinium salts,
  autooxidation of, 250
Binding site,
  $FMNH_2$, 372
  of firefly luciferase, 304
  of firefly luciferase for luciferin, ATP and MgATP, 297
  of firefly luciferase for dehydroluciferyl adenylate, 298
  of firefly luciferase for dehydroluciferin, 298
Bioluminescence,
  mechanisms, 5
  oxygen labelling of *Renilla*, *Cypridina*, 6

Biomass,
  determination of, 470
Biosynthesis of luciferins,
  model studies, 318
Blue fluorescent protein, 342
Born-Oppenheimer approximation,
  for diatomic system, 8
Boron, 411
Bromine, 431,484
Bromine fluoride, 481
Bromoform, 486

C

Calcium ion,
  specificity of in *Aequorea* luminescence, 472
Calcium requirement,
  in bioluminescence, 388
Calculation,
  statistical mechanical, 78
Cameron bands, 487
Campanularia, 362
Carbon dioxide, 338,339,341,399, 403,487
Carbon dioxide production,
  oxidative mechanism of during sea pansy luminescence, 353
Carbon dioxide,
  oxygen-18 incorporation in *Cypridina* luminescence, 349
  oxygen-18 incorporation in firefly luminescence, 349
  oxygen-18 incorporation sea pansy luminescence, 349
  production of during firefly luminescence, 290
  source of oxygen in during bioluminescence, 351
  source of oxygen in during chemiluminescence, 352
Carbon disulfide, 416
Carbon monoxide, 398,402,411, 418,486,487
Carbon monoxide,
  spectra, 118

SUBJECT INDEX

Catalysis,
  of luminol chemilumines-
    cence, 427
Catalytic sites,
  of firefly luciferase, 304
Cavernularia, 362,367
Cetyltrimethylammonium bromide,
  318,331
Charge transfer, 123
Chemiluminescence,
  concept of electronic states
    in, 2
  role of ionic intermediates
    in, 3
  role of symmetry in, 3
  mechanisms in solution, 4
Chemiluminescence-gas phase, 7
  in three-atom systems, 17
  orbital degeneracy-induced, 40
  in terms of electron transfer,
    40
Chemiluminescence,
  infrared, 43
  low level, 474
  of monoacylhydrazides, 235
  of diazaquinones, 236
  of luminol, 237
  O+NO, 83
Chemiluminescence rate,
  effect of water on, 352
Chemiluminescence yields,
  of triplet-triplet annihila-
    tion, 171
  in radical ion recombination
    reactions, 186,187
  total of hetero-excimer emis-
    sion, 172
  in radical ion recombination,
    172
Chemiluminescence,
  in the atmosphere, 412,416
Chemiluminescence intensity,
  corrections in ECL, 198
Chlorophyll, 498
Chromium (III),
  catalyst of luminol CL, 427,
    432
  analysis for, 436,432
Chromium (II), 431,439

Cis-Diethoxydioxetane, 266,271
Cis-Diethoxyethylene,
  as catalysts of dioxetane
    decomposition, 271
Clytia, 362
Cobalt (II),
  in luminol reaction, 427,432,
    436,442
Coefficients,
  spontaneous emission, 47
Coelenterate, 361,362,367
Complexation,
  effect of on luminol CL, 430,
    436
Control mechanisms,
  in bioluminescence, 391
Copper (II),
  in luminol reaction, 427,433,
    435,442
Coronene, 178
Cross section,
  energy dependence at threshold
    for collision excitation,
    36,37
Ctenophore, 495
Cyalume,
  chemical light curve, 457
2-cyano-6-chlorobenzothiazole,
  inactivation of firefly
    luciferase by, 305
Cyclic dioxetanones, 315
Cyclobutane,
  dissociation of, 23
Cyclopentadiene, 211
Cypridina luciferin, 328,337,
  339,340,341,343,363,365
Cypridina oxyluciferin, 328,329,
  337,338,339,343

D

Dark current, 467
Deactivations,
  dual, 67
1,4-dehydro-5-amino-phthalazine-
  1,4-dione, 209
Dehydroluciferin,
  binding of to luciferase, 287
  firefly, 286

Dehydroluciferyl adenylate,
  formation of, 290
Diazabicyclooctane,
  as catalysts in dioxetane
    decomposition, 271
Diazoquinones, 209,210
9,10-dibromoanthracene,
  as fluorescor, 271
Dicarboxylperoxide, 138,139
Diethylamine, 217,218,271
4-diethylamino phthalic hydrazide, 213
1,4-dihydro(2,3g)naphtho-phthal-
  azine-1,4-dione,
  chemiluminescence of, 214, 222
8,9-dihydropyrazino[1,2-a:3,4a']
  [1,1']-biisoquinolinium
  dibromide monohydrate,
  250,252,255,256,261
5,5 dimethyl oxyluciferin, 325
Dimethyl phthalate, 217
Dinoflagellates, 495
Dioxetane, 6,22,23,237,249,258,
  263,265-278,492,493
Dioxetanedione, 138
Dioxyluciferin, 325
Diphenylanthracene(DPA),
  138,193,226,235,479,493
Diphyes, 362
Diradicals, 269
Doppler width, 105

E

Equilibrium dialysis,
  for determination of dehydro-
    luciferin binding sites in
    firefly luciferase, 299
N-ethyl-maleimide,
  labeling of active sulfhydryls
    in firefly luciferase, 304
Electrochemiluminescence, 5
  CL efficiency, 197,198,200, 204
Electrolysis, 132
Electron exchange,
  singlet-triplet conversion
    via, 41

Electron spin resonance, 490
Electron transfer,
  in hydrazide chemiluminescence, 234
  in chemiluminescence of
    monoacylhydrazides, 235
  in bacterial bioluminescence, 385
Electronic excitation, 119
Electronic states,
  time dependence, 12
  adiabatic, 13
  diabatic, 13
Emission,
  Vegard-Kaplan band, 74
  fully allowed, 78
Emission spectrum,
  bacteria, 373
Emitter,
  in firefly luminescence, 289
Encounter complex,
  between solvated radical ions, 170
  competing reactions of, 170
  spin relaxation time, 173
Energy balance, 480
Energy exchange,
  vibrational-vibrational, 47, 49
Energy gap, 123
Energy pooling, 483
Energy transfer, 481,482,483,
  485,489
  in bioluminescence, 387,391
  vibrational, 91
  from vibration in NaN_2, 36
  vibration-to-electronic in K, 38
  energetics of, for K, Na, Br, 39
Endogenous rhythm,
  of spontaneous bioluminescence, 475
Enthalpy, 480
  redox processes in ECL, 194
Ethylene, 399
Etioluciferin, 337,338,339,343
Excimers, 159
Excitation,

vibrational, 54
Extraterrestrial life,
  detection of, 471

F

Fall-off,
  ratios, 88
Firefly dehydroluciferin, 327, 328
Firefly luciferin analogues, 288
Firefly luminescence, 285, 453, 493, 494
Firefly oxyluciferin, 325, 327
Flash apparatus,
  vacuum U.V., 62
Flavin cation, 374, 378
  as bacterial emitter, 377, 378, 382
Fluorescein,
  iodide quenching of fluorescence, 246
Fluorescers, 136
Fluorescence,
  of flavin, 373, 381
  of aminophthalate, 242
  from K*, 38
  from collisionally excited molecular states, 34
Flow system,
  for BIQI++ luminescence, 257
FMN,
  specificity of in bacterial bioluminescence, 471
$FMNH_2$,
  non-enzymatic oxidation of, 369
  reaction of with bacterial luciferase, 382
Free energy, 173, 174, 461
Franck-Condon factor, 119, 484

G

Gas,
  non-reactive, 48
Green fluorescent protein, 391

H

Hammett reaction constants, 316
Hemin, 217, 218, 220
Hetero-excimer,
  chemiluminescence yield of, 181
  dissociation, 169, 172
  formation of, 169, 181
  free enthalpy of, 176, 177
  thermal dissociation, 190
Hexamethylphosphoramido(HMPA), 133
Homonuclear diatomics, 116
Hydrazide, 223, 231
Hydrogen,
  atomic, 398, 401, 403, 417
  molecular, 401, 409, 412
Hydrogen peroxide, 211, 217, 218
  excitation energy, 131
  in luminol reaction, 427
Hydroxylation,
  mechanisms of in bacterial bioluminescence, 385
Hydrophobicity,
  of firefly luciferase, 300
Hydrozoa,
  bioluminescence of, 388
Hypochlorite, 431, 447

I

Illumination,
  emergency, 451
Infra-red,
  chemiluminescence in, 1, 45
Intensities,
  total emission, 65
Interferometer,
  Fourier, 90
Intermediate II,
  bacteria, 374
Internuclear distance,
  of the nitrogen molecule, 106
Intersystem crossing, 154
Intrinsic efficiency,
  of phototubes, 463
Inverse predissociation, 77
Iodine, 483

radical cation spectra, 252
EPR signal of I^{++}, 251
reaction with luminol, 431, 441,443
in CL titrations, 445
Iodine fluoride, 481
Ion diffusion,
in ECL, 193
Ion exchange, 439,441
Ion pairs,
collision formation of free, 34
cross sections for forming K,SO$_2$, 35
state in energy transfer, 40
Iron (II),
as catalyst of luminol CL, 243,428
analysis for, 427,439
Iron-pentacarbonyl, 415
IsoATP,
effect on emission color in firefly luminescence, 296

J

Jellyfish,
bioluminescence of, 363

K

α-keto-β-methyl-n-valeric acid, 337,338,339
α-ketoperoxide,
intermediate in sea pansy luminescence, 350
Kirkwood-Onsager continuum model,
equivalent sphere radius from, 177

L

Lasers,
chemical, 44,49,64,486,487
Latia,
bioluminescence of, 341,342
Lovenella,
bioluminescence of, 362

Lead,
tetramethyl, 416
Levels,
high vibrational, 70,79
Lifetimes,
synthetic dioxetanes, 266
$O_2(^1\Delta_g)$, 124
Light capacity,
for CL systems, 452
Light emission,
factors affecting color of in firefly luminescence, 293
Light inhibition,
in bioluminescence, 495
Lighting,
chemical systems, 451,456, 458
Lightstick, 456
Limit,
predissociation, 77
Linear peroxide,
in firefly luminescence, 293
Lipids, 497
Liquid scintillation counter, 494
Lithium phosphide, 488
Local symmetry,
effect of in NO + O$_3$, 27
Lophine dimer, 333
Luciferase, 391,493,494,495,496
Luciferase structure,
effect on emission color in firefly luminescence, 294
Luciferin, 493,495
binding sites on luciferase, 305
dimer, 331
firefly, 286
quantum yield and CL, 470
Luciferin sulfokinase, 362
Luciferyl adenylate chemiluminescence,
oxygen-18 studies, 350
Luciferyl adenylate,
formation of, 290
Luciferyl sulfate, 362
Luminol,
chemiluminescence of 209,225,

SUBJECT INDEX

226,231,235,238,245,427-432,492
Luminol diazaquinone, 211,213
Luminol,
  light capacity, 453
Lumisome,
  cellular site of bioluminescence, 387
  composition of, 391
  emission kinetics, 389
  emission spectra, 389
Lyman-Birge-Hopfield bands, 485

M

M-dependence, 89
M-effect,
  the ratio of, 85
  finding of, 87
Magnetic field effects,
  in ECL, 194
Malate, 494
  determination of, 472
Manganese (II), 429
Maxwell-Boltzmann Distribution,
  the excited state population, 105
P-mercuribenzoate,
  reaction with sulfhydryls in firefly luciferase, 302
Mercury,
  lowest triplet level, 107
Metal vapor,
  light emission from, 101
Methoxide ion, 254
Methods,
  low-pressure, 61
  in IRCL, 44
N-methylacridone,
  rate of formation of, 317
Methyl glyoxal, 363
5-methyl oxyluciferin, 325
2-methyl-5-phenylimidazo[1,2-a]-pyrazin-3-one, 329
Methylene blue, 489
Methyne, 486
Micelles,
  chemiluminescence in, 317
Microsomal chemiluminescence, 497

Microwave discharge, 102
Migration,
  H atom, 67
Mnemiopsis,
  bioluminescence of, 362
Model compounds,
  in the study of bioluminescence, 313
Model,
  vertical transition, 78
Modified oxidative mechanism,
  in firefly luminescence, 292
Molecular beams,
  fast alkali, 31
  crossed Na and N_2, 34
  "seeded-jet" technique, 36

N

Naphthalene-2,3-dicarboxylic hydrazide, 212
Nickel (II), 428,433,435,439,442
Nitrogen,
  afterglow, 107,485,487
  electronically excited, 74
  atomic, 398,403
  molecular, 398,404,406,412
  radiative recombination, 12, 15
  theoretical description, 12
Non-exponential decay,
  fluorescence of complex NO_2, 94
Non-radiative processes, 111
Non-resonant transfer,
  from a thermalized distribution, 105
Nitrogen dioxide (NO_2), 90,395, 400,402,404,407,413,416
  fluorescence spectrum, 93
  electronic quenching, 85
  photodissociation of, 94
  pressure dependence of, 84
  quantum calculations, 93
  radiative lifetime of, 87
  state correlation diagram for, 19,27
Nitrogen oxide (NO), 395,402,

407,412,416,486,487
spectra, 118
radiative recombination with
O, 18

O

Obelia,
　bioluminescence of, 362,367
Orbital symmetry,
　in O+NO+M reaction, 21
Orbiting resonances,
　in radiative recombination, 11
Overlap, 119
Overton bands, 56
Oxaloacetate,
　determination of, 472,494
Oxalic acid,
　oxidation, 140
Oxidative mechanism,
　in firefly luminescence, 291
Oxygen,
　atomic, 395,408,411
　molecular, 395,401,406,411, 415
　in chemiluminescence, 265
　in luminol CL, 430,441
　radical anion, 223
　requirement for in lumisome bioluminescence, 389
Oxygen-18, 337,338,339,367
　in firefly luminescence, 292
　methodology, in luminescence, 345,346,347,348
　use of in mechanism studies in chemi- and bioluminescence, 345
Oxyluciferin,
　firefly, 289
Oxyluminesce, olefin, 249
Ozone, 396,405,411,417,481,482

P

Parazoanthus,
　bioluminescence of, 362,367
Pentacene, 178
Pelagia,
　bioluminescence of, 362
Perhydroxyl-radicals, 131
　reaction with luminol, 430 434,447
Peroxalic acid, 137
Peroxy carbonates, 238
Peroxydisulfate, 211
Peroxylactones, 268,493
α-peroxylactone intermediate,
　in firefly luminescence, 354
Phenylalanine, 365
Phialidium,
　bioluminescence of, 362
Phosphorus, 402,411
Phosphides, 488
Photochemiluminescence, 481, 488,490,492
Photo-decomposition, 270
Photodissociation,
　of H_2O, 19
Photo-oxidation, 488
Photinus pyralis,
　bioluminescence of, 285
Photoprotein, 342,362,387,391, 495
Photopic factor, 453
Phthalic hydrazide, 210,232
Phthaloyl peroxide,
　sensitized chemiluminescence of, 225
pK_a,
　firefly luciferin, 287
Polar solvent,
　quenching in, 124
Poly(ethylene 2,6 naphthalene-dicarboxylate), 488
Poisson distribution, 465,466
Potassium iodide, 246
Potassium iodide quenching, 245
Preannihilation, 479
Predissociation, 13,14,79,485
Probabilities,
　exchange, 54
　transition, 55
Process,
　allowed, 129
Processes,
　spin forbidden, 77
Product analysis, 69

Ptilosarcus,
  bioluminescence of, 362,367
Pumping,
  treanor type, 70
Purification,
  of lumisomes, 390
Pyrazine nucleus, 255,365
Pyrophosphate, 494

Q

Quantum efficiency, 233
Quantum yields,
  of aldehyde and $FMNH_2$ in bacterial bioluminescence, 381,383
  in hydrazide luminescence, 233
  in $BIQI^{++}$ luminescence, 257, 259
  in firefly luminescence, 453
  of fluorescence, 462
  in luminol luminescence, 232
Quenching,
  in ECL, 194
  hydrogen, 116
  in laser emissions, 65
  in IR chemiluminescence, 52
  mechanism of, 121
  physical, 112
  by SO_2, 66
Quinones, 490,491

R

Radiation,
  magnetic and electrical dipole, 78
Radiative recombination,
  of atoms, 8
  on a single potential curve, 8-11
  probability of radiation, 9
  emission rate coefficient, 9
  classical approach, 9
  in He, 10
  in Ar+O, 11
  in C+H, 11
  via curve crossing, 11-15
  in N+O, 12
  phenomenological description, 14
  as resonance scattering process, 14
  in N, 15
  in triatomic systems, 16-20
  in H+NO, 17
  in O+NO, 18
  in five-atom systems, 20
  in O+NO+M, 20
Radicals,
  carbonate, 139
Radical ions, 143
  electrochemically generated, 193
  recombination of, 96,144,179, 184,185
  temperature dependence of CL, 189
Raschig rings, 134
Rates,
  in population of vibrational levels, 44
  of $BIQI^{++}$ luminescence decay 258
  in formation of CO, 46
  in vibrational deactivation, 44
Rate coefficient, 52
Rate constants,
  quenching of, 112
Reactors,
  fast-flow, 395,396
Reactions,
  photodissociation, 63
Recombination processes,
  two body and three body, 74, 80,135
Red light emitter,
  structure of in firefly luminescence, 293
Reflectivity,
  from ECL electrodes, 203
Relaxation,
  of HCl and HF, 54
  radiative, 47
  rotational and vibrational rates, 78
  vibrational, 52,67

Renilla,
  bioluminescence of, 341,361,
    362,367
  auto-oxidized luciferin, 362,
    363
Renilla-like luciferase, 367
  luciferin, 367
Renilla luciferin, 361,362,363,
    364,365,366
Renilla oxyluciferin, 367
Renilla reniformis,
  bioluminescence of, 361,387
Rhodamine-B, 406
Rotating ring-disk electrode,
  ECL, 194
Rubrene, 178,225

S

S-route,
  ECL, 194
Salcomine, 329
Schumann-Runge bands, 403,404
Scattering,
  resonance, 79
Scotopic, 470
Sea pansy,
  bioluminescence of, 361,493
Second-order rate constant,
  for O+NO reaction, 96
Selection rules,
  harmonic oscillator, 52
Semiquinone, 491
Shock tube, 400
Singlet oxygen, 265,481,482,483,
    489,491,497,498
Solvent,
  absorption, 124
  aprotic, 132
  quenching, 123
Spin change, 127
Spin-forbidden, 129
Stoichiometry,
  in bacterial bioluminescence,
    382
State distributions,
  importance of, 44
States,
  excited electronic, 74
Steady state, analysis, 46

analysis, 46
Stern-Volmer,
  quenching, 493
  relationship, 48
  model, 91
Stylatula, bioluminescence of,
    362,367
Subunits,
  bacterial luciferase, 370
  firefly luciferase, 301
Sucrose gradient densities,
  of lumisomes, 390
Sulfhydryl groups,
  role of in firefly luciferase,
    302
Sulfur, 403,409
Sulfur dioxide, 396,405,409,416,
    447,486
Sulfur oxide, 396,416
Superoxide anion, 132,265
Superoxide dismutase, 497,498
Systems,
  discharge flow, 74

T

Tetracene,
  cation-anion annihilation
    rate constants, 160
Tetramethoxy BIQ,
  preparation of, 255
Tetramethyl-p-phenylenediamine
    (TMPD),
  in radical ion recombination,
    183,184,185,186
Temperature dependence, 89
Temperature,
  pH and metal ions, effect on
    emission color in firefly
    luminescence, 295
  rotational-translational, 66
Thioglycolate, 325
Thiol groups,
  modification of on bacterial
    luciferase, 372
Threshold energy, 67
TMAE, 259
TPCK,
  inhibitory effect on firefly
    luciferase, 302

Trans-azodicarboxylates,
  chemiluminescence of, 225
Transitions,
  collision induced, 52,80,118
  in O_2, 127
  triplet-singlet, 139
Transition state energy,
  effect of substituents on, 234
Transvibronic reactions,
  in molecular beams, 29
  of alkali atoms, 30
Triplet acetone,
  involvement in dioxetane decomposition, 270
Triplet-singlet energy transfer, 232
Triplet-triplet annihilation, 161,169,175,194,480
Triethylamine,
  as catalyst of dioxetane decomposition, 271
Tris(p-dimethylaminophenyl)amine (TPDA),
  in radical ion recombination systems, 183,184,185,187
Trimethyl aluminum, 414
2,6,8-triphenylimidazo[1,2-a]-pyrazin-3-one, 333
Tri-p-tolylaminium perchlorate,
  yield and energies of some radical ion recombination reactions with, 178
Trace element analysis, 427
Trace elements, 473
Triplet state emission, 489
Tryptophan, 365
Tyrosine, 365

U

Unpaired electron, 119

V

Vanadium, 428,431,439,447
V.E.R.,
  modes of transformation, 92
Vesicular nature,
  of lumisomes, 391
Vibraluminescence, 90
Vibrational excitation, 45,119
Vibrational frequency, 116
Vibronic states,
  in triatomic system, 16
  density of in five-atom system, 20

W

Water,
  as reactant in bioluminescence and chemiluminescence, 349, 350
  effect on reaction rate in chemiluminescence, 352,353

Y

Yellow-green emitter,
  structure of in firefly luminescence, 294

Z

Zero point energy,
  of NO_2, 95

5/21   P2653
27.50   33.3